21世纪高等学校计算机教育实用规

新编计算机操作系统双语教程

朱天翔　王溪波　编著

清华大学出版社
北京

内容简介

本教材的创新点是采用双语制，提供中英文教学素材，适应高等院校所提倡的双语教学模式，响应国际型人才培养战略的要求。在内容的编排上，每章后面将本章的主要概念、原理和算法附上英文教学内容。既可作为高等院校计算机相关专业的计算机操作系统课程的双语教材，也可供广大师生自学之用。

本书介绍了计算机操作系统的基本概念、原理和相关的技术。从计算技术的产生到操作系统的发展，从单机批处理操作到多道程序系统的实现，由浅入深、循序渐进，构成计算机操作系统的整体架构。全书共分 8 章，分别介绍计算机操作系统的基本概念；讲述处理机的管理内容，包括进程管理、进程同步、进程通信等；介绍内存管理和虚拟存储器的实现；阐述文件管理、设备管理的相关知识。

图书在版编目（CIP）数据

新编计算机操作系统双语教程：汉、英/朱天翔等编著. --北京：清华大学出版社，2016
21 世纪高等学校计算机教育实用规划教材
ISBN 978-7-302-43821-2

Ⅰ. ①新…　Ⅱ. ①朱…　Ⅲ. ①操作系统－双语教学－高等学校－教材－汉、英　Ⅳ. ①TP316

中国版本图书馆 CIP 数据核字（2016）第 119127 号

责任编辑：付弘宇　薛　阳
封面设计：常雪影
责任校对：焦丽丽
责任印制：沈　露

出版发行：清华大学出版社
网　　址：http://www.tup.com.cn，http://www.wqbook.com
地　　址：北京清华大学学研大厦 A 座　　邮　　编：100084
社 总 机：010-62770175　　邮　　购：010-62786544
投稿与读者服务：010-62776969，c-service@tup.tsinghua.edu.cn
质 量 反 馈：010-62772015，zhiliang@tup.tsinghua.edu.cn
课 件 下 载：http://www.tup.com.cn，010-62795954
印 装 者：清华大学印刷厂
经　　销：全国新华书店
开　　本：185mm×260mm　　印　　张：20.75　　字　　数：504 千字
版　　次：2016 年 7 月第 1 版　　印　　次：2016 年 7 月第1 次印刷
印　　数：1～2000
定　　价：44.50 元

产品编号：068488-01

前　言

1946 年世界上第一台计算机的面世开启了人类信息化文明的新时代。现今世界正在被以计算机技术为核心的信息化文明深深地影响和改变。

计算机是实现信息化的重要工具。操作系统是覆盖在计算机硬件之上的第一层系统软件。学习操作系统知识体系，对于计算机相关专业的本科生至关重要。

双语教学是目前各高等院校提倡的教学模式，是培养学生成为国际型人才的重要的教学手段。本教材采用双语制，每章后面将本章的主要概念、算法附上英文教学内容，为双语教学提供方便，有利于学生专业外语能力的提高。

学习计算机操作系统知识体系主要分为四个阶段：① 学习某一种具体的操作系统（如 Windows XP）的使用；② 学习计算机操作系统的基本原理；③ 通过学习研究某一种操作系统（如 Linux）的具体实现来体验计算机操作系统的基本原理；④ 操作系统的编程训练，在某一种具体的操作系统源代码中加入自己的个性化代码，培养开发大中型计算机软件所必备的编程能力和团队协作精神。

本书对应的是上述第二个阶段的教学环节，即为高等院校计算机及相关专业本科生的“计算机操作系统”课程提供双语授课教材。全书贯穿操作系统的核心概念、原理和各种算法，使学生了解计算机系统中硬件、软件的相互配合及高效率工作的原理。

全书共 8 章，第 1 章介绍计算机操作系统的基本概念；第 2~4 章主要讲述处理机的管理，分别阐述进程管理、进程同步、进程通信等内容；第 5、6 章讲述内存管理和虚拟存储器的实现；第 7 章阐述文件管理的相关内容；第 8 章介绍设备管理的相关内容。

本教材的编写过程中，得到了沈阳工业大学和清华大学出版社的大力支持，在此表示衷心的感谢！此外，朱琪、李康泰、常欣、王传鹰等同志在本教材的编撰、整理和绘图等工作中，都付出了许多艰辛的劳动，为本教材的出版做出了许多贡献，谨向上述各位表示衷心的感谢！

本教材难免会有疏漏及不当之处，恳请读者批评指正。

编　者

2016 年 5 月

目　录

第1章　操作系统引论

操作系统是控制和管理计算软硬件资源，合理地组织计算机工作流程，方便用户使用的系统软件。操作系统是配置在计算机硬件上的第一层软件，是硬件系统功能的首次扩充。它在计算机系统中占据了非常重要的地位，人们常把计算机的操作系统称为“人机接口”。

1.1　计算机的基本工作原理

1.1.1　自动计算

人类在蒙昧时代就已具有识别事物多寡的能力。原始人在采集、狩猎等生产活动中首先注意到一只羊与许多羊、一头狼与整群狼在数量上的差异。通过一只羊与许多羊、一头狼与整群狼的比较，逐渐认识到它们之间存在着某种共通的东西（即它们的单位性）。当对数的认识变得越来越明确时，人们感到有必要以某种方式来表达事物的这一属性，于是导致了记数。由记数到算数运算，逐步发展成为当今庞大的学科——数学。

数学的发展解决人们生产实践过程中的许多问题，从简单的买菜到复杂的导弹弹道的计算。数学已经渗透到人们生活的各个领域，为人类的发展做出了巨大贡献。

对于一些非常复杂的计算人们可以找到求解的方法，但是计算过程可能需要极大的工作量。例如导弹弹道的计算，其计算工作量可能需要几千人年。在这种情况下，人们就梦想着有一种机器，能够帮助人们自动计算。它能按照人们的求解方法，不厌其烦地、快速地计算，求出问题的解。

计算机的诞生，实现了人们自动计算的梦想。计算机（Computer）是一种能够按照事先存储的程序，自动、高速地进行大量数值计算和各种信息处理的现代化智能电子设备。它不仅具有计算的功能，还有逻辑判断、高速运算、大容量储存、记忆、输入、输出及处理等与人脑类似的功能。现代的计算机已不能与我国古老的算盘和早期的机械、机电计算机同日而语，称其为电脑可谓名符其实。1945 年底，世界上第一台电子计算机在美国宾夕法尼亚大学诞生。这台电子计算机安装有 18000 多个电子管，重 30 多吨，运算速度是每秒 5000 次。

1.1.2　计算机基础

1．进位计数制

所谓进位计数制是指按进位的原则进行计数。目前常用的进位计数制有十进制、二进制、八进制、十六进制等。

十进制中的数包括 0、1、2、3、4、5、6、7、8、9，其进位规则为逢十进一。

二进制中的数包括 0、1，其进位规则为逢二进一。

八进制中的数包括 0、1、2、3、4、5、6、7，其进位规则为逢八进一。

十六进制中的数包括 0、1、2、3、4、5、6、7、8、9、A、B、C、D、E、F，其进位规则为逢十六进一。

基数：某种数制中使用的数字的个数。例如：十进制数的基数是十；二进制数的基数是二；八进制数的基数是八；十六进制数的基数是十六。

数位：在某种数制中，数字在一个数中所处的位置称为数位。例如十进制数中包含个位、十位、百位、千位等。

位值：位值也叫权（位权），任何一个数都是由一串数字（符号）表示的，其中每一位所表示的值除其本身的数值外，还与它所处的位置有关，由位置决定的值就叫权。

不同进制中的权是不一样的，例如：十进制数中 10^0、10^1、10^2、…；二进制数中 2^0、2^1、2^2、…；八进制数中 8^0、8^1、8^2、…；十六进制中 16^0、16^1、16^2、…。

例 1-1 一个二进制数$(11011.101)_2$，求这个数各位权的表示。

解：

数	1	1	0	1	1	0	1	0	1
数位	4	3	2	1	0	–1	–2	–3	–4
权	2^4	2^3	2^2	2^1	2^0	2^{-1}	2^{-2}	2^{-3}	2^{-4}

即 $(11011.101)_2=1\times2^4+1\times2^3+0\times2^2+1\times2^1+1\times2^0+1\times2^{-1}+0\times2^{-2}+1\times2^{-3}$。由此我们可以得出一个结论：对于 M 位进制，整数的权为 M^i，从右向左，i=0，1，2，3…；小数的权为 M^I，从左向右，I = –1，–2，–3…。

2．二进制

二进制是计算机技术中广泛采用的一种数制，由 18 世纪德国数理哲学大师莱布尼兹首先使用。二进制数据用 0 和 1 两个数码来表示，它的基数为 2，进位规则是“逢二进一”，借位规则是“借一当二”，当前计算机系统使用的基本上是二进制系统。数值计算可以采用任意进制，即二进制同十进制的计算都可以得到正确的结果。

二进制系统具有以下特点：

（1）二进制数容易用物理器件实现，低电平和高电平这两个物理状态就可以分别代表 0 和 1；

（2）二进制数具有良好的可靠性，因为只有两个物理状态，数据传输和运算过程中，不容易因为干扰而发生错误；

（3）二进制运算法则简单；

（4）二进制中使用的 1 和 0，可分别用来代表逻辑运算中的“真”和“假”，可以很方便地实现逻辑运算。

3．布尔代数

布尔（Boole George）是英国数学家及逻辑学家。他是鞋匠之子，十六岁时在私立学校教数学，到 1835 年他自己开办了一所中等学校。在这个时期，他对数学产生了深厚的兴趣，一边教书，一边自修高等数学。1849 年（尽管他没有学位）被任命为科克的女王学院的数学教授。1854 年，他出版了《思维规律的研究》一书，其中完满地讨论了这个主题并奠定了现在所谓的符号逻辑的基础。在布尔代数里，布尔构思出一个关于 0 和 1 的代数系统，用基础的逻辑符号系统描述物体和概念。这种代数为今后数字计算机开关电路设计提供了最重要数学方法。

在数学上可以证明，任何复杂的逻辑关系，都可以由布尔代数表达的“与”“或”“非”三种基本逻辑组合而成。二进制、布尔代数为计算机的产生打下了坚实的基础。一个复杂的逻辑关系的电路，可以化简为“与”“或”“非”三种基本的开关电路实现，数字逻辑作为一门新的研究科目出现了。一个复杂的计算机的基本功能的逻辑可以通过数字逻辑电路得以实现。

4．计算机体系结构

数学、逻辑学、电子学理论以及工程技术的飞速发展，使电子计算机的研制成为可能。那么究竟应该采用什么样的模式，什么样的体系结构来实现电子计算呢？冯•诺依曼提出了一套可行的计算机体系结构的设计。冯•诺依曼理论的要点是：数字计算机的数制采用二进制；计算机应该按照程序顺序执行。人们把冯•诺依曼的这个理论称为冯•诺依曼体系结构。从世界上第一台电子计算机（ENIAC）到当前最先进的计算机都是采用的冯•诺依曼体系结构。

根据冯•诺依曼体系结构构成的计算机，必须具有如下功能：把需要的程序和数据送至计算机中；必须具有长期记忆程序、数据、中间结果及最终运算结果的能力；能够完成各种算术、逻辑运算和数据传送等数据加工处理的能力；能够根据需要控制程序走向，并能根据指令控制机器的各部件协调操作；能够按照要求将处理结果输出给用户。为了完成上述的功能，计算机必须具备五大基本组成部件，包括：输入数据和程序的输入设备、记忆程序和数据的存储器、完成数据加工处理的运算器、控制程序执行的控制器、输出程序处理结果的输出设备。

虽然计算机的制造技术从计算机出现到今天已经发生了极大的变化，但在基本的体系结构一直沿袭着冯•诺依曼的传统结构，即计算机硬件系统由运算器、控制器、存储器、输入设备、输出设备五大部件构成。计算机系统体系结构如图 1-1 所示。图 1-1 中实线代表数据流，虚线代表指令流，计算机各部件之间的联系就是通过这两股信息流动来实现的。原始数据和程序通过输入设备送入存储器，在运算处理过程中，数据从存储器读入运算器进行运算，运算的结果存入存储器，必要时再经输出设备输出，指令也以数据形式存于存储器中，运算时指令由存储器送入控制器，由控制器控制各部件分析处理。

冯•诺依曼体系结构具有如下基本特点：

（1）计算机由运算器、控制器、存储器、输入设备和输出设备五部分组成；

（2）采用存储程序的方式，程序和数据放在同一个存储器中，指令和数据一样可以送到运算器运算，即由指令组成的程序是可以修改的；

（3）数据以二进制码表示；

（4）指令由操作码和地址码组成；

（5）指令在存储器中按执行顺序存放，由指令计数器(即程序计数器 PC)指明要执行的指令所在的单元地址，一般按顺序递增，但可按运算结果或外界条件而改变；

（6）机器以控制器为中心，输入输出设备与存储器间的数据传送都通过控制器实现。

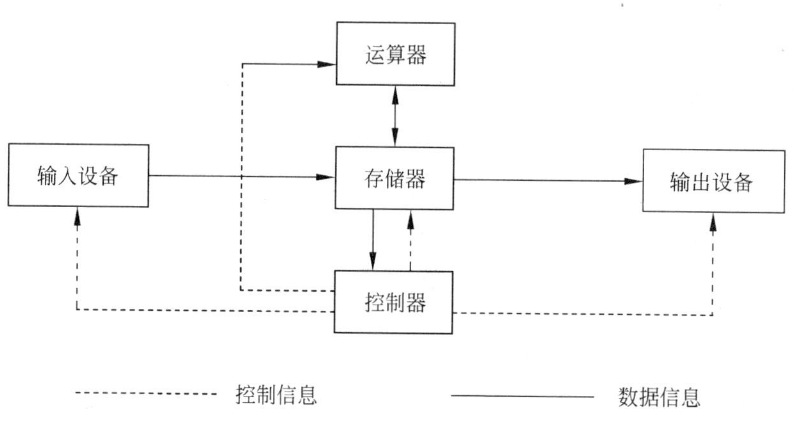

图 1-1　计算机体系结构

5．计算机系统

一个完整的计算机系统包括硬件系统和软件系统两大部分，如图 1-2 所示。

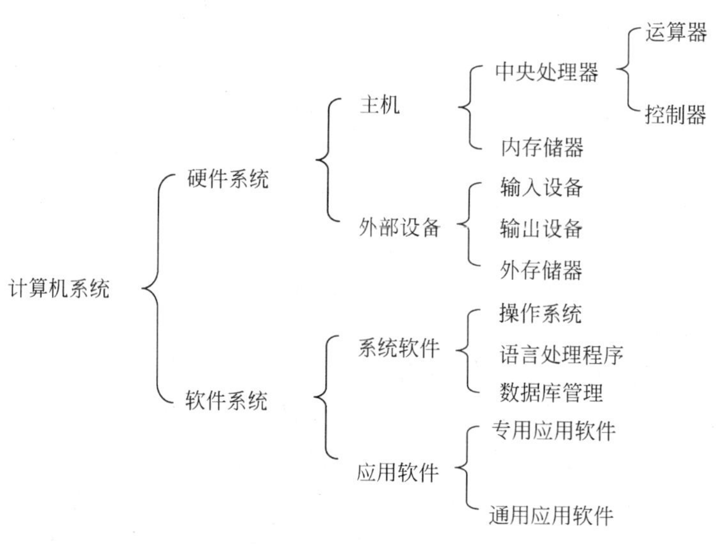

图 1-2　计算机系统

计算机硬件系统是指构成计算机的所有实体部件的集合，通常这些部件由电路（电子元件）、机械等物理部件组成。直观地看，计算机硬件是一大堆设备，它们都是看得见、摸得着的，是计算机进行工作的物质基础，也是计算机软件发挥作用、施展其技能的舞台。

计算机软件是指在硬件设备上运行的各种程序以及有关资料。所谓程序实际上是用户用于指挥计算机执行各种动作以便完成指定任务的指令的集合。用户要让计算机做的工作可能是很复杂的，因而指挥计算机工作的程序也可能是庞大而复杂的，有时还可能要对程序进行修改与完善。因此，为了便于阅读和修改，必须对程序作必要的说明或整理出有关的资料。这些说明或资料（称之为文档）在计算机执行过程中可能是不需要的，但对于用户阅读、修改、维护、交流，这些程序却是必不可少的。因此，也有人简单地用一个公式来说明包括其基本内容：软件=程序+文档。

通常，人们把不装备任何软件的计算机称为硬件计算机或裸机。裸机由于不装备任何软件，所以只能运行机器语言程序，这样的计算机，它的功能显然不会得到充分有效的发挥。普通用户面对的一般不是裸机，而是在裸机之上配置若干软件之后构成的计算机系统。有了软件，就把一台实实在在的物理机器（有人称为实机器）变成了一台具有抽象概念的逻辑机器（有人称为虚机器），从而使人们不必更多地了解机器本身就可以使用计算机，软件在计算机和计算机使用者之间架起了桥梁。正是由于软件的丰富多彩，可以出色地完成各种不同的任务，才使得计算机的应用领域日益广泛。当然，计算机硬件是支撑计算机软件工作的基础，没有足够的硬件支持，软件也就无法正常工作。实际上，在计算机技术的发展进程中，计算机软件随硬件技术的迅速发展而发展；反过来，软件的不断发展与完善又促进了硬件的新发展，两者的发展密切地交织着，缺一不可。

计算机硬件的基本功能是接受计算机程序的控制来实现数据输入、运算、数据输出等一系列根本性的操作。输入设备负责把用户的信息（包括程序和数据）输入到计算机中；输出设备负责将计算机中的信息（包括程序和数据）传送到外部媒介，供用户查看或保存；存储器负责存储数据和程序，并根据控制命令提供这些数据和程序，它包括内存（储器）和外存（储器）；运算器负责对数据进行算术运算和逻辑运算（即对数据进行加工处理）；控制器负责对程序所规定的指令进行分析，控制并协调输入、输出操作或对内存的访问。

中央处理器简称 CPU（Central Processing Unit），它是计算机系统的核心，中央处理器包括运算器和控制器两个部件。

计算机所发生的全部动作都是 CPU 来控制。其中，运算器主要完成各种算术运算和逻辑运算，是对信息加工和处理的部件。控制器是对计算机发布命令的“决策机构”，用来协调和指挥整个计算机系统的操作，它本身不具有运算功能，而是通过读取各种指令，并对其进行翻译、分析，而后对各部件做出相应的控制。它主要由指令寄存器、译码器、程序计数器、操作控制器等组成。中央处理器是计算机的心脏，CPU 品质的高低直接决定了计算机系统的档次。

1.2 操作系统的产生

计算机产生之后，并不是马上就具有了操作系统。从第一台计算机诞生(1945 年)到 20 世纪 50 年代中期的计算机，属于第一代，这时还未出现计算机操作系统。这时计算机的操作是由程序员采用人工操作方式直接使用计算机硬件系统。

1.2.1 早期计算机的使用

程序员为了在计算机上算一道题，先要预约登记一段机时，到时他将预先准备好的表示指令和数据的插接板带到机房，由操作员将其插入计算机，并设置好计算机上的各种控制开关，启动计算机运行。假如程序员设计的程序是正确的，并且计算机也没有发生故障，若干小时后他就能获得计算结果，否则将前功尽弃，再约定下次上机时间。

汇编语言和高级语言的问世，以及程序和数据可以通过穿孔纸带或卡片装入计算机，改善了软件的开发环境，但计算机的操作方式并没有多大的改进。程序员首先将记有程序和数据的纸带或卡片装到输入设备上，拨动开关，将程序和数据装入内存；接着，程序员要启动汇编或编译程序，将源程序翻译成目标代码；假如程序中不出现语法错误，下一步程序员就可通过控制台按键设定程序执行的起始地址，并启动程序的执行。

在程序的执行期间，程序员要观察控制台上的各种指示灯以监视程序的运行情况。如果发现错误，并且还未用完所预约的上机时间，就可通过指示灯检查存储器中的内容，直接在控制台上进行调试和排错。如果程序运行正常，最终将结果在电传打字机等输出设备上打印出来。当程序运行完毕并取走计算结果后，才让下一个用户上机。

总之，在早期的计算机系统中，每一次独立的运行都需要很多的人工干预，操作过程烦琐，占用机时多，也很容易产生错误。在一个程序的运行过程中，要独占系统的全部硬件资源，设备利用率很低。

早期的计算机不具备操作系统，这种人工操作方式有以下两方面的缺点：

（1）用户独占全机；

（2）CPU 等待人工操作。

1.2.2 批处理系统

计算机的人工操作方式，费时、费力，远远不能发挥计算机处理机的高速运算能力。计算机设备作为一种高速的、昂贵的设备，其运用的效率是人们关注的热点，为了提高计算机的利用率，方便人们使用计算机，人们开始研究帮助人们使用计算机的操作系统。

早期的计算机操作系统大多是批处理系统。这种系统中，把用户的计算任务按“作业（Job）”进行管理。所谓作业，是用户定义的、由计算机完成的工作单位。它通常包括一组计算机程序、文件和操作系统的控制语句。逻辑上，一个作业可由若干有序的步骤组成。由作业控制语句明确标识的计算机程序的执行过程称为作业步，一个作业可以指定若干要执行的作业步。例如编译作业步、装配作业步、运行作业步、出错处理作业步等。

批处理（Batch Processing）就是将作业按照它们的性质分组（或分批），然后再成组（或成批）地提交给计算机系统，由计算机自动完成后再输出结果，从而减少作业建立和结束过程中的时间浪费。早期的批处理系统属于单道批处理系统，其目的是减少作业间转换时的人工操作，从而减少 CPU 的等待时间。单道批处理系统的工作流程如图 1-3 所示。

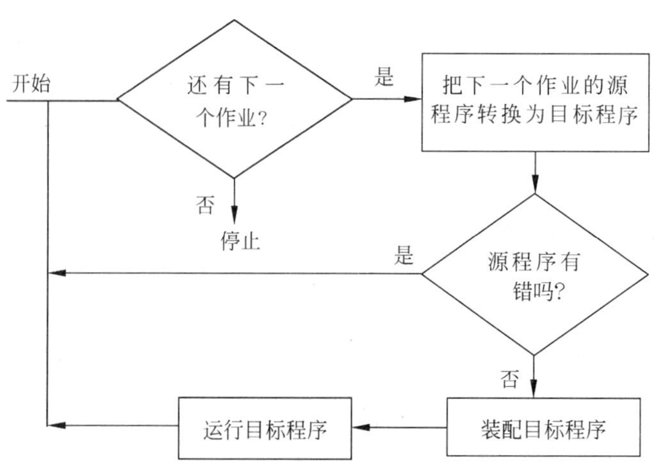

图 1-3　批处理系统的工作流程

单道批处理系统的特征是内存中只允许存放一个作业，即当前正在运行的作业才能驻留内存，作业的执行顺序是先进先出，即按顺序执行。一个作业单独进入内存并独占系统资源，直到运行结束后下一个作业才能进入内存，当作业进行 I/O 操作时，CPU 只能处于等待状态，因此，CPU 利用率较低，尤其是对于 I/O 操作时间较长的作业。

在单道批处理系统中，内存中仅有一道作业，它无法充分利用系统中的所有资源，致使系统性能较差。为了进一步提高系统资源的利用率和系统吞吐量，在 20 世纪 60 年代中期又引入了多道程序设计（Multiprogramming）技术，由此而形成了多道批处理系统（Multiprogrammed Batch Processing System）。多道程序设计的基本思想是在内存里同时存放若干道程序，它们可以交替地运行。作业执行的次序与进入内存的次序无严格的对应关系，用户提交的作业都先存放在外存上并排成一个队列，称为“后备队列”；然后，由作业调度程序按一定的算法从后备队列中选择若干个作业调入内存，使它们共享 CPU 和系统中的各种资源。作业通过进程调度来使用 CPU，一个作业在等待 I/O 处理时，调度另外一个作业使用 CPU，这样处理机得到了比较充分的利用。CPU 的利用率显著地提高了。多道批处理系统的工作流程如图 1-4 所示。

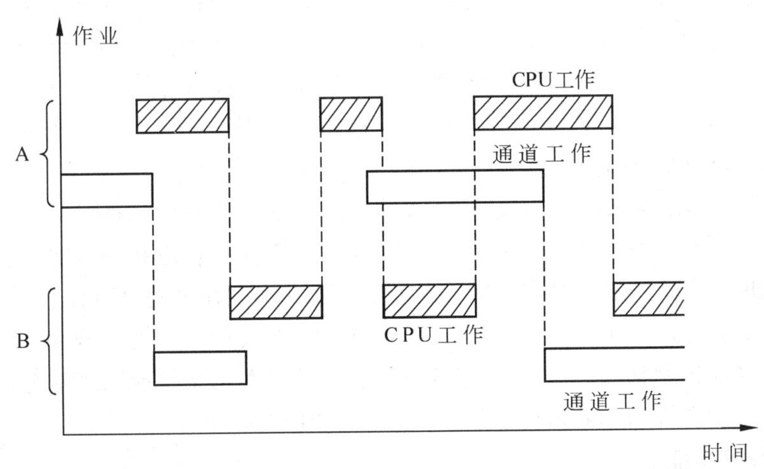

图 1-4　多道批处理系统的工作流程

作业 A 和作业 B 就是交替运行的，当作业 A 执行通道操作，不使用 CPU 时，作业 B 执行程序，使用 CPU。通道是专门负责 I/O 操作的设备，可以独立完成程序的 I/O 操作。当一个作业利用通道做 I/O 操作的同时，另一个作业可以使用 CPU 执行程序，提高 CPU 的利用率。

多道处理系统的优点是系统资源为多个作业所共享，其工作方式是作业之间自动调度执行。在运行过程中用户不干预自己的作业，从而大大提高了系统资源的利用率和作业吞吐量。其缺点是无交互性，用户一旦提交作业就失去了对其运行的控制能力，而且是批处理的，作业周转时间长，用户使用不方便。

批处理系统是最早出现的一种操作系统，严格地说，它只能算作是操作系统的前身而并非是现在人们所理解的计算机操作系统。尽管如此，该系统比起人工操作方式的系统已有很大进步。

1.2.3 分时系统

如果说，推动多道批处理系统形成和发展的主要动力是提高资源利用率和系统吞吐量，其实分时系统的形成也是为了更好地提高资源利用率和系统吞吐量。

1．分时

分时是对时间的共享，是为提高资源利用率采用的并行操作技术。如 CPU 和通道并行操作、通道与通道并行操作、通道与 I/O 设备并行操作。这些已成为现代计算机系统的基本特征。与这三种并行操作相应的有三种对内存访问的分时：CPU 与通道对内存访问的分时，通道与通道对 CPU 和内存的分时，同一通道中的 I/O 设备对内存和通道的分时。

2．时间片

分时系统将 CPU 的时间划分成若干个片段，称为时间片。操作系统以时间片为单位，轮流为每个终端用户服务。每个用户轮流使用时间片，而使每个用户并不感到有别的用户存在。

3．时间片轮转调度

时间片轮转调度是一种最古老、最简单、最公平且使用最广的算法。每个进程被分配一个时间段，也就是时间片，即该进程允许运行的时间。如果在时间片结束时进程还在运行，则 CPU 将被剥夺并分配给另一个进程。如果进程在时间片结束前阻塞或结束，则 CPU 当即进行切换。调度程序所要做的就是维护一张就绪进程队列列表，当进程用完它的时间片后，它被移到队列的末尾。

时间片轮转调度中需要关注的是时间片的长度。从一个进程切换到另一个进程是需要一定时间的——保存和装入寄存器值及内存映像，更新各种表格和队列等。假如进程切换（Process Switch）需要 5 ms，再假设时间片设为 20 ms，则在做完 20 ms 有用的工作之后，CPU 将花费 5 ms 来进行进程切换。CPU 时间的 20%被浪费在了管理开销上。时间片设得太短会导致过多的进程切换，降低了 CPU 效率；而设得太长又可能引起对短的交互请求的响应变差。将时间片设为 100 ms 通常是一个比较合理的折衷。

4．分时系统特征

分时系统具有多路性、交互性、“独占”性和及时性的特征。

多路性：同时有多个用户使用一台计算机，宏观上看是多个人同时使用一个 CPU，微观上是多个人在不同时刻轮流使用 CPU。

交互性：用户根据系统响应结果进一步提出新请求（用户直接干预每一步）。

“独占”性：用户感觉不到计算机为其他人服务，就像整个系统为他所独占。

及时性：系统对用户提出的请求及时响应。

为实现分时系统，其中，最关键的问题是如何使用户能与自己的作业进行交互，即当用户在自己的终端上键入命令时，系统应能及时接收并及时处理该命令，再将结果返回给用户，此后，用户可继续键入下一条命令。这种过程即人机交互。应强调指出，即使有多个用户同时通过自己的键盘键入命令，系统也应能全部地及时接收并处理。

分时系统具有的许多优点促使它迅速发展，其优点主要是：

（1）为用户提供了友好的接口，即用户能在较短时间内得到响应，能以对话方式完成对其程序的编写、调试、修改、运行和得到运算结果；

（2）促进了计算机的普遍应用，一个分时系统可带多台终端，可同时为多个远近用户使用，这给教学和办公自动化提供很大方便；

（3）便于资源共享和交换信息，为软件开发和工程设计提供了良好的环境。

分时系统具有现代操作系统的特征，是比较成熟的操作系统设计理念，现代的大多数操作系统都是基于分时系统的设计思想设计的。

5．特权指令

操作系统特权指令是指具有特殊权限的指令。这类指令只用于操作系统或其他系统软件，一般不直接提供给用户使用。在多用户、多任务的计算机系统中特权指令必不可少。它主要用于系统资源的分配和管理，包括改变系统工作方式，检测用户的访问权限，修改虚拟存储器管理的段表、页表，完成任务的创建和切换等。

常见的特权指令有以下几种：

（1）有关对 I/O 设备使用的指令，如启动 I/O 设备指令、测试 I/O 设备工作状态和控制 I/O 设备动作的指令等；

（2）有关访问程序状态的指令，如读程序状态字（PSW）的指令等；

（3）存取特殊寄存器指令 如存取中断寄存器、时钟寄存器等指令；

（4）其他指令。

CPU 执行特权指令的状态称为管态，又称为特权状态、系统态或核心态。CPU 执行用户指令的状态称为目态，又称为用户态。

1.2.4 实时系统

与分时系统相对应的是实时操作系统，实时操作系统是保证在一定时间限制内完成特定功能的操作系统。例如，可以为确保生产线上的机器人能获取某个物体而设计一个操作系统。一些实时操作系统是为特定的应用设计的，另一些是通用的。

强实时系统（Hard Real-Time）：在航空航天、军事、核工业等一些关键领域中，应用时间需求应能够得到完全满足，否则就造成如飞机失事等重大的安全事故，造成重大地生命财产损失和生态破坏。因此，在这类系统的设计和实现过程中，应采用各种分析、模拟及形式化验证方法对系统进行严格的检验，以保证在各种情况下应用的时间需求和功能需求都能够得到满足。

弱实时系统（Soft Real-Time）：某些应用虽然提出了时间需求，但实时任务偶尔违反这种需求对系统的运行以及环境不会造成严重影响，例如视频点播（Video-On-Demand，VOD）系统、信息采集与检索系统就是典型的弱实时系统。在 VOD 系统中，系统只需保证绝大多数情况下视频数据能够及时传输给用户即可，偶尔的数据传输延迟对用户不会造成很大损失，也不会造成像飞机失事一样严重的后果。

1.3 操作系统的概念

1.3.1 操作系统的定义

计算机操作系统的定义：控制和管理计算机软硬件资源、合理地组织计算机工作流程，方便用户使用计算机的系统软件。

操作系统是配置在计算机硬件上的第一层软件，是硬件系统功能的首次扩充。它在计算机系统中占据了非常重要的地位，人们常把计算机的操作系统称为“人机接口”。

为了深入理解操作系统的定义，我们应注意以下几点：

（1）操作系统是系统软件，而且是裸机之上的第一层软件；

（2）操作系统的基本职能是控制和管理系统内的各种资源。计算机系统的基本资源包括硬件（如处理机、内存、各种设备等）、软件（系统软件和应用软件）和数据；

（3）合理地、有效地组织多道程序的运行；

（4）设置操作系统的另一个目的是扩充机器功能，方便用户使用计算机。

计算机的资源由操作系统来管理，用户不必理会系统内存如何分配、如何保护，不必理会多个程序之间如何协调工作，也不必理会硬盘数据如何存储、数据怎样导入内存和系统设备如何管理。这些工作都由操作系统完成，用户使用计算机变得非常方便。

1.3.2 操作系统与计算机其他软件及硬件的关系

计算机操作系统作为用户与计算机硬件系统之间接口的含义是：操作系统处于用户与计算机硬件系统之间，用户通过操作系统来使用计算机系统。或者说，用户在操作系统帮助下，能够方便、快捷、安全、可靠地操纵计算机硬件和运行自己的程序。应当注意，操作系统是一个系统软件，因而这种接口是软件接口。操作系统与计算机软硬件的关系如图 1-5 所示。

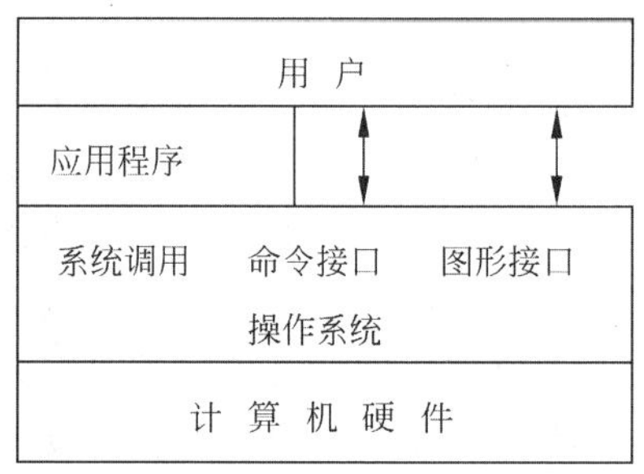

图 1-5　操作系统与计算机软硬件的关系

计算机操作系统的主要功能也正是针对四类资源进行有效的管理，即：处理机管理，用于分配和控制处理机；存储器管理，主要负责内存的分配与回收；I/O 设备管理，负责 I/O 设备的分配与操纵；文件管理，负责文件的存取、共享和保护。

对于一台完全无软件的计算机系统（即裸机），即使其功能再强，也必定是难于使用的。如果我们在裸机上面覆盖上一层 I/O 设备管理软件，用户便可利用它所提供的 I/O 命令来操纵计算机，进行数据输入和打印输出。此时用户所看到的机器，将是一台比裸机功能更强、使用更方便的机器。通常把覆盖了软件的机器称为扩充机器或虚机器。如果我们又在第一层软件上再覆盖上面一层文件管理软件，则用户可利用该软件提供的文件存取命令，来进行文件的存取。此时，用户所看到的这台机器是一台功能更强的虚机器。如果我们又在文件管理软件上面再覆盖一层面向用户的窗口软件，则用户便可在窗口环境下方便地使用计算机，形成一台功能更加强大的虚机器。

1.3.3 操作系统的使用

1．命令接口

联机用户接口是为联机用户提供的，它由一组键盘操作命令及命令解释程序所组成。当用户在终端或控制台上每键入一条命令后，系统便立即转入命令解释程序，对该命令加以解释并执行。在完成指定功能后，控制又返回到终端或控制台上，等待用户键入下一条命令。用户可通过先后键入不同命令的方式，来实现对作业的控制，直至作业完成。

脱机用户接口是为批处理作业的用户提供的，故也称为批处理用户接口。该接口由一组作业控制语言（Job Control Language，JCL）组成。批处理作业的用户不能直接与自己的作业交互作用，只能委托系统代替用户对作业进行控制和干预。这里的 JCL 便是提供给批处理作业用户的、为实现所需功能而委托系统代为控制的一种语言。用户用 JCL 把需要对作业进行的控制和干预事先写在作业说明书上，然后将作业连同作业说明书一起提供给系统。当系统调度到该作业运行时，又调用命令解释程序，对作业说明书上的命令，逐条地解释执行。如果作业在执行过程中出现异常现象，系统也将根据作业说明书上的指示进行干预。这样，作业一直在作业说明书的控制下运行，直至遇到作业结束语句时，系统才停止该作业的运行。

2．系统调用

系统调用也叫程序接口，是为用户程序在执行中访问系统资源而设置的，是用户程序取得操作系统服务的唯一途径。它是由一组系统调用组成的，每一个系统调用都是一个能完成特定功能的子程序，每当应用程序要求操作系统提供某种服务（功能）时，便调用具有相应功能的系统调用。早期的系统调用都是用汇编语言提供的，只有在用汇编语言书写的程序中，才能直接使用系统调用。但在高级语言以及 C 语言中，往往提供了与各系统调用一一对应的库函数，这样，应用程序便可通过调用对应的库函数来使用系统调用。

3．图形接口

用户虽然可以通过联机用户接口来取得操作系统的服务，但这时要求用户能熟记各种命令的名字和格式，并严格按照规定的格式输入命令，这既不方便又花时间，于是，图形用户接口便应运而生。图形用户接口采用了图形化的操作界面，用非常容易识别的各种图标（Icon）来将系统的各项功能、各种应用程序和文件，直观、逼真地表示出来。用户可用鼠标或通过菜单和对话框，来完成对应用程序和文件的操作。此时用户已完全不必像使用命令接口那样去记住命令名及格式，从而把用户从烦琐且单调的操作中解脱出来。

1.4 操作系统的引导

操作系统是帮助我们使用计算机运行程序的工具。而计算机操作系统本身又是一套系统程序，那么，操作系统程序又由谁来运行呢？

为了运行操作系统程序，需要定义一套机制，做一些约定，来执行操作系统程序，这就是操作系统的引导。

首先约定，计算机加电后，CPU 到一个固定的内存地址去读取指令，执行程序，这段程序是负责加载操作系统程序的程序。这段程序掉电不能消失，所以存放这段程序的内存必须是只读存储器（ROM）。然后，只读存储器的加载程序负责读取外存储器（即安装了某种操作系统的外存储器）的第一个存储区到内存储器，在这个存储区中包含引导操作系统的程序。加载程序把引导程序读到内存后，将程序的执行权交给这个引导程序。引导操作系统的程序由操作系统程序制作的厂家提供。

以常用的 PC 引导 Windows 操作系统为例，简单说明计算机系统启动过程。PC 的启动过程包括两个方面：① 硬件加电自检；② 操作系统引导并启动。目前的计算机硬件加电自检基本相似，系统启动过程不同。Windows 操作系统的引导流程如图 1-6 所示。

硬件系统加电自检过程，主板 CMOS 芯片上的 BIOS（Basic Input/Output System）程序检测与主板相连的硬件：CPU、内存、光驱、硬盘、软驱键盘等。如果出现错误则发出相应的声响，或文字信息提示错误。如：内存与主板插槽接触不良（灰尘、潮湿导致的），就会不断发出一声长鸣。硬件一切正常，BIOS 读取硬盘 MBR（Main Boot Record，主引导记录）中的代码。MBR 代码程序可以读取分区表，加载系统卷中的引导扇区。MBR 和引导扇区的内容是安装系统时写在存储介质上的。MBR 和引导扇区位置是固定的。MBR 是硬盘的第 0 个盘片的第 0 个面的第 0 磁道的 0 扇区（计算机中，常常“第一个”用 0 来表示）。

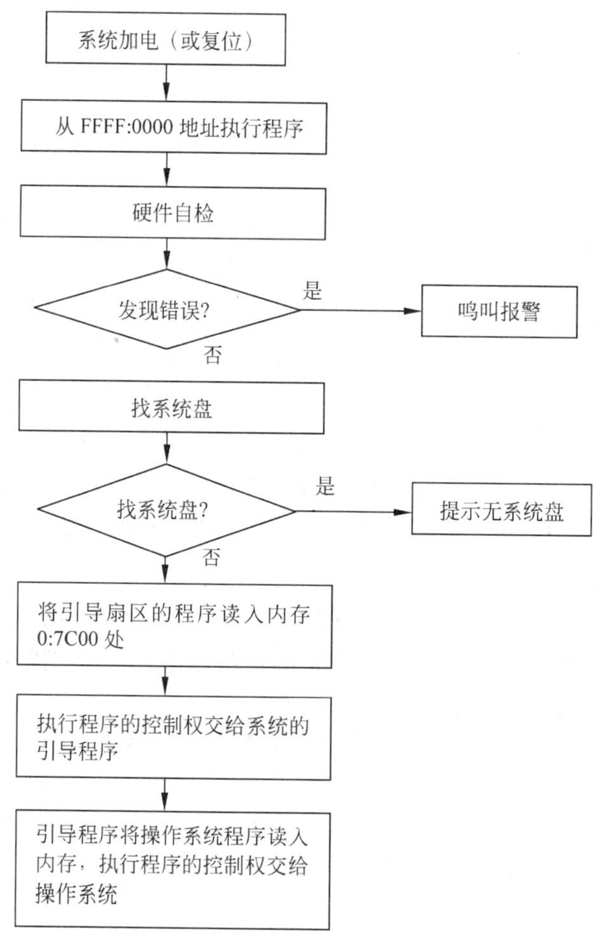

图 1-6 Windows 操作系统的引导过程

操作系统 Windows XP Professional 启动过程，主引导扇区存储的是 Windows XP 引导代码。读取系统卷根目录，并加载 NTLDR(NT Loader)程序。NTLDR 是 Windows XP 的装载程序，存储在主引导扇区。NTLDR 读取并分析 boot.ini 文件：如果 boot.ini 有多个条目，NTLDR 显示选择菜单；如果在系统目录下有 hiberfil.sys 文件且有效，NTLDR 将读取该文件里的信息并让系统恢复到休眠以前的状态；完成 boot.ini 引导选择后，用户可以按 F8 键进入高级启动选项；NTLDR 执行 Ntdetect.com 进行硬件和 BIOS 信息检测；加载注册表 HKLM\System，引导驱动程序 Ntoskrnl.exe，hal.dll 等，加载了 NTLDR 后（屏幕黑屏），BIOS 把控制权交给了操作系统 Ntoskrnl.exe 的入口函数。Ntoskrnl 接收到控制权时，屏幕显示 Windows 启动微标。Ntoskrnl 创建会话管理器进程（Session Manager）。\windows\system32\smss.exe 中的 smss.exe 是第一个用户态进程。smss.exe 程序初始化 paging files 和其余的注册表项，启动 Csrss.exe 和 Winlogon.exe 程序，显示 Windows 登录窗口。Winlogin.exe 加载 GINA（Graphical Identification adn Authentication）并等待用户登录。启动 Services.exe，启动所有标识为自动启动的 Win32 服务程序。Windows XP 整个启动过程完成。

1.5 操作系统特征

1．并发

并行性和并发性是既相似又有区别的两个概念。并行性是指两个或多个事件在同一时刻发生；而并发性是指两个或多个事件在同一时间间隔内发生。在多道程序环境下，并发性是指宏观上在一段时间内多道程序同时运行，但在单处理器系统中，每一时刻仅能执行一道程序，故微观上这些程序是在交替执行的。

2．共享

所谓共享是指系统中的资源可供主存中多个并发执行的进程共同使用。由于资源的属性不同，故多个进程对资源的共享方式也不同，可分为两种资源共享方式。

1）互斥共享方式

系统中的某些资源（如打印机），虽然它们可以提供给多个进程使用，但在一段时间内却只允许一个进程访问该资源。当一个进程正在访问该资源时，其他欲访问该资源的进程必须等待，仅当该进程访问完并释放该资源后，才允许另一进程对该资源进行访问。我们把在一段时间内只允许一个进程访问的资源称为临界资源，许多物理设备以及某些变量、表格都属于临界资源，它们要求互斥地被共享。

2）同时访问方式

系统中还有一类资源，允许在一段时间内多个进程同时对它进行访问。这里所谓的“同时”往往是宏观上的。而在微观上，这些进程可能是交替地对该资源进行访问。典型的可供多个进程同时访问的资源是磁盘，一些用重入码编写的文件也可同时共享。

并发和共享是操作系统的两个最基本的特征，它们又互为存在条件。一方面，资源共享是以程序（进程）的并发执行为条件的；若系统不允许程序并发执行，自然就不存在资源共享问题。另一方面，若系统不能对资源共享实施有效管理，则必将影响到程序的并发执行，甚至根本无法并发执行。

3．虚拟

操作系统中的所谓“虚拟”是指通过某种技术把一个物理实体变成若干个逻辑上的对应物。物理实体（前者）是实的，即实际存在的；而后者是虚的，是用户感觉上的东西。例如，在多道分时系统中，虽然只有一个 CPU，但每个终端用户却都认为有一个 CPU 在专门为它服务，亦即利用多道程序技术和分时技术可以把一台物理 CPU 虚拟为多台逻辑上的 CPU，也称为虚处理器。类似地，也可以把一台物理 I/O 设备虚拟为多台逻辑上的 I/O 设备。

4．异步性

在多道程序环境下，允许多个进程并发执行，但由于受资源等因素的限制，通常进程的执行并非“一气呵成”，而是以“走走停停”的方式运行。主存中的每个进程在何时执行，何时暂停，以怎样的速度向前推进，每道程序总共需多少时间才能完成，都是不可预知的。很可能是先进入主存的作业后完成，后进入主存的作业先完成。或者说，进程是以异步方式运行的。尽管如此，只要运行环境相同，作业经多次运行，都会获得完全相同的结果，因此，异步运行方式是允许的。进程的异步性，是操作系统的一个重要特征。

1.6 操作系统组成

根据操作系统的功能划分，操作系统一般由四大组成部分，即处理机管理、内存管理、文件管理和设备管理。

1.6.1 处理机管理

1．进程控制

在传统的多道程序环境下，要使作业运行，必须先为它创建一个或几个进程，并为之分配必要的资源。当进程运行结束时，立即撤销该进程，以便能及时回收该进程所占用的各类资源。进程控制的主要功能是为作业创建进程、撤销已结束的进程，以及控制进程在运行过程中的状态转换。在现代操作系统中，进程控制还应具有为一个进程创建若干个线程的功能和撤销（终止）已完成任务的线程的功能。

2．进程调度

在后备队列上等待的每个作业，通常都要经过调度才能执行。在传统的操作系统中，包括作业调度和进程调度两步。作业调度的基本任务，是从后备队列中按照一定的算法，选择出若干个作业，为它们分配其必需的资源（首先是分配内存）。在将它们调入内存后，便分别为它们建立进程，使它们都成为可能获得处理机的就绪进程，并按照一定的算法将它们插入就绪队列。而进程调度的任务，则是从进程的就绪队列中选出一新进程，把处理机分配给它，并为它设置运行环境，使进程投入执行。值得提出的是，在多线程操作系统中，通常是把线程作为独立运行和分配处理机的基本单位，为此，须把就绪线程排成一个队列，每次调度时，是从就绪线程队列中选出一个线程，把处理机分配给它。

3．进程同步

为使多个进程能有条不紊地运行，系统中必须设置进程同步机制。进程同步的主要任务是为多个进程（含线程）的运行进行协调。有两种协调方式：① 进程互斥方式，这是指诸进程（线程）在对临界资源进行访问时，应采用互斥方式；② 进程同步方式，指在相互合作去完成共同任务的诸进程（线程）间，由同步机构对它们的执行次序加以协调。

为了实现进程同步，系统中必须设置进程同步机制。最简单的用于实现进程互斥的机制，是为每一个临界资源配置一把锁，当锁打开时，进程（线程）可以对该临界资源进行访问；而当锁锁上时，则禁止进程（线程）访问该临界资源。

4．进程通信

在多道程序环境下，为了加速应用程序的运行，应在系统中建立多个进程，并且再为一个进程建立若干个线程，由这些进程（线程）相互合作去完成一个共同的任务。而在这些进程（线程）之间，又往往需要交换信息。例如，有三个相互合作的进程，它们是输入进程、计算进程和打印进程。输入进程负责将所输入的数据传送给计算进程；计算进程利用输入数据进行计算，并把计算结果传送给打印进程；最后，由打印进程把计算结果打印出来。进程通信的任务就是用来实现在相互合作的进程之间的信息交换。

当相互合作的进程（线程）处于同一计算机系统时，通常在它们之前是采用直接通信方式，即由源进程利用发送命令直接将消息（Message）挂到目标进程的消息队列上，以后由目标进程利用接收命令从其消息队列中取出消息。

1.6.2 内存管理

1．内存分配

计算机操作系统在实现内存分配时，可采取静态和动态两种方式。在静态分配方式中，每个作业的内存空间是在作业装入时确定的；在作业装入后的整个运行期间，不允许该作业再申请新的内存空间，也不允许作业在内存中“移动”；在动态分配方式中，每个作业所要求的基本内存空间，也是在装入时确定的，但允许作业在运行过程中，继续申请新的附加内存空间，以适应程序和数据的动态增涨，也允许作业在内存中“移动”。

为了实现内存分配，在内存分配的机制中应具有以下的结构和功能：

（1）内存分配数据结构，该结构用于记录内存空间的使用情况，作为内存分配的依据；

（2）内存分配功能，系统按照一定的内存分配算法，为用户程序分配内存空间；

（3）内存回收功能，系统对于用户不再需要的内存，通过用户的释放请求，去完成系统的回收功能。

2．内存保护

内存保护的主要任务，是确保每道用户程序都只在自己的内存空间内运行，彼此互不干扰。为了确保每道程序都只在自己的内存区中运行，必须设置内存保护机制。一种比较简单的内存保护机制，是设置两个界限寄存器，分别用于存放正在执行程序的上界和下界。系统须对每条指令所要访问的地址进行检查，如果发生越界，便发出越界中断请求，以停止该程序的执行。如果这种检查完全用软件实现，则每执行一条指令，便须增加若干条指令去进行越界检查，这将显著降低程序的运行速度。因此，越界检查都由硬件实现。当然，对发生越界后的处理，还须与软件配合来完成。

3．地址映射

一个应用程序（源程序）经编译后，通常会形成若干个目标程序；这些目标程序再经过链接便形成了可装入程序。这些程序的地址都是从 0 开始的，程序中的其他地址都是相对于起始地址计算的；由这些地址所形成的地址范围称为地址空间，其中的地址称为逻辑地址或相对地址。此外，由内存中的一系列单元所限定的地址范围称为内存空间，其中的地址称为物理地址。

在多道程序环境下，每道程序不可能都从 0 地址开始装入（内存），这就致使地址空间内的逻辑地址和内存空间中的物理地址不相一致。使程序能正确运行，存储器管理必须提供地址映射功能，以将地址空间中的逻辑地址转换为内存空间中与之对应的物理地址。该功能应在硬件的支持下完成。

4．内存扩充

存储器管理中的内存扩充任务，并非是去扩大物理内存的容量，而是借助于虚拟存储技术，从逻辑上去扩充内存容量，使用户所感觉到的内存容量比实际内存容量大得多；或者是让更多的用户程序能并发运行。这样，既满足了用户的需要，改善了系统的性能，又基本上不增加硬件投资。为了能在逻辑上扩充内存，系统必须具有内存扩充机制，用于实现下述各功能：

（1）请求调入功能；

（2）置换功能。

1.6.3 文件管理

1．文件存储空间的管理

由文件系统对诸多文件及文件的存储空间，实施统一的管理。其主要任务是为每个文件分配必要的外存空间，提高外存的利用率，并能有助于提高文件系统的运行速度。

为此，系统应设置相应的数据结构，用于记录文件存储空间的使用情况，以供分配存储空间时参考；系统还应具有对存储空间进行分配和回收的功能。为了提高存储空间的利用率，对存储空间的分配，通常是采用离散分配方式，以减少外存零头，并以盘块为基本分配单位。盘块的大小通常为 512B~8KB。

2．目录管理

为了使用户能方便地在外存上找到自己所需的文件，通常由系统为每个文件建立一个目录项。目录项包括文件名、文件属性和文件在磁盘上的物理位置等。由若干个目录项又可构成一个目录文件。目录管理的主要任务，是为每个文件建立其目录项，并对众多的目录项加以有效地组织，以实现方便地按名存取。用户只需提供文件名，即可对该文件进行存取。其次，目录管理还应能实现文件共享，此外，还应能提供快速的目录查询手段，以提高对文件的检索速度。

3．文件的读/写管理和保护

该功能是根据用户的请求，从外存中读取数据；或将数据写入外存。在进行文件读（写）时，系统先根据用户给出的文件名，去检索文件目录，从中获得文件在外存中的位置。然后，利用文件读（写）指针，对文件进行读（写）。一旦读（写）完成，便修改读（写）指针，为下一次读（写）做好准备。由于读和写操作不会同时进行，故可合用一个读/写指针。文件保护要做到：

（1）防止未经核准的用户存取文件；

（2）防止冒名顶替存取文件；

（3）防止以不正确的方式使用文件。

1.6.4 设备管理

设备管理用于管理计算机系统中所有的外围设备，而设备管理的主要任务是，完成用户进程提出的 I/O 请求；为用户进程分配其所需的 I/O 设备；提高 CPU 和 I/O 设备的利用率；提高 I/O 速度；方便用户使用 I/O 设备。为实现上述任务，设备管理应具有缓冲管理、设备分配和设备处理，以及虚拟设备等功能。

1．缓冲管理

CPU 运行的高速性和 I/O 低速性间的矛盾自计算机诞生时起便已存在。而随着 CPU 速度大幅度提高，使得此矛盾更为突出，严重降低了 CPU 的利用率。如果在 I/O 设备和 CPU 之间引入缓冲，则可有效地缓和 CPU 和 I/O 设备速度不匹配的矛盾，提高 CPU 的利用率，进而提高系统吞吐量。因此，在现代计算机系统中，都毫无例外地在内存中设置了缓冲区，而且还可通过增加缓冲区容量的方法，来改善系统的性能。

最常见的缓冲区机制有单缓冲机制、能实现双向同时传送数据的双缓冲机制，以及能供多个设备同时使用的公用缓冲池机制。

2．设备分配

设备分配的基本任务，是根据用户进程的 I/O 请求、系统的现有资源情况以及按照某种设备分配策略，为之分配其所需的设备。如果在 I/O 设备和 CPU 之间，还存在着设备控制器和 I/O 通道时，还须为分配出去的设备分配相应的控制器和通道。

为了实现设备分配，系统中应设置设备控制表、控制器控制表等数据结构，用于记录设备及控制器的标识符和状态。据这些表格可以了解指定设备当前是否可用，是否忙碌，以供进行设备分配时参考。设备使用完后，应立即由系统回收。

3．设备处理

设备处理程序又称为设备驱动程序。其基本任务是用于实现 CPU 和设备控制器之间的通信，即由 CPU 向设备控制器发出 I/O 命令，要求它完成指定的 I/O 操作；反之由 CPU 接收从控制器发来的中断请求，并给予迅速的响应和相应的处理。

处理过程是设备处理程序首先检查 I/O 请求的合法性，了解设备状态是否是空闲的，了解有关的传递参数及设置设备的工作方式。然后，便向设备控制器发出 I/O 命令，启动 I/O 设备去完成指定的 I/O 操作。设备驱动程序还应能及时响应由控制器发来的中断请求，并根据该中断请求的类型，调用相应的中断处理程序进行处理。对于设置了通道的计算机系统，设备处理程序还应能根据用户的 I/O 请求，自动地构成通道程序。

Chapter 1 Introduction to Computer

An operating system is a program that manages the computer hardware. It also provides a basis for application programs and acts as an intermediary between the computer user and the computer hardware. An amazing aspect of operating systems is how varied they are in accomplishing these tasks. Mainframe operating systems are designed primarily to optimize utilization of hardware. The operating systems of the personal computer (PCs) support complex games, business applications, and everything in between. Operating systems for handheld computers are designed to provide an environment in which a user can easily interface with the computer to execute programs. Thus, some operating systems are designed to be convenient, others to be efficient, and others some combination of the two.

1.1 The Basic Principle of Computer

1.1.1 Computer Architecture

All computers share the same basic architecture, whether it is a mainframe or a Palm Pilot. All have memory, an I/O system, and arithmetic/logic unit, and a control unit, shoun in Figure1.1. This type of architecture is named Von Neumann's Architecture after the mathematician, John Von Neumann who conceived of the design. John Von Neumann began his "Preliminary Discussion" with a broad description of the general-purpose computing machine containing four basic components. These are known as relating to arithmetic, memory, control, and connection with the human operator. In other words, the arithmetic/logical unit (ALU), the memory, the control unit (CU), and the input-output (I/O) devices that we see in the classical model of what a computer looks like.

1. Memory

Computer Memory is the subsystem that serves as temporary storage for all program instructions and data that are being executed by the computer. It is typically called RAM (Random Access Memory). Memory is divided into cells, each cell has a unique address so that the data can be fetched.

2. Arithmetic/logical unit (ALU)

This subsystem is to perform all arithmetic operations and comparisons for equality. In the Von Neumann's architecture, the arithmetic/logical unit (ALU) and the Control Unit (CU) are separate components, but in modern systems they are integrated into the processor. The ALU has 3 sections, the register, the ALU circuitry, and the pathways in between. The register is basically a storage cell that works like RAM and holds the results of the calculations. It is mush faster than RAM and is addresses differently. The ALU circuitry is that actually performs the calculations, and it is designed from AND, OR, and NOT gates just as any chip.

3. Control Unit (CU)

The control unit is responsible for fetching the next program instruction to be run from memory, decoding it to determine what needs to be done, and then issuing the proper command to the ALU, memory and I/O controllers to get the job done. These steps are done continuously until the last line of a program is done, which is usually QUIT or STOP.

4. Input-Output (I/O) devices

This is the subsystem that allows the computer to interact with other devices and communicate to the outside world. It is also responsible for program storage, such as hard drive control.

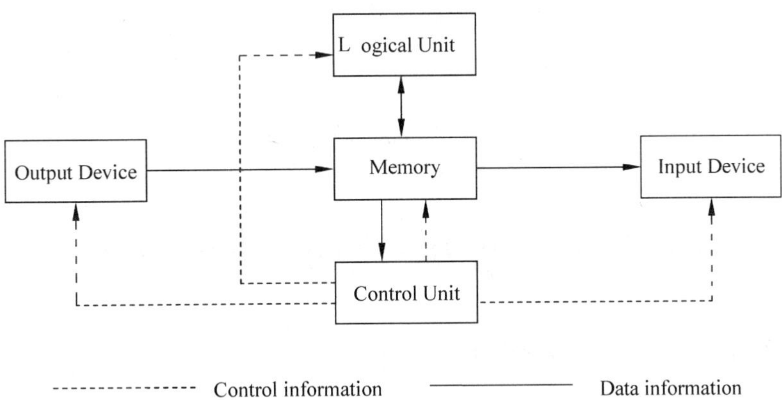

Figure 1.1 Computer Architecture

1.1.2 Computer System

Computer System in shown in Figure1.2.

Hardware includes every computer-related object that you can physically touch and handle like disks, screens, keyboards, printers, chips, wires, central processing unit, floppies, USB ports, pen drives etc.

Software is a general term used to describe a collection of computer programs, procedures, and documentation that perform some task on a computer system. Practical computer systems divide software systems into three major classes: system software, programming software, and application software, although the distinction is arbitrary and often blurred. Software is an ordered sequence of instructions for changing the state of the computer hardware in a particular sequence. It is usually written in high-level programming languages that are easier and more efficient for humans to use (closer to natural language) than machine language. High-level languages are compiled or interpreted into machine language object code. Software may also be written in an assembly language, essentially, a mnemonic representation of a machine language using a natural language alphabet.

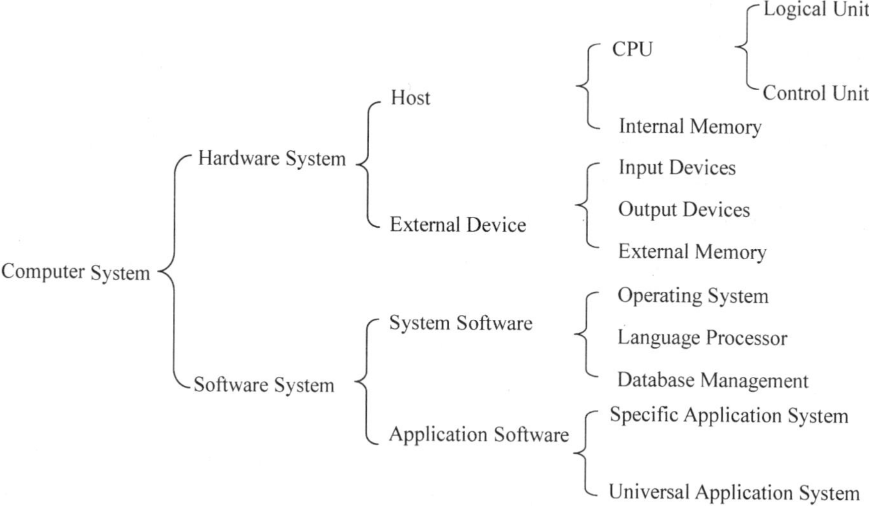

Figure 1.2 Computer System

1.2 Operating System Concepts

1.2.1 Operating System Definitions

The user's view of the computer varies according to the interface being used. Most computer users sit in front of a PC, consisting of a monitor, keyboard, mouse and system unit. Such a system is designed for one user to monopolize its resources. The goal is to maximize the work (or play) that the user is performing. In this case, the operating system is designed mostly for ease of use, with some attention paid to performance and none paid to resource utilization—how various hardware and software resources are shared. Performance is, of course, important to the user; but rather than resource utilization, such systems are optimized for the single-user experience.

From the computer's point of view, the operating system is the program most intimately involved with the hardware. In this context, we can view an operating system as a resource allocator. A computer system has many resources that may be required to solve a problem: CPU time, memory space, file-storage space, I/O devices and so on. The operating system acts as the manager of these resources. Facing numerous and possibly conflicting requests for resources, the operating system must decide how to allocate them to specific programs and users so that it can operate the computer system efficiently and fairly. As we have seen, resource allocation is especially important where many users access the same mainframe or minicomputer.

A slightly different view of an operating system emphasizes the need to control the various I/O devices and user programs. An operating system is a control program. A control program manages the execution of user programs to prevent errors and improper use of the computer. It is especially concerned with the operation and control of I/O devices.

1.2.2 Use of the Operating System

1. Command Interpreter

The main function of the command interpreter is to get and execute the next user-specified command. Many of the commands given at this level manipulate files: create, delete, list, print, copy, execute and so on. The MS-DOS and UNIX shells operate in this way. There are two general ways in which these commands can be implemented.

In one approach, the command interpreter itself contains the code to execute the command. For example, a command to delete a file may cause the command interpreter to jump to a section of its code that sets up the parameters and makes the appropriate system call. In this case, the number of commands that can be given determines the size of the command interpreter, since each command requires its own implementing code.

2. System calls

The system-call level must provide the basic functions, such as process control and file and device manipulation. Higher-level requests, satisfied by the command interpreter or system programs, are translated into a sequence of system calls. System services can be classified into several categories: program control, status requests and I/O requests. Program errors can be considered implicit requests for service.

3. Graphical User Interface

Rather than having users directly enter commands via a command-line interface, a GUI provides a mouse-based window-and-menu system as an interface. A GUI provides a desktop metaphor where the mouse is moved to position its pointer on images, or icons, on the screen (the desktop) that represent programs, files, directories, and system functions. Depending on the mouse pointer's location, clicking a button on the mouse can invoke a program, select a file or directory—known as a folder—or pull down a menu that contains commands.

Graphical interfaces became more widespread with the advent of Apple Macintosh computers in the 1980s. The user interface to the Macintosh operating system (Mac OS) has undergone various changes over the years, the most significant being the adoption of the Aqua interface that appeared with Mac OS X. Microsoft's first version of Windows—version 1.0—was based upon a GUI interface to the MS-DOS operating system. The various versions of Windows systems proceeding this initial version have made cosmetic changes to the appearance of the GUI and several enhancements to its functionality, including the Windows Explorer.

1.2.3 Operating System Several Related Concepts

One of the most important aspects of operating systems is the ability to multiprogram. A single user cannot, in general, keep either the CPU or the I/O devices busy at all times. Multiprogramming increases CPU utilization by organizing jobs (code and data) so that the CPU always has one to execute.

The idea is as follows: The operating system keeps several jobs in memory simultaneously. This set of jobs can be a subset of the jobs kept in the job pool—which contains all jobs that enter the system—since the number of jobs that can be kept simultaneously in memory is usually smaller than the number of jobs that can be kept in the job pool. The operating system picks and begins to execute one of the jobs in memory. Eventually, the job may have to wait for some task, such as an I/O operation, to complete. In a non-multiprogrammed system, the CPU would sit idle. In a multiprogrammed system, the operating system simply switches to, and executes, another job. When that job needs to wait, the CPU is switched to another job, and so on. Eventually, the first job finishes waiting and gets the CPU back. As long as at least one job needs to execute, the CPU is never idle.

Multiprogrammed systems provide an environment in which the various system resources (for cxample, CPU, memory and peripheral devices) are utilized effectively, but they do not provide for user interaction with the computer system. Time sharing (or multitasking) is a logical extension of multiprogramming. In time-sharing systems, the CPU executes multiple jobs by switching among them, but the switches occur so frequently that the users can interact with each program while it is running.

A time-shared operating system allows many users to share the computer simultaneously. Since each action or command in a time-shared system tends to be short, only a little CPU time is needed for each user. As the system switches rapidly from one user to the next, each user is given the impression that the entire computer system is dedicated to his use, even though it is being shared among many users.

Time-sharing and multiprogramming require several jobs to be kept simultaneously in memory. Since in general main memory is too small to accommodate all jobs, the jobs are kept initially on the disk in the job pool. This pool consists of all processes residing on disk awaiting allocation of main memory. If several jobs are ready to be brought into memory, and if there is not enough room for all of them, then the system must choose among them. When the operating system selects a job from the job pool, it loads that job into memory for execution. Having several programs in memory at the same time requires some form of memory management. In addition, if several jobs are ready to run at the same time, the system must choose among them. Making this decision is CPU scheduling. Finally, running multiple jobs concurrently requires that their ability to affect one another be limited in all phases of the operating system, including process scheduling, disk storage, and memory management. These considerations are discussed throughout the text.

In a time-sharing system, the operating system must ensure reasonable response time, which is sometimes accomplished through swapping, where processes are swapped in and out of main memory to the disk. A more common method for achieving this goal is virtual memory, a technique that allows the execution of a process that is not completely in memory. The main advantage of the virtual-memory scheme is that it enables users to run programs that are larger than actual physical memory. Further, it abstracts main memory into a large, uniform array of storage, separating logical memory as viewed by the user from physical memory. This arrangement frees programmers from concern over memory-storage limitations.

We must ensure that the operating system maintains control over the CPU. We must prevent a user program from getting stuck in an infinite loop or not calling system services and never returning control to the operating system. To accomplish this goal, a timer can be used. A timer can be set to interrupt the computer after a specified period. The period may be fixed (for example, 1/60 second) or variable (for example, from 1 millisecond to 1 second). A variable timer is generally implemented by a fixed-rate clock and a counter. The operating system sets the counter. Every time the clock ticks, the counter is decremented. When the counter reaches 0, an interrupt occurs. For instance, a 10-bit counter with a 1-millisecond clock allows interrupts at intervals from 1 millisecond to 1， 024 milliseconds, in steps of 1 millisecond.

Before turning over control to the user, the operating system ensures that the timer is set to interrupt. If the timer interrupts, control transfers automatically to the operating system, which may treat the interrupt as a fatal error or may give the program more time. Clearly, instructions that modify the content of the timer are privileged.

Thus, we can use the timer to prevent a user program from running too long. A simple technique is to initialize a counter with the amount of time that a program is allowed to run. A program with a 7-minute time limit, for example, would have its counter initialized to 420. Every second, the timer interrupts and the counter is decremented by 1. As long as the counter is positive, control is returned to the user program. When the counter becomes negative, the operating system terminates the program for exceeding the assigned time limit.

1.3 System Components

Even though not all systems have the same structure, many modern operating systems share the same goal of supporting the following types of system components.

1.3.1 Process Management

A program does nothing unless its instructions are executed by a CPU. A program in execution, as mentioned, is a process. A time-shared user program such as a compiler is a process. A word-processing program being run by an individual user on a PC is a process. A system task, such as sending output to a printer, can also be a process (or at least part of one).

A process needs certain resources—including CPU time, memory, files, and I/O devices—to accomplish its task. These resources are either given to the process when it is created or allocated to it while it is running. In addition to the various physical and logical resources that a process obtains when it is created, various initialization data (input) may be passed along. For example, consider a process whose function is to display the status of a file on the screen of a terminal. The process will be given as an input the name of the file and will execute the appropriate instructions and system calls to obtain and display on the terminal the desired information. When the process terminates, the operating system will reclaim any reusable resources.

A program by itself is not a process; a program is a passive entity, such as the contents of a file stored on disk, whereas a process is an active entity. A single-threaded process has one program counter specifying the next instruction to execute. The execution of such a process must be sequential. The CPU executes one instruction of the process after another, until the process completes. Further, at any time, one instruction at most is executed on behalf of the process. Thus, although two processes may be associated with the same program, they are nevertheless considered two separate execution sequences. A multithreaded process has multiple program counters, each pointing to the next instruction to execute for a given thread.

A process is the unit of work in a system. Such a system consists of a collection of processes, some of which are operating-system processes (those that execute system code) and the rest of which are user processes (those that execute user code). All these processes can potentially execute concurrently by multiplexing the CPU among them on a single CPU, for example.

The operating system is responsible for the following activities in connection with process management:

(1) Creating and deleting both user and system processes;

(2) Suspending and resuming processes;

(3) Providing mechanisms for process synchronization;

(4) Providing mechanisms for process communication;

(5) Providing mechanisms for deadlock handling.

1.3.2 Memory Management

Main memory is central to the operation of a modern computer system. Main memory is a large array of words or bytes ranging in size from hundreds of thousands to billions. Each word or byte has its own address. Main memory is a repository of quickly accessible data shared by the CPU and I/O devices. The central processor reads instructions from main memory during the instruction-fetch cycle and both reads and writes data from main memory during the data-fetch cycle (on a Von Neumann architecture). The main memory is generally the only large storage device that the CPU is able to address and access directly. For example, for the CPU to process data from disk, those data must first be transferred to main memory by CPU-generated I/O calls.

In the same way, instructions must be in memory for the CPU to execute them.

For a program to be executed, it must be mapped to absolute addresses and loaded into memory. As the program executes, it accesses program instructions and data from memory by generating these absolute addresses. Eventually, the program terminates, its memory space is declared available, and the next program can be loaded and executed.

To improve both the utilization of the CPU and the speed of the computer's response to its users, general-purpose computers must keep several programs in memory, creating a need for memory management. Many different memory-management schemes are used. These schemes reflect various approaches, and the effectiveness of any given algorithm depends on the situation. In selecting a memory-management scheme for a specific system, we must take into account many factors—especially on the hardware design of the system. Each algorithm requires its own hardware support.

The operating system is responsible for the following activities in connection with memory management:

(1) Keeping track of which parts of memory are currently being used and by whom;

(2) Deciding which processes (or parts thereof) and data to move into and out of memory.

1.3.3 File-System Management

File management is one of the most visible components of an operating system. Computers can store information on several different types of physical media. Magnetic disk, optical disk, and magnetic tape are the most common. Each of these media has its own characteristics and physical organization. Each medium is controlled by a device, such as a disk drive or tape drive, which also has its own unique characteristics. These properties include access peed, capacity, data-transfer rate, and access method (sequential or random).

A file is a collection of related information defined by its creator. Commonly, files represent programs (both source and object forms) and data. Data files may be numeric, alphabetic, alphanumeric, or binary. Files may be free-form (for example, text files), or they may be formatted rigidly (for example, fixed fields). Clearly, the concept of a file is an extremely general one.

The operating system implements the abstract concept of a file by managing mass storage media, such as tapes and disks, and the devices that control them. Also, files are normally organized into directories to make them easier to use. Finally, when multiple users have access to files, it may be desirable to control by whom and in what ways (for example, read, write, append) files may be accessed.

The operating system is responsible for the following activities in connection with file management:

(1) Creating and deleting files;

(2) Creating and deleting directories to organize files;

(3) Supporting primitives for manipulating files and directories;

(4) Mapping files onto secondary storage.

1.3.4 I/O Systems

One of the purposes of an operating system is to hide the peculiarities of specific hardware devices from the user. For example, in UNIX, the peculiarities of I/O devices are hidden from the bulk of the operating system itself by the I/O subsystem. Buffers are often used in conjunction with I/O to hardware, such as disk drives, sending or receiving data to or from a network, or playing sound on a speaker. A line to a rollercoaster in an amusement park shares many similarities. People who ride the coaster come in at an unknown and often variable pace, but the roller coaster will be able to load people in bursts (as a coaster arrives and is loaded). The queue area acts as a buffer: a temporary space where those wishing to ride wait until the ride is available. Buffers are usually used in a FIFO (first in, first out) method, outputting data in the order it arrived. The I/O subsystem consists of several components:

(1) A memory-management component that includes buffering, caching, and spooling;

(2) A general device-driver interface;

(3) Drivers for specific hardware devices.

Only the device driver knows the peculiarities of the specific device to which it is assigned.

习　　题

一、选择题

1．操作系统是一种________。

A．通用软件　　B．系统软件　　C．应用软件　　D．软件包

2．操作系统是对________进行管理的软件。

A．软件　　B．硬件　　C．计算机资源　　D．应用程序

3．从用户的观点看，操作系统是________。

A．用户与计算机之间的接口

B．控制和管理计算机资源的软件

C．合理地组织计算机工作流程的软件

D．由若干层次的程序按一定的结构组成的有机体

4．操作系统是现代计算机系统不可缺少的组成部分，是为了提高计算机的________，方便用户使用计算机而配备的一种系统软件。

A．速度　　B．利用率　　C．灵活性　　D．兼容性

5．若把操作系统看作计算机系统资源的管理者，下列________不属于操作系统所管理的资源。

A．程序　　B．内存　　C．CPU　　D．中断

6．在下列操作系统的各个功能组成部分中，________不需要硬件的支持。

A．进程调度　　B．时钟管理　　C．地址映射　　D．中断系统

7．操作系统中采用多道程序设计技术提高 CPU 和外部设备的________。

A．利用率　　B．可靠性　　C．稳定性　　D．兼容性

8．操作系统的基本类型主要有________。

A．批处理系统、分时系统及多任务系统

B．实时系统、批处理系统及分时操作系统

C．单用户系统、多用户系统及批处理系统

D．实时系统、分时系统、多用户系统

9．________操作系统允许在一台主机上同时连接多台终端，多个用户可以通过各自的终端同时交互地使用计算机。

A．网络　　B．分布式　　C．分时　　D．实时

10．如果分时操作系统的时间片一定，那么________，则响应时间越长。

A．用户数越少　　B．用户数越多　　C．内存越少　　D．内存越多

11．分时操作系统通常采用________策略为用户服务。

A．可靠性和灵活性　　B．时间片轮转

C．时间片加权分配　　D．短作业优先

12．________系统允许用户把若干个作业提交给计算机系统。

A．单用户　　B．分布式　　C．批处理　　D．监督

13．在________操作系统控制下，计算机系统能及时处理由过程控制反馈的数据并作出响应。

A．实时　　B．分时　　C．分布式　　D．单用户

14．操作系统的________管理部分负责对进程进行调度。

A．主存储器　　B．控制器　　C．运算器　　D．处理机

15．在计算机系统中，操作系统是________。

A．一般应用软件　　B．数据库管理软件

C．核心系统软件　　D．程序编译软件

16．现代操作系统的两个基本特征是________和资源共享。

A．多道程序设计　　B．中断处理

C．程序的并发执行　　D．实现分时与实时处理

17．在批处理系统中，作业调度程序从后备作业队列中选出若干作业，使其进入________。

A．高速缓存　　B．内存　　C．外存　　D．存储器

18．操作系统是________的接口。

A．主机和外设　　B．系统软件和应用软件

C．用户和计算机　　D．高级语言和机器语言

19．操作系统的作用是________。

A．把源程序编译成目标程序　　B．便于进行目录管理

C．控制和管理系统资源的使用　　D．把高级语言变成机器语言

20．引入多道程序设计的目的是________。

A．增强系统的用户友好性　　B．提高系统实用性

C．充分利用 CPU　　D．扩充内存容量

21．在精确制导导弹中使用的操作系统应属于________。

A．批处理操作系统　　B．个人计算机操作系统

C．实时操作系统　　D．网络操作系

22．操作系统是一种______________软件。

A．系统　　B．编辑　　C．应用　　D．实用

二、填空题

1．计算机系统是由________系统和____________系统两部分组成的。

2．所谓是________指将一个以上的作业放入主存，并且同时处于运行状态，这些作业共享处理机的时间和外围设备等其他资源。

3．分时操作系统的主要特征有_________、_________、_________、_________。

4．采用__________技术能充分发挥 CPU 与外设的并行工作的能力。

5．操作系统的基本功能包括处理机管理、存储器管理、设备管理、__________。

6．操作系统是控制和管理计算机________，合理组织计算机工作流程，方便用户使用的_____________。

第2章 进程管理

进程是程序在一个数据集合上运行的过程，它是系统进行资源分配和调度的一个基本单位。进程控制块是系统描述进程信息的一套数据结构，是进程存在的标志。进程控制块是进程实体的一部分。进程具有就绪、执行、阻塞三种基本状态。

2.1 计算机程序的执行

2.1.1 计算机程序执行

计算机指令是计算机能够识别的代码，是要计算机执行某种操作的命令。计算机程序是计算机指令序列，是人们为了解决某种问题用计算机指令编排的一系列的加工步骤，通过这些指令来指挥计算机做什么、怎么做。

计算机程序的执行过程是：计算机程序被首先加载入内存，计算机的控制器根据指令计数器（PC）的值，从内存中读取一条指令，执行这条指令，指令计数器的值自动增加，指向下一条指令，指令序列被顺序执行，如图 2-1 所示。如果执行一条跳转指令，则修改指令计数器的值，跳转到其他的内存地址去执行指令；在正常执行一条指令后，如果出现中断，则跳转到中断服务程序地址，执行中断服务程序。

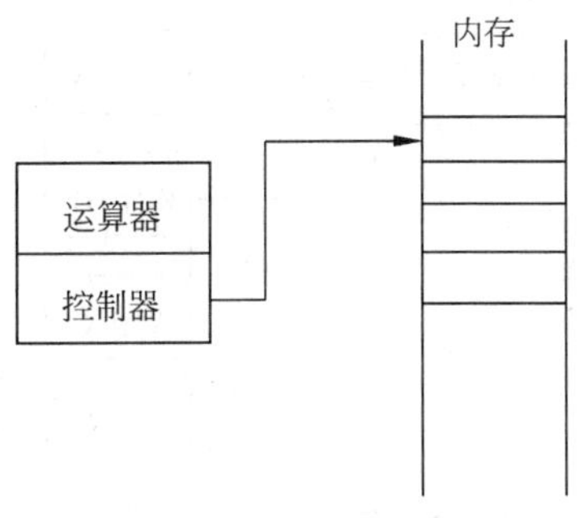

图 2-1　计算机程序的执行

只有很好地理解计算机程序的执行过程，才能够很好地理解计算机操作系统是怎样调度和管理系统内的进程的。理解计算机程序的执行过程，需要注意下面三点。

（1）计算机指令的执行，必须在内存中。计算机的控制器只能按照内存的地址读取指令，执行指令，所以需要执行的指令，一定在内存中。

计算机同硬盘或其他外设交换数据，控制外部设备，就需要把外部设备的接口寄存器编址到计算机的内存，占用一定量的内存地址单元，控制器才能同它们交换数据，指挥它们工作。

在一般情况下，运行的程序应在内存中。有时为了提高内存的利用率，同时运行更多的程序，可以采用虚拟存储器技术，将程序中马上要运行的一部分指令放到内存中，其他指令可以放在硬盘上，这种虚拟存储技术的细节，在后续的章节中会做详细的介绍。

（2）指令逐条执行。一般情况下，控制器执行一条指令，指令计数器会自动指向下一条指令，这样，指令被顺序的逐条执行，除非遇到跳转指令或者发生中断。

（3）操作系统调度和管理系统进程，是通过中断机制实现的。操作系统本身也是计算机程序，它的各条指令也需要放在内存才能运行。被操作系统管理的用户程序也是程序，它要执行的指令也需要放在内存中。操作系统程序同应用程序的切换是靠系统的中断机制实现的。

2.1.2 多道程序的执行

基于上述计算机程序执行的原理，当多道程序在计算机上运行时，每个程序正在执行的指令是必须放在内存中的。如图 2-2 所示。操作系统（OS）控制用户程序的运行。程序 A、程序 B、程序 C 以时间片为单位，交替运行。程序间的调度和切换由操作系统程序控制，当正在运行的程序因 I/O 而暂停执行时，操作系统可调度另一道程序运行，从而保持了 CPU 处于忙碌状态。

现代的操作系统大多数采用分时处理方式，基于时间片轮转调度。在早期的时间片轮转法中，系统将所有的就绪进程按先来先服务的原则排成一个队列，每次调度时，把 CPU 分配给队首进程，并令其执行一个时间片，时间片的大小从几 ms 到几百 ms，当执行的时间片用完时，由一个计时器发出时钟中断请求，中断服务程序控制转去执行调度程序来停止该进程的执行，并将它送往就绪队列的末尾；然后，再把处理机分配给就绪队列中新的队首进程，同时也让它执行一个时间片，这样就可以保证就绪队列中的所有进程，在给定的时间内，均能获得一个时间片的处理机执行时间。

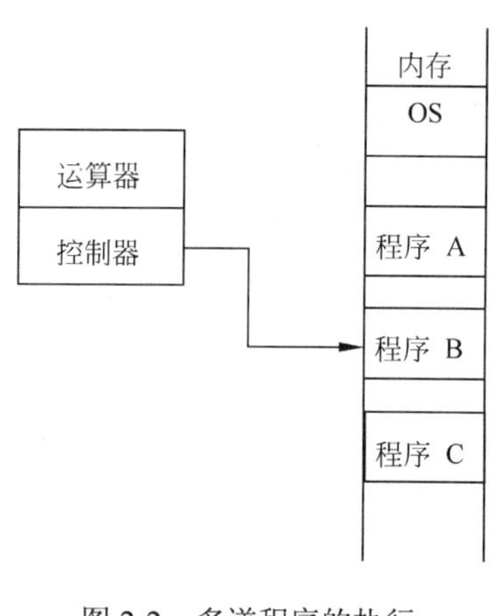

图 2-2　多道程序的执行

随着操作系统的发展，调度算法越来越复杂，由单队列的调度发展成基于优先级的多级反馈队列调度。这种调度算法设置多个就绪队列，并为各个队列赋予不同的优先级，第一个队列的优先级最高，第二个队列次之，其余各队列的优先权逐个降低，各队列中进程执行时间片的大小各不相同，优先权愈高的队列为每个进程所规定的执行时间片就愈越长。

2.2 进程的概念

2.2.1 进程概念的引入

多道程序执行时，每个运行的应用程序都需要放入内存，在运行过程中，它们需要共享系统资源，从而导致各程序在执行过程中出现相互制约的关系，程序的执行表现出间断性的特征。这些特征都是在程序的执行过程中发生的，是动态的过程。

程序本身是一组指令的集合，是一个静态的概念，无法描述程序在内存中的执行过程的含义，程序这个静态概念已不能准确地反映程序执行过程的特征。操作系统控制、管理、调度这些处于运行状态的应用程序，这时程序的意义已经发生变化，处于运行状态的程序，是操作系统管理调度的对象，是系统分配资源的基本单位。为了深刻描述程序动态执行过程的性质，人们引入进程（Process）概念。

应用程序提交，被操作系统接纳，操作系统就为其创建一个进程，操作系统为这个运行过程中的程序分配它所需要的系统资源，控制它的运行过程直到程序运行结束，操作系统撤销进程。所以说进程有一个从被创建产生到运行结束而消亡的过程，进程是一个动态的概念。

如果一个程序被提交多次，对于操作系统而言，它对应多个进程，如图 2-3 所示，程序 A 被提交两次，它分别对应进程 1、进程 3。在这种情况下，程序的概念已经不能满足操作系统控制和管理系统中运行程序的需要，因此引入进程的概念是必需的。

进程是操作系统的核心，所有基于多道程序设计的操作系统都是建立在进程的概念之上。目前的计算机操作系统均提供了多任务并行环境。无论是应用程序还是系统程序。都需要针对每一个任务创建相应的进程。

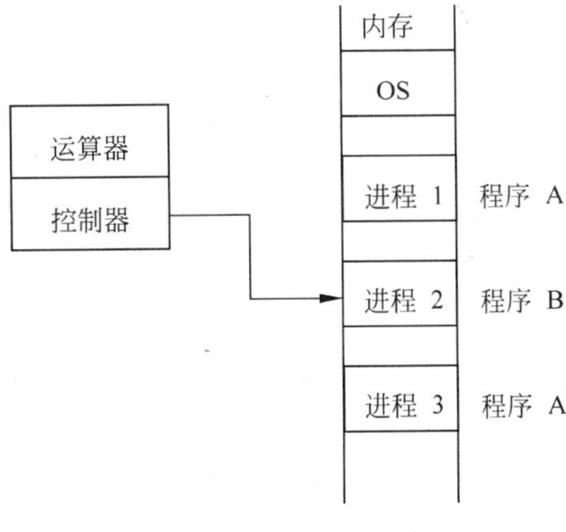

图 2-3　程序与进程

2.2.2　进程的概念

进程是一个具有独立功能的程序关于某个数据集合的一次运行活动。它可以申请和拥有系统资源，是一个动态的概念，是一个活动的实体。它不只是程序的代码，还包括当前的活动，通过程序计数器的值和寄存器的内容来表示。

进程是程序在一个数据集合上的一次运行活动，是操作系统进行资源分配和调度的一个基本单位。

学习进程的概念要注意以下三点。① 进程是一个实体。每一个进程都有它自己的地址空间，一般情况下，包括文本区域（Text Region)、数据区域（Data Region）和堆栈（Stack Region)。文本区域存储处理器执行的代码；数据区域存储变量和进程执行期间使用的动态分配的内存；堆栈区域存储着活动过程调用的指令和本地变量。② 进程是一个“执行中的程序”。程序是一个没有生命的实体，只有处理器赋予程序生命时，它才能成为一个活动的实体，称为进程。③ 它是操作系统管理、调度和分配资源的基本单位。

进程是操作系统中最基本、重要的概念。是多道程序系统出现后，为了刻画系统内部出现的动态情况，描述系统内部各道程序的活动规律引进的一个概念，所有多道程序设计操作系统都建立在进程的基础上。

2.2.3　进程与程序的关系

程序是有序代码的集合，通常对应着文件，可以复制，其本身没有任何运行的含义，是一个静态的概念。而进程是程序在处理机上的一次执行过程，它是一个动态的概念。

进程有被创建、生存到退出消亡的过程，是有一定生命周期的。程序是静态的，可以作为一种软件资料长期存在，是永久的，无生命的。

进程更能真实地描述并发，而程序不能。

进程是由进程控制块、程序段、数据段三部分组成的。

进程具有创建其他进程的功能，而程序没有。

同一程序同时运行于若干个数据集合上，它将属于若干个不同的进程。也就是说同一程序可以对应多个进程。通过调用关系，一个进程也可以包括多个程序。

在操作系统中，程序并不能独立运行，作为资源分配、调度和独立运行的基本单位都是进程。

2.2.4 进程的特征

结构特征：进程由程序、数据和进程控制块三部分组成。

动态性：进程的实质是程序在多道程序系统中的一次执行过程，进程是动态产生、动态消亡的。

并发性：任何进程都可以同其他进程一起并发执行。

独立性：进程是一个能独立运行的基本单位，也是系统分配资源和调度的独立单位。

异步性：由于进程间的相互制约，使进程具有执行的间断性，即进程按各自独立的、不可预知的速度向前推进。

2.3 进程控制块

2.3.1 进程控制块概述

编写程序、应用计算机运行程序的过程就是对数据进行加工的过程，加工首先需要清楚地了解被加工的对象，就是通过一套数据结构来描述加工对象的各种信息，其次需要知道加工处理的过程，就是我们常说的算法。所以说程序就是数据结构加算法。

操作系统管理、调度系统中的进程就是操作系统程序加工的对象。针对这个加工对象需要一套清楚的、全面的数据结构描述，这套数据结构就是进程控制块（PCB），即详细描述系统进程信息的数据结构叫做进程控制块（PCB）。

2.3.2 进程控制块的内容

在不同的操作系统中对进程的控制和管理机制不同，PCB 中的信息多少也不一样，通常 PCB 应包含如下一些信息。

（1）进程标识符 name：每个进程都必须有一个唯一的标识符，可以是字符串，也可以是一个数字。

（2）进程当前状态 status：说明进程当前所处的状态。为了管理方便，系统设计时会将相同状态的进程组成一个队列，如就绪进程队列。等待进程则要根据等待的事件组成多个等待队列，如等待打印机队列、等待磁盘 I/O 完成队列等。

（3）进程相应的程序和数据地址：把 PCB 与其程序和数据联系起来。

（4）进程资源清单：列出所拥有的除 CPU 外的资源记录，如拥有的 I/O 设备，打开的文件列表等。

（5）进程优先级（Priority）：进程的优先级反映进程的紧迫程度，通常由用户指定和系统设置。

（6）CPU 现场保护区（Cpustatus）：当进程因某种原因不能继续占用 CPU 时（如等待打印机），释放 CPU，这时就要将 CPU 的各种状态信息保护起来，为将来再次得到处理机恢复 CPU 的各种状态，继续运行。

（7）用于实现进程间互斥、同步和通信所需的信号量。

（8）进程所在队列 PCB 的链接字，它指出该进程所在队列中下一个进程 PCB 的首地址。

（9）与进程有关的其他信息。如进程记账信息，进程占用 CPU 的时间等。

2.3.3 Linux 的进程控制块

在 Linux 中每一个进程都由 task_struct 数据结构来定义。task_struct 就是我们所说的 PCB。

```
struct task_struct {
   long state;  /*任务运行状态（-1不可运行，0可运行(就绪)，>0已停止）*/
   long counter; /*运行时间片计数器(递减)*/
   long priority; /*优先级*/
   long signal;  /*信号*/
   struct sigaction sigaction[32];/ *信号执行属性结构，对应信号将要执行的操作和标
                                    息*/
   long blocked; /* bitmap of masked signals */
   int exit_code; /*任务执行停止的退出码*/
   unsigned long start_code,end_code,end_data,brk,start_stack;
/*代码段地址 代码长度 代码长度+数据长度 总长度 堆栈段地址*/
    long pid,father,pgrp,session,leader;/*进程标识号(进程号) 父进程号 父进程组
                                          会话号 会话首领*/
   unsigned short uid,euid,suid;/ *用户标识号（id） 有效用户id 保存的用户id*/
   unsigned short gid,egid,sgid; /*组标识号（组id） 有效组id 保存的组id*/
   long alarm;  /*报警定时值*/
   long utime,stime,cutime,cstime,start_time;/ * 运行时间：用户态内核态 子进程
                                    用户态，子进程内核态 及进程开始运行时刻*/
   unsigned short used_math;/ *标志：是否使用协处理器*/
   int tty;     /* -1 if no tty, so it must be signed */
   unsigned short umask; /*文件创建属性屏蔽位*/
   struct m_inode * pwd; /*当前工作目录i 节点结构*/
   struct m_inode * root; /*根目录i节点结构*/
   struct m_inode * executable;  /*执行文件i节点结构*/
   unsigned long close_on_exec; /*执行时关闭文件句柄位图标志*/
   struct file * filp[NR_OPEN];  /*进程使用的文件表结构*/
   struct desc_struct ldt[3];     /*本任务的局部描述符表。*/
   struct tss_struct tss;        /*本进程的任务状态段信息结构*/
};
```

2.3.4 进程控制块的组织方式

1．线性表方式

不论进程的状态如何，将所有的 PCB 连续地存放在内存的系统区。这种方式适用于系统中进程数目不多的情况，如图 2-4 所示。

1	PCB1
2	PCB2
3	PCB3
4	PCB4
5	PCB5
6	PCB6
7	PCB7

图 2-4　线性表方式

2．索引表方式

该方式是线性表方式的改进，系统按照进程的状态分别建立就绪索引表、阻塞索引表等。并把各索引表在内存的首地址记录在内存的一些专用单元中。在每个索引表的表目中记录具有相应状态的某个 PCB 在 PCB 表中的地址。如图 2-5 所示，采用索引表方式的 PCB 块的组织结构。

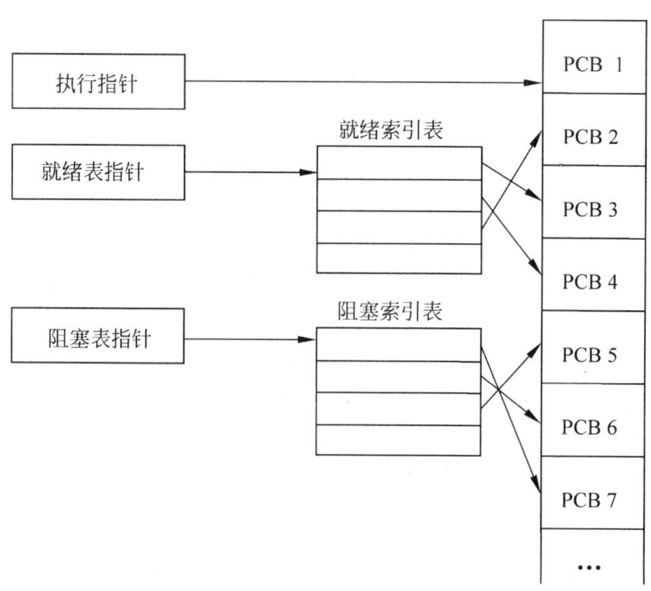

图 2-5　索引表方式

3．链接表方式

系统按照进程的状态将进程的 PCB 组成队列，从而形成就绪队列、阻塞队列、空白队列、运行进程指针等。图 2-6 给出了一种链接表方式的 PCB 组织方式。

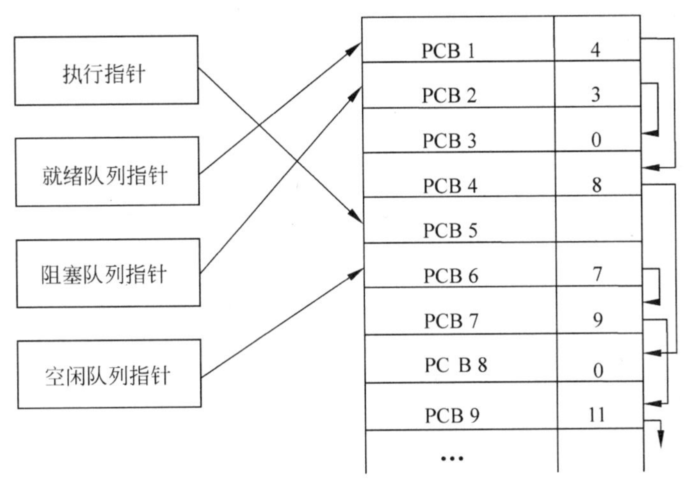

图 2-6　链接表方式

其中的就绪队列通常按进程优先级的高低排列，把高优先级的进程的 PCB 排在队列前面。另外，阻塞队列也可以根据阻塞原因，把阻塞的进程排在不同的队列，例如分成等待 I/O 的队列和等待分配内存的队列等。

2.4　进程的状态

在分时系统中，多道程序被提交给系统接纳运行，这种运行是宏观上的运行。在微观上，它们以时间片为单位，轮流使用 CPU，使各进程向前推进执行。由于这种轮转较快，在宏观上，每个用户都会看到自己的程序在连续地推进执行，感到好像自己独占 CPU 一样，可实际上，它们是走走停停，交替前进的。

2.4.1　进程基本状态

一个进程的生命期可以划分为一组状态，这些状态刻划了整个进程。进程基本状态如图 2-7 所示。系统根据 PCB 结构中的状态值控制进程。进程的基本状态有三种，即就绪状态、执行状态、阻塞状态。

1．就绪状态（Ready）

进程已获得除处理器外的所需资源，等待分配处理器资源；只要分配了处理器进程就可执行。就绪进程可以按多个优先级来划分队列。例如，当一个进程由于时间片用完而进入就绪状态时，排入低优先级队列；当进程由 I/O 操作完成而进入就绪状态时，排入高优先级队列。

2．运行状态（Running）

进程占用处理器资源；处于此状态的进程的数目小于等于处理器的数目。在没有其他进程可以执行时(如所有进程都在阻塞状态)，通常会自动执行系统的空闲进程。

3．阻塞状态（Blocked）

由于进程等待某种条件（如 I/O 操作或进程同步），在条件满足之前无法继续执行。该事件发生前即使把处理机分配给该进程，也无法运行。

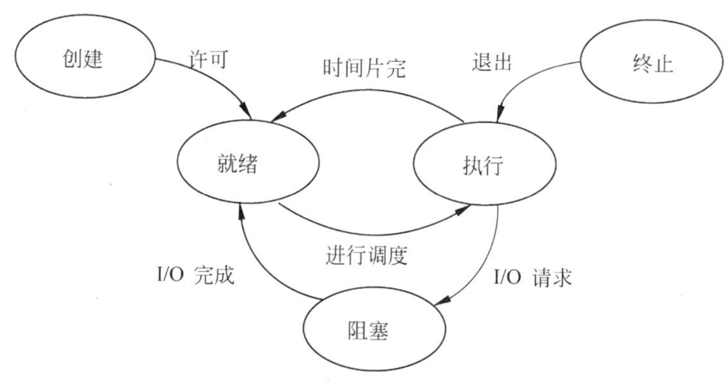

图 2-7　进程基本状态

2.4.2　进程基本状态的转换

进程的状态反映进程执行进程的变化。这些状态随着进程的执行和外界条件发生变化和转换。进程被创建后，已经具备了运行的条件，只要获得 CPU 就可以运行，所以它首先进入的是就绪状态，等待进程调度程序调度；一旦被调度到（分配 CPU），获得 CPU，就可以执行程序，处于执行状态；在执行状态，如果时间片用完，操作系统通过一个时钟中断，使它停下来，把它的状态再置为就绪状态；在执行状态如果进程申请 I/O 操作，或者需要等待某种资源，才能继续向前推进，这时进程转入阻塞状态，不参与进程的调度；当它等待的 I/O 操作完成了，或者它等待的资源具备了，进程转入就绪状态。

进程状态的转换，需要调用操作系统中最基本的功能操作，这样操作是不可被中断的，这就引入了一个新的概念，原子操作——指不会被更高级的中断机制打断的操作。这种操作一旦开始，就一直运行到结束，中间不会有任何进程打断。我们把操作系统的这种原子操作。也叫做原语。进程调度程序阻塞和唤醒进程是通过阻塞原语 block( )和唤醒原语 wakeup( )实现的。

1．进程阻塞过程

正在执行的进程，当发现上述某事件时，由于无法继续执行，于是进程便通过调用阻塞原语 block 把自己阻塞。可见，进程的阻塞是进程自身的一种主动行为。进入 block 过程后，由于此时该进程还处于执行状态，所以应先立即停止执行，把进程控制块中的现行状态由执行改为阻塞，并将 PCB 插入阻塞队列。如果系统中设置了因不同事件而阻塞的多个阻塞队列，则应将本进程插入到具有相同事件的阻塞（等待）队列。最后，转调度程序进行重新调度，将处理机分配给另一就绪进程，并进行切换，亦即，保留被阻塞进程的处理机状态（在 PCB 中），再按新进程的 PCB 中的处理机状态设置 CPU 的环境。

2．进程唤醒过程

当被阻塞进程所期待的事件出现时，如 I/O 完成或其所期待的数据已经到达，则由有关进程（例如，用完并释放了该 I/O 设备的进程）调用唤醒原语 wakeup()，将等待该事件的进程唤醒。唤醒原语执行的过程是：首先把被阻塞的进程从等待该事件的阻塞队列中移出，将其 PCB 中的现行状态由阻塞改为就绪，然后再将该 PCB 插入到就绪队列中。

2.4.3 带挂起的进程状态

进程的阻塞是因为需要等待资源或者等待某一事件，是客观的等待。有时我们需要人为地要求进程暂时停下来，有时操作系统由于某种原因也可能要求某一进程暂停。这种主动地让进程停下，不在参与进程调度的状态，我们称作挂起状态。

引起挂起状态的原因有如下几方面。

（1）终端用户的请求。当终端用户在自己的程序运行期间发现有可疑问题时，希望暂停使自己的程序静止下来。亦即，使正在执行的进程暂停执行；若此时用户进程正处于就绪状态而未执行，则该进程暂不接受调度，以便用户研究其执行情况或对程序进行修改。我们把这种静止状态成为“挂起状态”。

（2）父进程的请求。有时父进程希望挂起自己的某个子进程，以便考察和修改子进程，或者协调各子进程间的活动。

（3）负荷调节的需要。当实时系统中的工作负荷较重，已可能影响到对实时任务的控制时，可由系统把一些不重要的进程挂起，以保证系统能正常运行。

（4）操作系统的需要。操作系统有时希望挂起某些进程，以便检查运行中的资源使用情况或进行记账。

（5）对换的需要。为了缓和内存紧张的情况，将内存中处于阻塞状态的进程换至外存。

被挂起的进程可能是就绪的，也可能是阻塞的。处于就绪状态，并且是挂起的，我们称为静止就绪状态，原来就绪状态可以称为动态就绪状态；处于阻塞状态，并且是挂起的，我们称为静止阻塞状态，原来阻塞状态可以称为动态阻塞状态；带挂起的进程状态转换如图 2-8 所示。

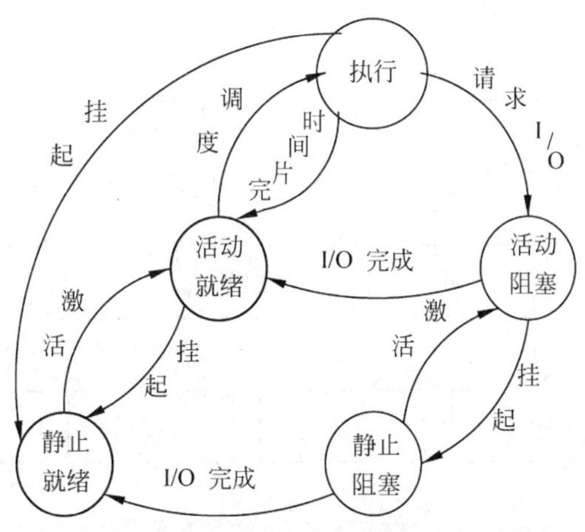

图 2-8　带挂起的进程状态

1．进程的挂起

当出现了引起进程挂起的事件时，例如，用户进程请求将自己挂起，或父进程请求将自己的某个子进程挂起，系统将利用挂起原语 suspend()将指定进程或处于阻塞状态的进程挂起。挂起原语的执行过程是：首先检查被挂起进程的状态，若处于活动就绪状态，便将其改为静止就绪；对于活动阻塞状态的进程，则将之改为静止阻塞。为了方便用户或父进程考查该进程的运行情况而把该进程的 PCB 复制到某指定的内存区域。最后，若被挂起的进程正在执行，则转向调度程序重新调度。

2．进程的激活过程

当发生激活进程的事件时，例如，父进程或用户进程请求激活指定进程，若该进程驻留在外存而内存中有足够的空间时，则可将在外存上处于静止就绪状态的进程换入内存。这时，系统将利用激活原语 active()将指定进程激活。激活原语先将进程从外存调入内存，检查该进程的现行状态，若是静止就绪，便将之改为活动就绪；若为静止阻塞便将之改为活动阻塞。假如采用的是抢占调度策略，则每当有新进程进入就绪队列时，应检查是否要重新调度，由调度程序将被激活进程与当前进程进行优先级的比较，如果被激活进程的优先级更低，就不必重新调度；否则，立即剥夺当前进程的运行，把处理机分配给刚被激活的进程。

2.5 进 程 控 制

2.5.1 进程的创建

1．进程树

进程的创建过程是操作系统重要的处理过程之一。在系统中运行的进程都是由进程创建出来的，创建进程的进程与被创建的进程之间构成父子关系，子进程还可以再创建它的子进程，由此可以构成一棵进程树，如图 2-9 所示。

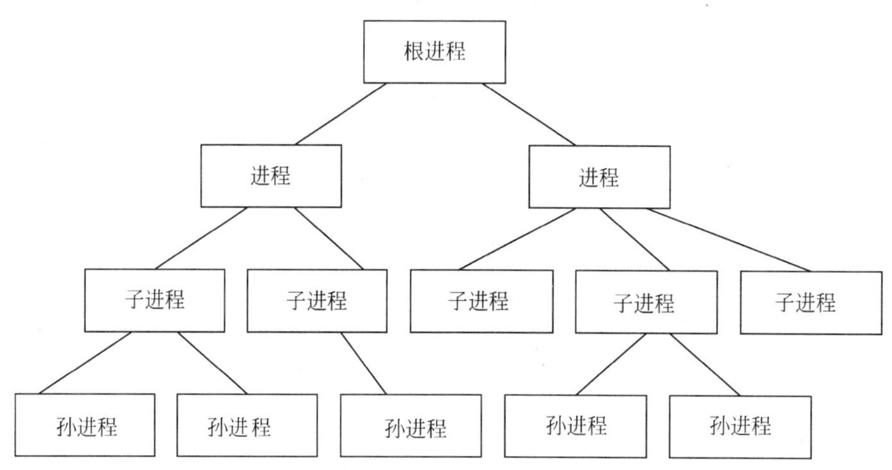

图 2-9　进程树

操作系统中的作业管理程序，可以接受用户提交的作业，经过判断，如果可以接纳，就为其创建一个进程，使用户程序得以运行，这个用户程序对应的进程是一个用户进程，它是操作系统作业管理进程的子进程。用户进程也可以通过系统调用创建再创建自己的子进程，所以创建进程的进程可以是操作系统进程也可以是用户进程。

在 Linux 系统中，可以通过 fork()（或 clone()）系统调用可用来创建子进程，调用 fork()

函数，系统找到一个空闲进程资源（find_empty_process），然后复制（copy_process）父进程相关信息，再分配内存及其他资源，这个函数实现创建子进程是基于父进程的复制创建的，fork 创建子进程，除了子进程标识符和其 PCB 结构中的某些特性参数不同之外，子进程是父进程的精确复制，父、子进程的运行是无关的，所以运行顺序也不固定，若要求父子进程运行顺序一定，则要用到进程间的通信。另外一个系统调用是 exec()，通过调用 exec()，子进程可以拥有自己的可执行代码。即 exec()用一个新进程覆盖调用进程，它的参数包括新进程对应的文件和命令行参数，成功调用时，不再返回。在大多数程序中，系统调用 fork()和 exec()是结合在一起使用的，父进程生成一个子进程，然后通过调用 exec()覆盖该子进程。子进程在创建好后并不能立即执行，至少需要一次调度，调度到子进程时，执行任务的切换过程，子进程才能执行。

2．根进程

子进程的创建是基于父进程的，因此一直追溯上去，总有一个进程是原始的，是没有父进程的，这个进程是根进程。根进程没有可以复制和参考的对象，它所拥有的所有信息和资源都是强制设置的，不是复制的，这个过程称为手工设置，也就是说根进程是“纯手工打造”的，它是操作系统中“最原始”的一个进程，它是一个模子，后面的任何进程都是基于根进程生成的。

在 Linux 系统中的这个根进程的进程号是 0，我们也把它叫做 0 号进程。创造 0 号进程最主要包括两个部分：一是创建进程 0 运行时所需的所有信息，即填充 0 号进程；二是调度 0 号进程的执行，即让它“动”起来，只有动起来，才是真正意义上的进程，符合进程动态的概念。

3．引起创建进程的事件

在多道程序环境中，只有进程才能在系统中运行。因此，为使程序能运行，就必须为它创建进程。导致一个进程去创建另一个进程的典型事件，可以有以下四类。

1）用户登录

在分时系统中，用户在终端键入登录命令后，如果是合法用户，系统将为该终端建立一个进程，并把它插入到就绪队列中。

2）作业调度

在批处理系统中，当作业调度程序按照一定的算法调度到某作业时，便将该作业装入到内存，为它分配必要的资源，并立即为它创建进程，再插入到就绪队列中。

3）提供服务

当运行中的用户程序提出某种请求后，系统将专门创建一个进程来提供用户所需要的服务，例如，用户程序要求进行文件打印，操作系统将为它创建一个打印进程，这样，不仅可以使打印进程与该用户进程并发执行，而且还便于计算出为完成打印任务所花费的时间。

4）应用请求

在上述三种情况中，都是由系统内核为它创建一个新进程，而这一类事件则是基于应用进程的需求，由它创建一个新的进程，以便使新进程以并发的运行方式完成特定任务。

4．进程的创建过程（Creation of Progress）

一旦操作系统发现了要求创建新进程的事件后，便调用进程创建原语 creat()按下述步骤创建一个新进程。

（1）申请空白 PCB。为新进程申请获得唯一的数字标识符，并从 PCB 集合中索取一个空白 PCB。

（2）为新进程分配资源。为新进程的程序和数据以及用户栈分配必要的内存空间。显然，此时操作系统必须知道新进程所需要的内存大小。

（3）初始化进程控制块。PCB 的初始化包括：①初始化标识信息，将系统分配的标识符和父进程标识符，填入新的 PCB 中；②初始化处理机状态信息，使程序计数器指向程序的入口地址，使栈指针指向栈顶；③初始化处理机控制信息，将进程的状态设置为就绪状态或静止就绪状态，对于优先级，通常是将它设置为最低优先级，除非用户以显式的方式提出高优先级要求。

（4）将新进程插入就绪队列。如果进程就绪队列能够接纳新进程，便将新进程插入到就绪队列中。

2.5.2 进程的终止

1．引起进程终止（Termination of Process）的事件

1）正常结束

在任何计算机系统中，都应有一个用于表示进程已经运行完成的指示。例如，在批处理系统中，通常在程序的最后安排一条 HALT 指令或终止的系统调用。当程序运行到 HALT 指令时，将产生一个中断，去通知 OS 本进程已经完成。 在分时系统中，用户可利用 Logs off 去表示进程运行完毕， 此时同样可产生一个中断，去通知 OS 进程已运行完毕。

2）异常结束

在进程运行期间，由于出现某些错误和故障而迫使进程终止。这类异常事件很多，常见的有：①越界错误，这是指程序所访问的存储区，已越出该进程的区域；②保护错，进程试图去访问一个不允许访问的资源或文件，或者以不适当的方式进行访问，例如，进程试图去写一个只读文件；③非法指令，程序试图去执行一条不存在的指令，出现该错误的原因，可能是程序错误地转移到数据区，把数据当成了指令，④特权指令错，用户进程试图去执行一条只允许 OS 执行的指令，⑤运行超时，进程的执行时间超过了指定的最大值；⑥等待超时，进程等待某事件的时间，超过了规定的最大值；⑦算术运算错，进程试图去执行一个被禁止的运算，例如，被 0 除；⑧I/O 故障。

3）外界干预

外界干预并非指在本进程运行中出现了异常事件，而是指进程应外界的请求而终止运行。这些干预有：①操作员或操作系统干预。由于某种原因，例如，发生了死锁，由操作员或操作系统终止该进程；②父进程请求。由于父进程具有终止自己的任何子孙进程的权利，因而当父进程提出请求时，系统将终止该进程；③父进程终止。当父进程终止时，OS 也将他的所有子孙进程终止。

2．进程的终止过程

（1）根据被终止进程的标识符，从 PCB 集合中检索出该进程的 PCB，从中读出该进程的状态。

（2）若被终止进程正处于执行状态，应立即终止该进程的执行，并置调度标志为真，用于指示该进程被终止后应重新进行调度。

（3）若该进程还有子孙进程，还应将其所有子孙进程予以终止，以防他们成为不可控的进程。

（4）将被终止进程所拥有的全部资源，或者归还给其父进程，或者归还给系统。

（5）将被终止进程（它的 PCB）从所在队列（或链表）中移出，等待其他程序搜集信息。

2.6 处理机调度

2.6.1 处理机调度的层次

一个应用程序作为一个作业提交给操作系统，希望它在计算机上运行结束，它是怎样一步一步地被接收运行到结束退出的呢？

完成一个独立任务的程序及其所需的数据组成一个作业。作业管理就是对用户提交的诸多作业进行管理，包括作业的组织、控制和调度等。作业管理的目标就是尽可能高效地利用整个系统的资源。

为了管理和调度作业，在支持多道程序运行的操作系统中为每个作业设置了一个作业控制块（Job Control Block，JCB）。作业控制块是描述作业全部信息的数据结构。如同进程控制块是进程在系统中存在的标志一样，它是作业在系统中存在的标志，其中保存了系统对作业进行管理和调度所需的全部信息。

作业控制块是作业在系统中存在的唯一标志，即一个作业控制块对应一个作业。操作系统根据作业控制块了解作业的情况，同时又利用作业控制块控制作业的运行。

操作系统向用户提供一组作业控制语言（或者是操作系统命令），用户用这种语言书写作业说明书（用户提交操作系统命令），然后将程序、数据和作业说明书一齐交给系统操作员。作业说明书在系统中生成一个称为作业控制块的表格，从而，操作系统通过该表了解到作业要求，并分配资源和控制作业中程序和数据的编译、链接、装入和执行等。

作业的提交首先进入操作系统的后备作业队列，操作系统的作业调度进程负责处理、判断，符合条件的作业被接纳，为其创建进程，送入进程的就绪队列；进程在就绪队列中等待进程调度程序的调度，经历进程的各种状态转换，获得了足够的 CPU 时间后，程序执行完毕，进程管理程序收回分配给它的所有资源，注销进程，程序运行结束退出。

1．高级调度（High Scheduling）

作业调度与进程调度是相互配合来实现多道作业的并行执行的。两者的关系如图 2-10 所示。

在这里我们看到，作业至少经历了作业调度、进程调度两个层级的调度过程，才能在计算机中运行结束。我们把这里的作业调度称为高级调度，把进程调度称为低级调度。

高级调度指的是作业调度，即根据作业控制块中的信息，审查系统能否满足用户作业的资源需求，以及按照一定的策略、算法，从后备队列中选取某些作业调入内存，并为它们创建进程、分配必要的资源。然后再将新创建的进程插入就绪队列，准备执行。在每次执行作业调度时，都须做出以下两个决定：

（1）接纳多少个作业；

（2）接纳哪些作业。

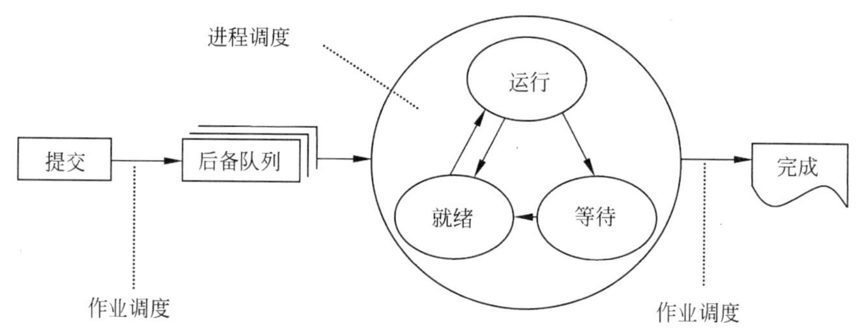

图 2-10　作业调度与进程调度

2．低级调度（Low Level Scheduling）

低级调度是指进程调度。被作业调度所接纳的进程，宏观上看都是处于运行状态了，但是 CPU 只有一个，这些进程是以时间片为单位轮流来使用 CPU 的。每一个时刻只能有一个进程使用 CPU，处于实际的执行状态。处于执行状态的进程怎样停下来，就绪状态的进程是怎样获得 CPU，执行它的指令的呢？这就是一个进程切换的过程。

进程切换就是从正在运行的进程中收回 CPU，然后再使就绪状态的进程来占用 CPU。收回 CPU，实质上就是把进程当前在 CPU 的寄存器中的中间数据找个地方存起来（保护现场），从而把 CPU 的寄存器腾出来让其他进程使用。那么被中止运行进程的中间数据存在进程的私有堆栈。

按照一定的调度算法，从就绪队列中选择一个进程来占用 CPU，实质上是把进程存放在私有堆栈中寄存器的数据（前一次本进程被中止时的中间数据）再恢复到 CPU 的寄存器中去（恢复现场），并把待运行进程的断点送入 CPU 的程序计数器 PC 中，于是这个进程就开始被 CPU 运行，也就是这个进程已经占有 CPU 的使用权了。

这就像多个同学要分时使用同一张课桌一样，所谓要收回正在使用课桌同学的课桌使用权，实质上就是让他把属于他的东西拿走；而赋予某个同学课桌使用权，只不过就是让他把他的东西放到课桌上罢了。

在切换时，一个进程存储在处理器各寄存器中的中间数据叫做进程的上下文，所以进程的切换实质上就是被中止运行进程与待运行进程上下文的切换。在进程未占用处理器时，进程的上下文是存储在进程的私有堆栈中的。

进程调度的功能主要包括下面三个方面。

（1）保存处理机的现场信息，记住进程的状态，如进程名称、指令计数器、程序状态寄存器以及所有通用寄存器等现场信息，将这些信息记录在进程控制块的私有堆栈中。

（2）按某种算法从就绪队列中选取进程，即根据一定的进程调度算法，决定哪个进程能获得 CPU，以及占用多长时间。

（3）进程切换，即正在执行的进程因为时间片用完或因为某种原因不能再执行的时候，保存该进程的现场，并收回 CPU，并把 CPU 分配给选中的进程。

进程调度中，很重要的一项就是根据一定调度算法，从就绪队列中选出一个进程占用 CPU 运行。算法是处理机调度的关键。

3．中级调度（Intermediate-Level Scheduling）

中级调度又称中程调度（Medium-Term Scheduling）。它是一种带有挂起功能的调度方式，引入中级调度的主要目的，是为了提高内存利用率和系统吞吐量，使那些暂时不能运行的进程不再占用宝贵的内存资源，而将它们调至外存上去等待（挂起进程），把此时的进程状态称为驻外存状态或挂起状态。当这些进程重新具备运行条件、且内存又稍有空闲时，由中级调度来决定把外存上的哪些具备运行条件的静态就绪的进程，重新调入内存，并修改其状态为活动就绪状态，挂在活动就绪队列上等待进程调度。三级级调度的调度模型如图 2-11 所示。

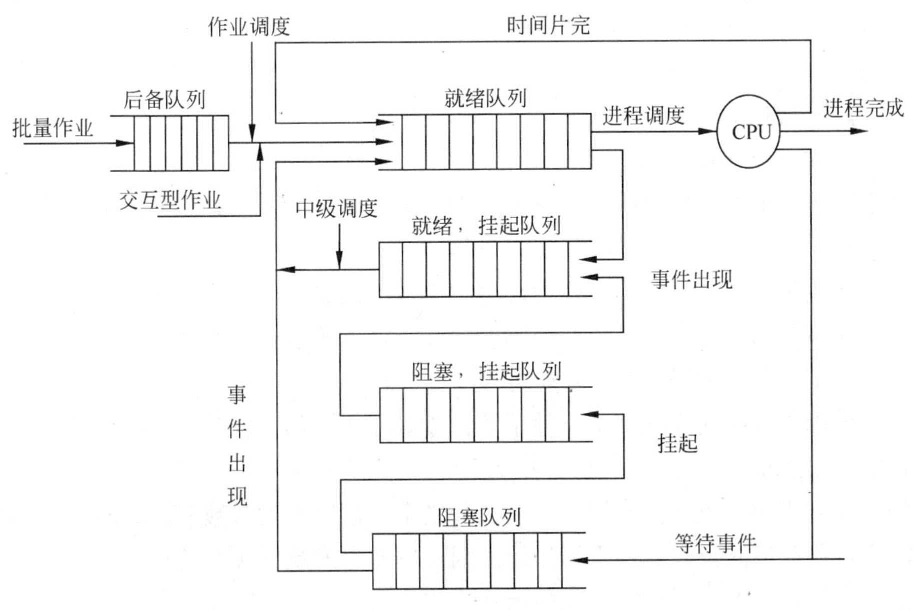

图 2-11　具有三级调度时的调度队列模型

2.6.2　进程调度的功能及实现方式

1．进程调度机制

进程调度模块是操作系统的核心模块，为实现进程的调度，操作系统需要设置如下三个基本的机制。

（1）排队，系统中就绪的进程可能有多个，就绪进程排成就绪队列可以方便调度程序调度。

（2）分派，进程调度程序按照一定的调度算法，从就绪队列中选择进程，选中的进程后，有分派程序把这个进程从就绪队列里取出来，做进程切换的准备。

（3）切换，进行进程切换，实质就是进程上下文的切换，操作系统先保护当前运行进程的上下文，进行现场保护，然后装入分派程序指定的进程城的上下文，使这个指定程序获得 CPU 控制权，并运行。

一个进程的上下文（Context）包括进程的状态、有关变量和数据结构的值、机器寄存器的值和 PCB 以及有关程序、数据等。一个进程的执行是在进程的上下文中执行。当正在执行的进程由于某种原因要让出处理机时，系统要做进程上下文切换，以使另一个进程得以执行。当进行上下文切换时点统要首先检查是否允许做上下文切换（在有些情况下，上下文切换是不允许的，例如系统正在执行某个不允许中断的原语时）。然后，系统要保留有关被切换进程的足够信息，以便以后切换回该进程时，顺利恢复该进程的执行。在系统保留了 CPU 现场之后，调度程序选择一个新的处于就绪状态的进程、并装配该进程的上下文，使 CPU 的控制权掌握在被选中进程手中。

进程调度程序实现这几种机制，需要上千条的指令，也是需要 CPU 时间的，所以，时间片太小，可能需要太多次的执行调度程序，CPU 的运行用户程序的效率就会降低，现代的计算中可以通过硬件的方法切换进程的上下文，节约进程调度时间，提高 CPU 运行用户程序的效率。

2．进程调度方式

1）非抢占方式（Non-preemptive Mode）

分派程序一旦把处理机分配给某进程后便让它一直运行下去，直到进程完成或发生某事件而阻塞时，才把处理机分配给另一个进程。

在采用非抢占调度方式时，可能引起进程调度的因素可归结为这样几个：

（1）正在执行的进程执行完毕，或因发生某事件而不能再继续执行；

（2）执行中的进程因提出 I/O 请求而暂停执行；

（3）在进程通信或同步过程中执行了某种原语操作，如 P 操作（wait 操作）、block 原语、wakeup 原语等。

这种调度方式的优点是实现简单、系统开销小，适用于大多数的批处理系统环境。但它难以满足紧急任务的要求——立即执行，因而可能造成难以预料的后果。显然，在要求比较严格的实时系统中，不宜采用这种调度方式。

2）抢占方式（Preemptive Mode）

当一个进程正在运行时，系统可以基于某种原则，剥夺已分配给它的处理机，将之分配给其他进程。剥夺原则有：优先权原则、短进程优先原则、时间片原则。

例如，有三个进程 P_1、P_2、P_3 先后到达，它们分别需要 20、4 和 2 个单位时间运行完毕。

假如它们按 P_1、P_2、P_3 的顺序执行，且不可剥夺，则三进程各自的周转时间分别为 20、24、26 个单位时间，平均周转时间是 23.33 个时间单位。

假如用时间片原则的剥夺调度方式，可得到：

可见：P_1、P_2、P_3的周转时间分别为26、10、6个单位时间(假设时间片为2个单位时间)，平均周转时间为14个单位时间。

2.7 调度算法

2.7.1 调度算法的性能评价准则

作业调度和进程调度性能的衡量方法可分为定性和定量两种。在定性衡量方面，首先是调度的可靠性。每一次调度不能引起数据结构的破坏等。这要求对调度时机的选择和保存CPU现场十分谨慎。另外，简洁性也是衡量调度的一个重要指标，如果调度程序过于繁琐和复杂，将会耗去较大的系统开销。这在用户进程调用系统调用较多的情况下，将会造成响应时间大幅度增加。

作业调度和进程调度的定量评价包括CPU的利用率评价、进程在就绪队列中的等待时间与执行时间的比率。定量分析可以从满足用户需求的角度定义指标，也可以从满足系统需求的角度定义指标。制定评价准则从用户的角度和从系统两个方面来讨论。

1．面向用户的准则

（1）周转时间。从用户角度看，作业被提交之后希望能够尽快地运行结束。我们就把一个作业提交给系统，系统接纳的时刻起，作业运行结束，这段时间间隔定义为周转时间。作业的周转时间越短，用户等待作业的时间就越短，用户的满意度就越高。所以我们可以把作业的周转时间作为调度算法的一个衡量指标，周转时间越短越好。

衡量系统的作业调度算法的好坏的第一个指标就可以采用作业的平均周转时间。设系统中有n个作业，第i（$1 \leqslant i \leqslant n$）个作业的周转时间为$T_i$，那么平均周转时间描述为：

$$T = \frac{1}{n}\left[\sum_{i=1}^{i} T_i\right]$$

（2）带权的周转时间。通过周转时间，可以看出用户等待时间的长短，但是只通过它来评价一个调度算法也不是很公平的。假设有两个作业，周转时间都为100秒，但作业1需要CPU的时间是99s，其他的等待时间是1s，作业2需要CPU的时间是1s，其他的等待时间是99s，这种情况下，好像对于作业2的用户并不公平。

因此，我们把作业的周转时间T与它需要CPU提供服务的时间T_s之比，即$W=T/T_s$，称为带权周转时间，而平均带权周转时间则可表示为：

$$W = \frac{1}{n}\left[\sum_{i=1}^{n} \frac{T_i}{T_{si}}\right]$$

（3）响应时间。指从作业提交到系统作出相应所经过的时间。在交互式系统中，作业的周转时间并不一定是最好的衡量准则，有时也使用另一种度量准则，即响应时间。从用户观点看，响应时间应该快一点好，但这常常要牺牲系统资源利用率为代价。

（4）优先权准则。作业周转时间和带权的作业周转时间可以反映用户的等待时间。但是，作业有的重要，有的不重要，有的着急有的不着急，轻重缓急各不相同，所以单纯地考虑等待时间的长短，也不见得是最好的方法。根据作业任务的轻重缓急，为每个作业设置不同的优先级别，作业调度算法能够尽量多地满足高优先级的作业要求，也是衡量作业调度算法优劣的评价准则。

2．面向系统的准则

设计作业调度算法，不单单看用户的要求，也应考虑系统的应用效率，从系统的角度出发，尽量使系统具有很高的工作效率。为满足系统的要求，主要应注意下面几个因素：

（1）CPU 利用率。CPU 是计算机系统中最重要的资源，所以应尽可能使 CPU 保持忙。

（2）吞吐量。工作量的大小以每单位时间所完成的作业数目来描述。

（3）各类资源的平衡利用。系统中不仅 CPU 的利用率要提高，也要考虑各类资源的平衡利用。使系统资源尽可能多的时间被利用，系统设备尽可能多的时间处于“忙”的状态，系统各种资源尽可能多地并行工作。

2.7.2 先来先服务调度算法

这是最简单、最朴素的处理机调度算法，其基本思想是按照进程进入就绪队列的先后顺序调度并分配处理机执行。先来先服务（FCFS）调度算法是一种不可抢占的算法，先进入就绪队列的进程，先分配处理机运行。一旦一个进程占有了处理机，它就一直运行下去，直到该进程完成工作或者因为等待某事件发生而不能继续运行时才释放处理机。

先来先服务算法简单，易于程序实现，但它性能较差，在实际运行的操作系统中，很少单独使用，它常常配合其他调度算法一起使用。

例 1 表 2-1 给出作业 1、2、3 的提交时间和运行时间。采用先来先服务调度算法，试问作业调度次序和平均周转时间各为多少？（时间单位：秒，以十进制进行计算。）

表 2-1 例 1 表

作业号	提交时间	运行时间
1	0.0	8.0
2	0.4	4.0
3	1.0	1.0

分析：解这样的题关键是要根据系统采用的调度算法，弄清系统中各道作业随时间的推进情况。用一个作业执行时间图来形象地表示作业的执行情况，帮助理解此题。

采用先来先服务调度算法，是按照作业提交的先后次序挑选作业，先进入的作业优先被挑选。然后按照“排队买票”的办法，依次选择作业。其作业执行时间如图 2-12 所示。

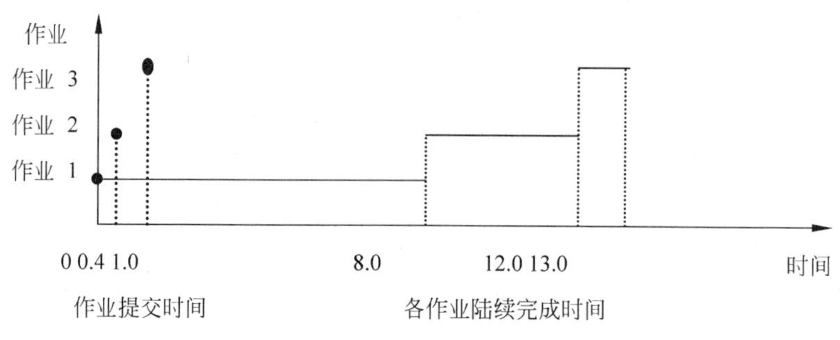

图 2-12 先来先服务调度算法

汇总这三个作业的开始运行时间、结束时间，统计它们的周转时间和带权的周转时间，计算它们的平均周转时间和带权的平均周转时间，如表 2-2 所示。

表 2-2　三个作业汇总各类时间表

作业	提交时间	运行时间	开始执行时间	完成时间	周转时间	带权的周转时间
作业 1	0	8.0	0	8.0	8.0	1
作业 2	0.4	4.0	8.0	12.0	11.6	2.9
作业 3	1.0	1.0	12.0	13.0	12.0	12.0
平均					10.53	5.3

2.7.3　短作业优先调度算法

先来先服务算法表面上看对所有作业都是公平的，并且一个作业的等待时间是可能预先估计的。但实际上这种算法是不利于小作业的，因为当一个大作业先进入就绪队列时，就会使其后的许多小作业等待很长的时间。这对小作业来说，等待时间可能要远远超出它运行的时间。使系统的吞吐量大为降低。

假设系统中有两个等待的作业，作业 1 需要 CPU 的时间是 99s，作业 2 需要 CPU 的时间是 1s，它们开始运行的次序不同，系统里总的周转时间的结果也不同。如表 2-3 所示。

表 2-3　作业 1、2 运行次序调换后的总周转时间

作业	开始时间	完成时间	周转时间	带权周转时间	作业	开始时间	完成时间	周转时间	带权周转时间
1	0	99	99	1	2	0	1	1	1
2	99	100	100	100	1	1	100	100	1.01
平均：			99.5	50.5	平均：			50.5	1.005

对于例 1 的三个作业，若采用短作业优先（SJF）调度算法，作业调度时根据作业的运行时间，优先选择计算时间短且资源能得满足的作业。其作业执行时间如图 2-13 所示。

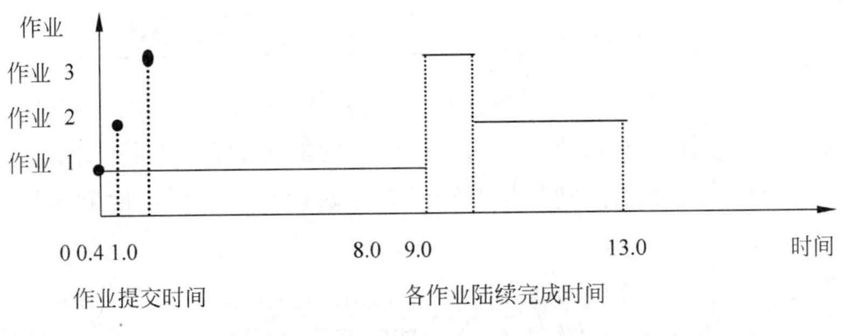

图 2-13　短作业优先调度算法

由于作业 1、2、3 是依次到来的，所以当开始时系统中只有作业 1，于是作业 1 先被选中。在 8.0 时刻，作业 1 运行完成，这时系统中有两道作业在等待调度，作业 2 和作业 3，按照短作业优先调度算法，作业 3 只要运行 1 个时间单位，而作业 2 要运行 4 个时间单位，于是作业 3 被优先选中，所以作业 3 先运行。待作业 3 运行完毕，最后运行作业 2。

作业调度的次序是 1，3，2。

汇总这三个作业的开始运行时间、结束时间，计算它们的平均周转时间和带权的平均周转时间，如表 2-4 所示。

表 2-4　三个作业汇总表

作业	提交时间	运行时间	开始执行时间	完成时间	周转时间	带权的周转时间
作业 1	0	8.0	0	8.0	8.0	1
作业 2	0.4	4.0	9.0	13.0	12.6	3.15
作业 3	1.0	1.0	8.0	9.0	8.0	8.0
平均					9.53	4.05

在系统的应用过程中，常常是先来先服务算法和短作业优先算法联合使用。一般情况下，按照先来先服务调度程序运行，当一个进程运行结束了，系统中有多个进程处于等待状态时，选择一个需要 CPU 时间较短的作业投入运行。

例 2　系统有 5 个进程 A、B、C、D、E，它们的到达时间分别是 0、1、2、3 和 4，它们需要的 CPU 服务时间分别是 4、3、5、2 和 4。采用先来先服务调度算法和短作业优先调度算法，试求系统的平均周转时间和带权的平均周转时间各为多少？

解：5 个进程的运行时间见表 2-5，求得平均周转时间分别为 9、8；带权的平均周转时间分别为 2.8 和 2.1。

表 2-5　例 2 表

调度算法 \ 作业情况	进程名	A	B	C	D	E	平　均
	到达时间	0	1	2	3	4	
	服务时间	4	3	5	2	4	
FCFS (a)	完成时间	4	7	12	14	18	
	周转时间	4	6	10	11	14	9
	带权周转时间	1	2	2	5.5	3.5	2.8
SJF (b)	完成时间	4	9	18	6	13	
	周转时间	4	8	16	3	9	8
	带权周转时间	1	2.67	3.1	1.5	2.25	2.1

短作业（进程）优先调度算法可以分别用于作业调度和进程调度。短作业优先的调度算法，是从后备队列中选择一个或若干个估计运行时间最短的作业，将它们调入内存运行。而短进程优先（SPF）调度算法，则是从就绪队列中选出一估计运行时间最短的进程，将处理机分配给它。

SJF 调度算法也存在不容忽视的缺点。

（1）该算法对长作业不利，如作业 C 的周转时间由 10 增至 16，其带权周转时间由 2 增至 3.1。更严重的是，如果有一长作业（进程）进入系统的后备队列（就绪队列），由于调度程序总是优先调度那些（即使是后进来的）短作业（进程），将导致长作业（进程）可能长期不被调度。

（2）该算法完全未考虑作业的紧迫程度，因而不能保证紧迫性作业（进程）会被及时处理。

（3）由于作业（进程）的长短只是根据用户所提供的估计执行时间而定的，而用户又可能会有意或无意地缩短其作业的估计运行时间，致使该算法不一定能真正做到短作业优先调度。

2.7.4 高优先权优先调度算法

1．优先权调度算法的类型

1）非抢占式优先权算法

在这种方式下，系统一旦把处理机分配给就绪队列中优先权最高的进程后，该进程便一直执行下去，直至完成；或因发生某事件使该进程放弃处理机时，系统方可再将处理机重新分配给另一优先权最高的进程。这种调度算法主要用于批处理系统中；也可用于某些对实时性要求不严的实时系统中。

2）抢占式优先权调度算法

在这种方式下，系统同样是把处理机分配给优先权最高的进程，使之执行。但在其执行期间，只要又出现了另一个其优先权更高的进程，进程调度程序就立即停止当前进程（原优先权最高的进程）的执行，重新将处理机分配给新到的优先权最高的进程。因此，在采用这种调度算法时，是每当系统中出现一个新的就绪进程 i 时，就将其优先权 P_i 与正在执行的进程 j 的优先权 P_j 进行比较。如果 $P_i \leq P_j$，则原进程 P_j 继续执行；但如果是 $P_i > P_j$，则立即停止 P_j 的执行，做进程切换，使 i 进程投入执行。显然，这种抢占式的优先权调度算法，能更好地满足紧迫作业的要求，故而常用于要求比较严格的实时系统中，以及对性能要求较高的批处理和分时系统中。

2．优先权的类型

1）静态优先权

静态优先权是在创建进程时确定的，且在进程的整个运行期间保持不变。一般地，优先权是利用某一范围内的一个整数来表示的，例如，0~7 或 0~255 中的某一整数，又把该整数称为优先数。只是具体用法各异：有的系统用 0 表示最高优先权，当数值愈大时，其优先权愈低；而有的系统恰恰相反。

确定进程优先权的依据有如下三个方面：

（1）进程类型；

（2）进程对资源的需求；

（3）用户要求。

2）动态优先权

动态优先权是指，在创建进程时所赋予的优先权，是可以随进程的推进或随其等待时间的增加而改变的，以便获得更好的调度性能。例如，我们可以规定，在就绪队列中的进程，随其等待时间的增长，其优先权以速率 a 提高。若所有的进程都具有相同的优先权初值，则显然是最先进入就绪队列的进程，将因其动态优先权变得最高而优先获得处理机，此即 FCFS 算法。若所有的就绪进程具有各不相同的优先权初值，那么，对于优先权初值低的进程，在等待了足够的时间后，其优先权便可能升为最高，从而可以获得处理机。当采用抢占式优先权调度算法时，如果再规定当前进程的优先权以速率 b 下降，则可防止一个长作业长期地垄断处理机。

3．高响应比优先调度算法

优先权的变化规律可描述为：

$$优先权 = \frac{等待时间 + 要求服务时间}{要求服务时间}$$

由于等待时间与服务时间之和，就是系统对该作业的响应时间，故该优先权又相当于响应比 RP。据此，又可表示为：

$$优先权 = \frac{等待时间 + 要求服务时间}{要求服务时间} = \frac{响应时间}{要求服务时间}$$

（1）如果作业的等待时间相同，则要求服务的时间愈短，其优先权愈高，因而该算法有利于短作业。

（2）当要求服务的时间相同时，作业的优先权决定于其等待时间，等待时间愈长，其优先权愈高，因而它实现的是先来先服务。

（3）对于长作业，作业的优先级可以随等待时间的增加而提高，当其等待时间足够长时，其优先级便可升到很高， 从而也可获得处理机。

2.7.5 基于时间片的轮转调度算法

1．时间片轮转法

时间片轮转（RR）调度是一种最古老，最简单，最公平且使用最广的算法。每个进程被分配一个时间段，称作它的时间片，即该进程允许运行的时间。如果在时间片结束时进程还在运行，则 CPU 将被剥夺并分配给另一个进程。如果进程在时间片结束前阻塞或结束，则 CPU 当即进行切换。

在早期的时间片轮转法中，系统将所有的就绪进程按先来先服务的原则，排成一个队列，每次调度时，把 CPU 分配给队首进程，并令其执行一个时间片。时间片的大小从几 ms 到几百 ms。当执行的时间片用完时，由一个计时器发出时钟中断请求，调度程序便据此信号来停止该进程的执行，并将它送往就绪队列的末尾；然后，再把处理机分配给就绪队列中新的队首进程，同时也让它执行一个时间片。这样就可以保证就绪队列中的所有进程，在一给定的时间内，均能获得一时间片的处理机执行时间。

时间片轮转调度中唯一有趣的一点是时间片的长度。从一个进程切换到另一个进程是需要一定时间的——保存和装入寄存器值及内存映像，更新各种表格和队列等。假如进程切换（Process Switch）——有时称为上下文切换（context switch），需要 5ms，再假设时间片设为 20ms，则在做完 20ms 有用的工作之后，CPU 将花费 5ms 来进行进程切换。CPU 时间的 20%被浪费在了管理开销上。

为了提高 CPU 效率，我们可以将时间片设为 500ms。这时浪费的时间只有 1%。但考虑在一个分时系统中，如果有 10 个交互用户几乎同时按下回车键，将发生什么情况？假设所有其他进程都用足它们的时间片的话，最后一个不幸的进程不得不等待 5s 钟才获得运行机会。多数用户无法忍受一条简短命令要 5s 钟才能做出响应。同样的问题在一台支持多道程序的个人计算机上也会发生。

结论可以归结如下：时间片设得太短会导致过多的进程切换，降低了 CPU 效率；而设得太长又可能引起对短的交互请求的响应变差。将时间片设为 100ms 通常是一个比较合理的折衷。

时间片轮转调度算法是一种既简单又有效的调度策略，它的基本思想是：对就绪队列中的每一进程分配一个时间片，时间片的长度 q 一般从 10~1100ms 不等。把就绪队列看成是一个环状结构，调度程序按时间片长度 q 轮流调度就绪队列中的每一进程，使每一进程都有机会获得相同长度的时间占用处理机运行。时间片轮转调度算法的性能极大的依赖于时间片长度 q 的取值，如果时间片过大。则 RR 算法就退化为 FIFO 算法了；反之，如果时间片过小，那么，处理机在各进程之间频繁转接，处理机时间开销变得很大，而提供给用户程序的时间将大大减少。

图 2-14 给出了时间片分别为 q=1 和 q=4 时，A、B、C、D、E 五个进程的运行情况。

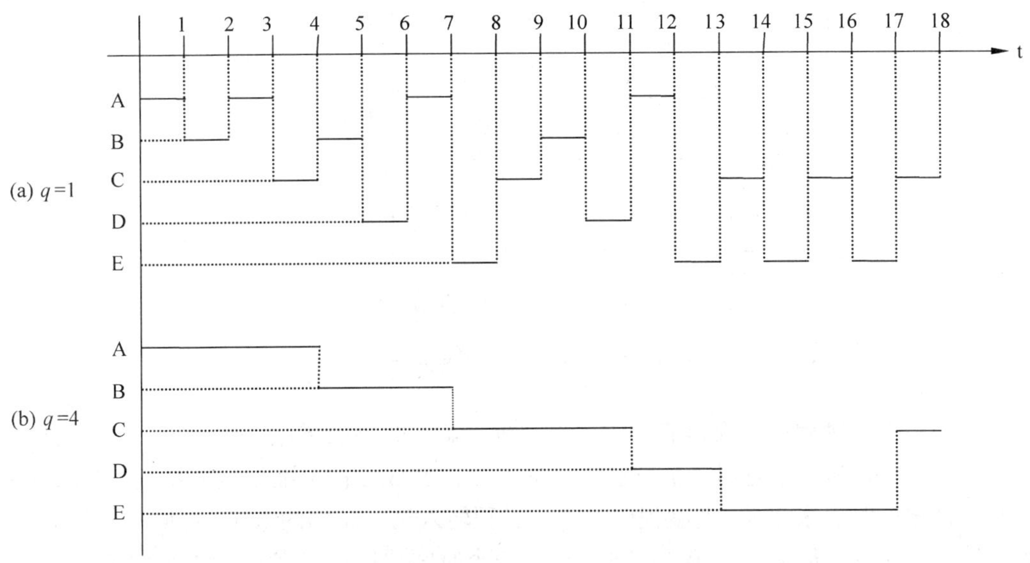

图 2-14　时间片轮转法 q=1 和 q=4 时进程运行情况

表 2-16 示出了时间片分别为 q=1 和 q=4 时，A、B、C、D、E 五个进程的平均周转时间和带权的平均周转时间。

表　2-6

作业情况 / 时间片	进程名	A	B	C	D	E	平均
	到达时间	0	1	2	3	4	
	服务时间	4	3	5	2	4	
RR　q=1	完成时间	12	10	18	11	17	
	周转时间	12	9	16	8	13	11.6
	带权周转时间	3	3	3.2	4	3.25	3.29
RR　q=4	完成时间	4	7	18	13	17	
	周转时间	4	6	16	10	13	9.8
	带权周转时间	1	2	3.2	5	3.25	2.89

2．多级反馈队列调度算法

（1）应设置多个就绪队列，并为各个队列赋予不同的优先级。 第一个队列的优先级最高，第二个队列次之，其余各队列的优先权逐个降低。该算法赋予各个队列中进程执行时间片的大小也各不相同，在优先权愈高的队列中，为每个进程所规定的执行时间片就愈小。例如，第二个队列的时间片要比第一个队列的时间片长一倍，……，第 i+1 个队列的时间片要比第 i 个队列的时间片长一倍。图 2-15 是多级反馈队列算法的示意。

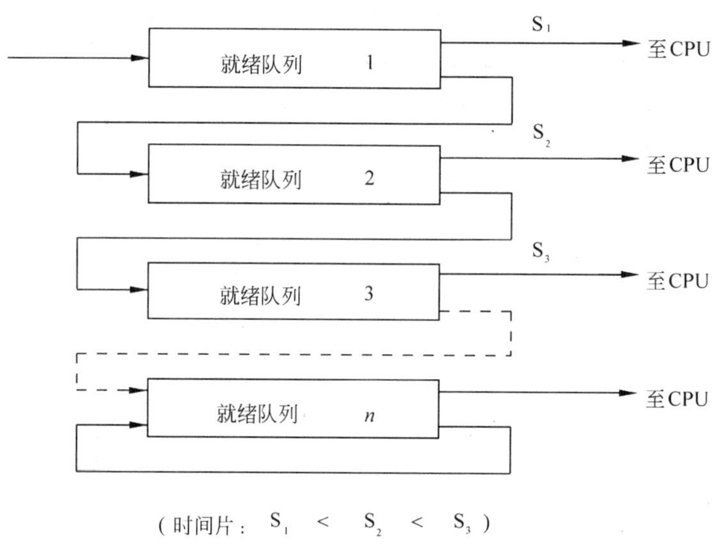

图 2-15 多级反馈队列调度算法

（2）当一个新进程进入内存后，首先将它放入第一队列的末尾，按 FCFS 原则排队等待调度。当轮到该进程执行时，如它能在该时间片内完成，便可准备撤离系统；如果它在一个时间片结束时尚未完成，调度程序便将该进程转入第二队列的末尾，再同样地按 FCFS 原则等待调度执行；如果它在第二队列中运行一个时间片后仍未完成，再依次将它放入第三队列，……如此下去，当一个长作业(进程)从第一队列依次降到第 n 队列后，在第 n 队列中便采取按时间片轮转的方式运行。

（3）仅当第一队列空闲时，调度程序才调度第二队列中的进程运行；仅当第 1~(i-1)队列均空时，才会调度第 i 队列中的进程运行。如果处理机正在第 i 队列中为某进程服务时，又有新进程进入优先权较高的队列（第 1~(i-1)中的任何一个队列），则此时新进程将抢占正在运行进程的处理机，即由调度程序把正在运行的进程放回到第 i 队列的末尾，把处理机分配给新到的高优先权进程。

2.8 实 时 调 度

实时调度是为了完成实时处理任务而分配计算机处理器的调度方法。实时处理任务要求计算机在用户允许的时限范围内给出计算机的响应信号。

实时系统根据其对于实时性要求的不同，可以分为软实时和硬实时两种类型。硬实时系统指要求计算机系统必须在用户给定的时限内完成，系统要有确保在最坏情况下的服务时间，即对于事件的响应时间的截止期限是无论如何都必须得到满足。例如航天中的宇宙飞船的控制等就是现实中这样的系统。

其他所有有实时特性的系统都可以称之为软实时系统。如果明确地来说，软实时系统就是那些从统计的角度来说，一个任务（在下面的论述中，我们将对任务和进程不作区分）能够得到有确保的处理时间，到达系统的事件也能够在截止期限到来之前得到处理，但违反截止期限并不会带来致命的错误，像实时多媒体系统就是一种软实时系统。

根据建立调度表和可调度性分析是脱机还是联机实现分为静态调度和动态调度，静态调度无论是单处理器调度还是分布式调度，一般是以单调速率调度（RMS）算法为基础；而动态调度则以最早截止时间优先算法（EDF）、最短空闲时间优先算法（LLF）为主。

单调速率调度（RMS）算法是单处理器下的最优静态调度算法。1973 年 Liu 和 Layland 发表文章首次提出了 RM 调度算法在静态调度中的最优性。它的一个特点是可通过对系统资源利用率的计算来进行任务可调度性分析，算法简单、有效，便于实现。不仅如此，他们还把系统的利用系数(Utilization Factor)和系统可调度性联系起来，推导出用 RM 调度所能达到的最小系统利用率公式。

最早截止时间优先算法（EDF）也称为截止时间驱动调度算法（DDS），是一种动态调度算法。EDF 指在调度时，任务的优先级根据任务的截止时间动态分配。截止时间越短，优先级越高。

最短空闲时间优先算法（LLF）也是一种动态调度算法。LLF 指在调度时刻，任务的优先级根据任务的空闲时间动态分配。空闲时间越短，优先级越高。空闲时间=deadline-任务剩余执行时间。LLF 可调度条件和 EDF 相同。

理论上，EDF 和 LLF 算法都是单处理器下的最优调度算法。但是由于 EDF 和 LLF 在每个调度时刻都要计算任务的截止时间或者空闲时间，并根据计算结果改变任务优先级，因此开销大、不易实现，其应用受到一定限制。

按系统分类，实时调度可以分为单处理器调度，集中式多处理器调度和分布式处理器调度。按任务是否可抢占又能分为抢占式调度和不可抢占式调度。

实时调度系统涉及几个基本概念。

（1）就绪时间：实时任务产生并可以开始处理的时间。

（2）开始截止时间：实时任务最迟开始处理的时间。

（3）处理时间：实时任务处理所需要的处理机的时间。

（4）完成截止时间：实时任务最迟完成时间。

（5）发生周期：周期性实时任务的发生间隔时间。

（6）优先级：实时任务相对紧迫程序。

2.8.1 实时调度的基本条件

1．系统处理能力

在实时系统中，通常都有着多个实时任务。若处理机的处理能力不够强，则有可能因处理机忙不过来而使某些实时任务不能得到及时处理，从而导致发生难以预料的后果。假定系统中有 m 个周期性的硬实时任务，它们的处理时间可表示为 C_i，周期时间表示为 P_i，则在单处理机情况下，必须满足下面的限制条件：

$$\sum_{i=1}^{m}\frac{C_i}{P_i}\leqslant 1$$

满足这个条件，系统才是可调度的。假如系统中有 6 个硬实时任务，它们的周期时间都是 50 毫秒，而每个任务的处理时间需要 10 毫秒，则不难算出，此时是不能满足上式的，因而系统是不可调度的。

解决的方法是提高系统的处理能力，其途径有二：其一仍是采用单处理机系统，但须增强其处理能力，以显著地减少对每一个任务的处理时间；其二是采用多处理机系统。假定系统中的处理机数为 N，则应将上述的限制条件改为：

$$\sum_{i=1}^{m}\frac{C_i}{P_i}\leqslant N$$

2．采用抢占式调度机制

当一个优先权更高的任务到达时，允许将当前任务暂时挂起，而令高优先权任务立即投入运行，这样便可满足该硬实时任务对截止时间的要求。但这种调度机制比较复杂。

对于一些小的实时系统，如果能预知任务的开始截止时间，则对实时任务的调度可采用非抢占调度机制，以简化调度程序和对任务调度时所花费的系统开销。但在设计这种调度机制时，应使所有的实时任务都比较小，并在执行完关键性程序和临界区后，能及时地将自己阻塞起来，以便释放出处理机，供调度程序去调度那种开始截止时间即将到达的任务。

3．具有快速切换机制

快速切换机制应具有如下两方面的能力。

（1）对外部中断的快速响应能力。为使在紧迫的外部事件请求中断时系统能及时响应，要求系统具有快速硬件中断机构，还应使禁止中断的时间间隔尽量短，以免耽误时机（其他紧迫任务）。

（2）快速的任务分派能力。在完成任务调度后，便应进行任务切换。为了提高分派程序进行任务切换时的速度，应使系统中的每个运行功能单位适当得小，以减少任务切换的时间开销。

4．实时调度算法的分类

1）非抢占式调度算法

非抢占式轮转调度算法为每一个被控对象建立一个实时任务并将它们排列成一轮转队列，调度程序每次选择队列中的第一个任务投入运行，该任务完成后便把它挂在轮转队列的队尾等待下次调度运行。

非抢占式优先调度算法。实时任务到达时，把它们安排在就绪队列的对首，等待当前任务自我终止或运行完成后才能被调度执行。

2）抢占式调度算法

基于时钟中断的抢占式优先权调度算法。实时任务到达后，如果该任务的优先级别高于当前任务的优先级并不立即抢占当前任务的处理机，而是等到时钟中断到来时，调度程序才剥夺当前任务的执行，将处理机分配给新到的高优先权任务。

立即抢占（Immediate Preemption）的优先权调度算法。在这种调度策略中，要求操作系统具有快速响应外部时间中断的能力。一旦出现外部中断，只要当前任务未处于临界区便立即剥夺当前任务的执行，把处理机分配给请求中断的紧迫任务。

非抢占式调度和抢占式的调度算法的调度方式如图 2-16 所示。

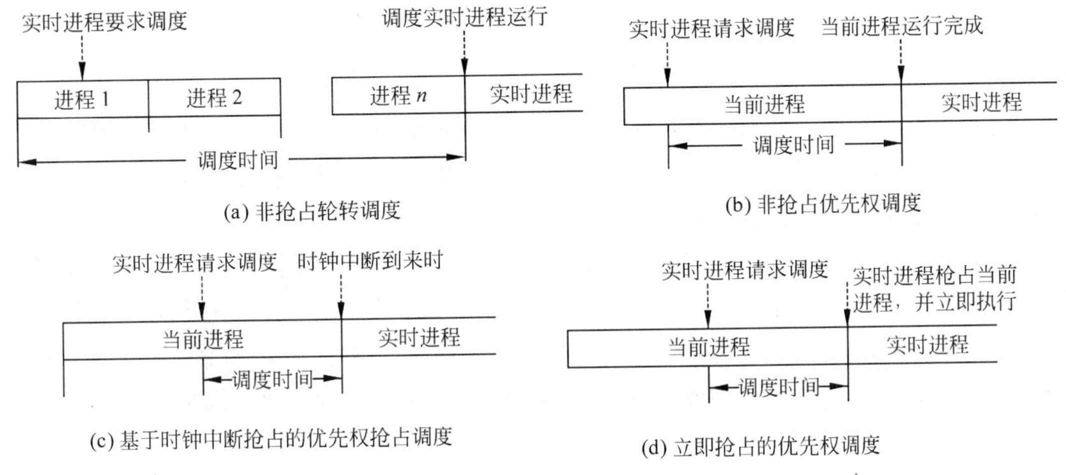

图 2-16　实时进程调度

2.8.2　实时调度算法

1．最早截止时间优先

最早截止时间优先算法（EDF）也称为截止时间驱动调度算法（DDS），即 EDF（Earliest Deadline First）算法 是一种动态调度算法。EDF 指在调度时，任务的优先级根据任务的截止时间动态分配。截止时间越短，优先级越高。

EDF 调度算法已被证明是动态最优调度，而且是充要条件。处理机利用率最大可达 100%，但瞬时过载时，系统行为不可预测，可能发生多米诺骨牌现象，一个任务丢失时会引起一连串的任务接连丢失。

如图 2-17 所示，EDF 算法用于非抢占调度方式，假设有 4 个实时任务，任务 1 首先到达，投入运行。然后任务 2 到达，任务 3 到达，任务 1 结束后，系统中有任务 2、任务 3 存在，任务 3 的最早截止时间比较早，所以调度任务 3 运行。运行期间任务 4 到达，任务 3 结束后，比较任务 2 与任务 4 的最早截止时间，任务 4 的最早截止时间较早，任务 4 先运行，任务 2 最后运行。

2．最低松弛度优先

该算法即 LLF（Least Laxity First）算法，也是一种动态调度算法，根据任务紧急（或松弛）的程度，来确定任务的优先级。在调度时刻，任务的优先级根据任务的空闲时间动态分配。空闲时间越短，优先级越高。空闲时间=截止时间–任务剩余执行时间。也就是说任务的紧急程度愈高，为该任务所赋予的优先级就愈高，以使之优先执行。

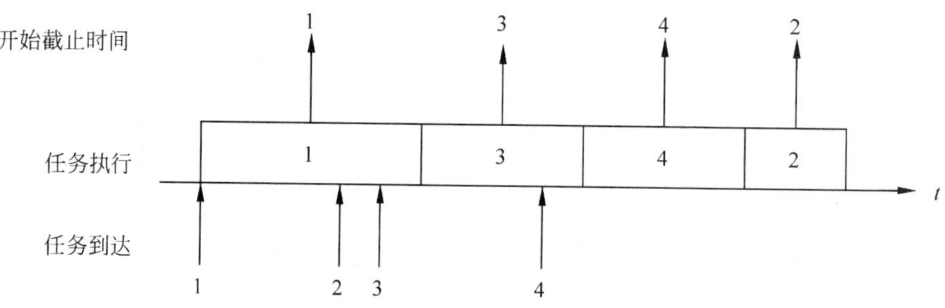

图 2-17　EDF 算法用于非抢占调度方式

例如，一个任务在 200ms 时必须完成，而它本身所需的运行时间是 100ms，因此，调度程序必须在 100ms 之前调度执行，该任务的紧急程度（松弛程度）为 100ms。另一任务在 400ms 时必须完成，它本身需要运行 150ms，则其松弛程度为 250ms。

在实现该算法时要求系统中有一个按松弛度排序的实时任务就绪队列，松弛度最低的任务排在队列最前面，调度程序总是选择就绪队列中的队首任务执行。该算法主要用于可抢占调度方式中。假如在一个实时系统中，有两个周期性实时任务 A 和 B，任务 A 要求每 20ms 执行一次，执行时间为 10ms；任务 B 只要求每 50ms 执行一次，执行时间为 25ms，如图 2-18 所示。

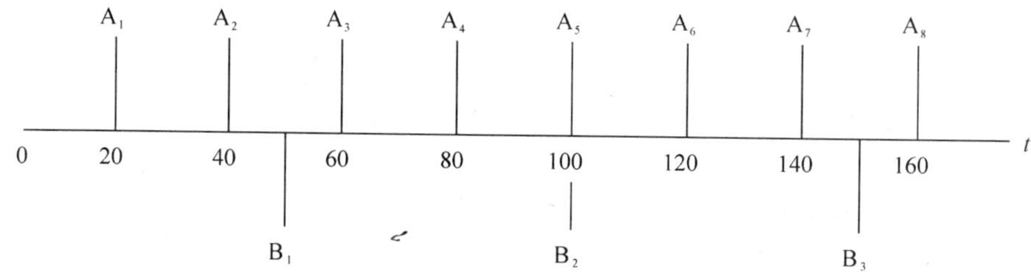

图 2-18　A 和 B 任务每次必须完成的时间

在刚开始时（t_1=0），A_1 必须在 20ms 时完成，而它本身运行又需 10ms，可算出 A_1 的松弛度为 10ms；B_1 必须在 50ms 时完成，而它本身运行就需 25ms，可算出 B_1 的松弛度为 25ms，故调度程序应先调度 A_1 执行。在 t_2=10 ms 时，A_2 的松弛度可按下式算出：

A_2 的松弛度=必须完成时间–其本身的运行时间–当前时间

=40ms–10ms–10ms

=20ms

类似地，可算出 B_1 的松弛度为 15ms，故调度程序应选择 B_1 运行。在 t_3=30ms 时，A_2 的松弛度已减为 0ms（即 40–10–30），而 B_1 的松弛度为 15ms（即 50–5–30），于是调度程序应抢占 B_1 的处理机而调度 A_2 运行。在 t_4=40ms 时，A_3 的松弛度为 10ms（即 60–10–40），

而 B_1 的松弛度仅为 5ms（即 50−5−40），故又应重新调度 B_1 执行。在 t_5=45ms 时，B_1 执行完成，而此时 A_3 的松弛度已减为 5ms（即 60−10−45），而 B_2 的松弛度为 30ms（即 100−25−45），于是又应调度 A_3 执行。在 t_6=55ms 时，任务 A 尚未进入第 4 周期，而任务 B 已进入第 2 周期，故再调度 B_2 执行。在 t_7=70ms 时，A_4 的松弛度已减至 0ms（即 80−10−70），而 B_2 的松弛度为 20ms（即 100−10−70），故此时调度又应抢占 B_2 的处理机而调度 A_4 执行，如图 2-19 所示。

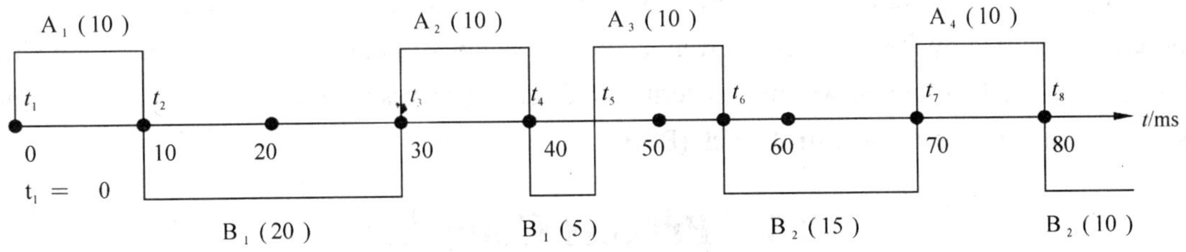

图 2-19 LLF 算法进行调度

Chapter 2 Process Management

A process is a program in execution. As a process executes, it changes state. The state of a process is defined by that process's current activity. Each process may be in one of the following states: new, ready, running, waiting, or terminated. Each process is represented in the operating system by its own process-control block (PCB).

2.1 Process Concepts

Early computer systems allowed only one program to be executed at a time. This program had complete control of the system and had access to all the system's resources. In contrast, current-day computer systems allow multiple programs to be loaded into memory and executed concurrently. This evolution required firmer control and more compartmentalization of the various programs; and these need resulted in the notion of a process, which is a program in execution. A process is the unit of work in a modern time-sharing system.

The more complex the operating system is, the more it is expected to do on behalf of its users. Although its main concern is the execution of user programs, it also needs to take care of various system tasks that are better left outside the kernel itself. A system therefore consists of a collection of processes: operating-system processes executing system code and user processes executing user code. Potentially all these processes can execute concurrently, with the CPU (or CPUs) multiplexed among them. By switching the CPU between processes, the operating system can make the computer more productive.

2.1.1 Process Concepts

A process is more than the program code, which is sometimes known as the text section. It also includes the current activity, as represented by the value of the program counter and the contents of the processor's registers. A process generally also includes the process stack, which contains temporary data (such as function parameters, return addresses, and local variables), and a data section, which contains global variables. A process may also include a heap, which is memory that is dynamically allocated during process run time. The structure of a process in memory is shown in Figure 2.1.

2.1.2 Process and Program

We emphasize that a program by itself is not a process; a program is a passive entity, such as a file containing a list of instructions stored on disk (often called an executable file), whereas a process is an active entity, with a program counter specifying the next instruction to execute and a set of associated resources. A program becomes a process when an executable file is loaded into memory. Two common techniques for loading executable files are double-clicking an icon representing the executable file and entering the name of the executable file on the command line (as in prog exe or a. out.)

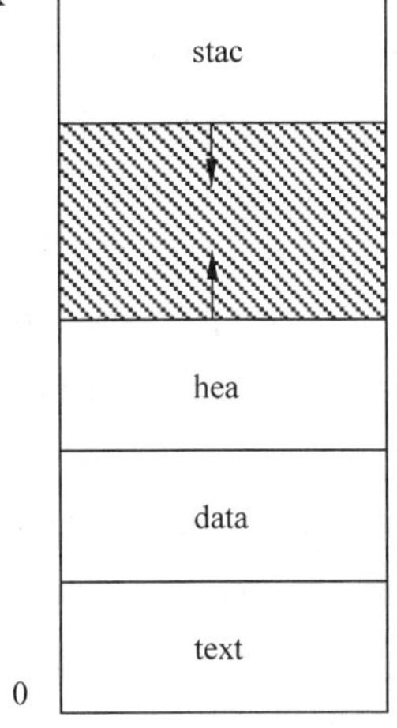

Figure 2.1 Process in memory

Although two processes may be associated with the same program, they are nevertheless considered two separate execution sequences. For instance, several users may be running different copies of the mail program, or the same user may invoke many copies of the web browser program. Each of these is a separate process; and although the text sections are equivalent, the data, heap, and stack sections vary. It is also common to have a process that spawns many processes as it runs.

2.2 Process Control Block

2.2.1 Process Control Block

If the mission of the operating system is to manage computing resources on behalf of processes, then it must be continuously informed about the status of each process and resource. The approach commonly followed to represent this information is to create and update status tables for each relevant entity, like memory, I/O devices, files and processes. Memory tables, for example, may contain information about the allocation of main and secondary (virtual) memory for each process, authorization attributes for accessing memory areas shared among different processes, etc. I/O tables may have entries stating the availability of a device or its assignment to a process, the status of I/O operations being executed, the location of memory buffers used for them, etc. File tables provide info about location and status of files.

Finally, process tables store the data the OS needs to manage processes. At least part of the process control data structure is always maintained in main memory, though its exact location and configuration varies with the OS and the memory management technique it uses. We'll refer by process image to the complete physical manifestation of a process, which includes instructions, program data areas (both static and dynamic - e.g. at least a stack for procedure calls and parameter passing) and the process management information. We'll call this last set the process control block (PCB).

2.2.2 Process Control Block Contents

The information in PCB is usually grouped into two categories: Process State Information and Process Control Information. Some of them are shown below.

(1) Process state. The state may be new, ready, running, waiting, halted, and so on.

(2) Program counter. The counter indicates the address of the next instruction to be executed for this process.

(3) CPU registers. The registers vary in number and type, depending on the computer architecture. They include accumulators, index registers, stack pointers, and general-purpose registers, plus any condition-code information.

(4) CPU-scheduling information. This information includes a process priority, pointers to scheduling queues, and any other scheduling parameters.

(5) Memory-management information. This information may include such information as the value of the base and limit registers, the page tables, or the segment tables, depending on the memory system used by the OS.

(6) Accounting information. This information includes the amount of CPU and real time used, time limits, account numbers, job or process numbers, and so on.

(7) I/O status information. This information includes the list of I/O devices allocated to the process, a list of open files, and so on.

2.3 Process State

2.3.1 Basic States of a Process

Basic States of a Process is shown in Figure 2.2.

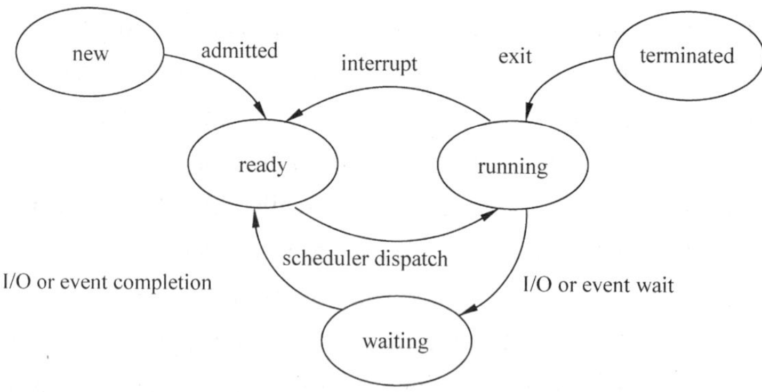

Figure 2.2 Basic states of a process

As a process executes, it changes state. The state of a process is defined in part by the current activity of that process. Each process may be in one of the following states.

(1) New. The process is being created.

(2) Running. Instructions are being executed.

(3) Waiting. The process is waiting for some event to occur (such as an I/O completion or reception of a signal).

(4) Ready. The process is waiting to be assigned to a processor.

(5) Terminated. The process has finished execution.

2.3.2 Process State Change

It's useful, at this point, to identify the possible transitions between process states. We can, for example draw a state transition diagram with four nodes, one for each of the so far identified states, plus one to point at a new process and one to a terminated process, and call admission the creation of a new process, which is putting into not-running state, dispatch its transition into running state, pause its transition from running into not-running, and exit the termination of a running process. This way of dealing with the two transitions which are not between running/not-running states, namely the admission and the exit, makes apparently sense, since a running process may be disposed immediately as a consequence of an error condition, without further need to pass it into not-running state; a new process must at least wait for the current process to be paused before gaining control of the CPU (here we obviously refer to a single processor architecture).

The new and terminated states are worth a bit of more explanation. The former refer to a process that has just been defined (e.g. because an user issued a command at a terminal), and for which the OS has performed the necessary housekeeping chores. The latter refers to a process whose task is not running anymore, but whose context is still being saved (e.g. because an user may want to inspect it using a debugger program).

A simple way to implement this process handling model in a multiprogramming OS would be to maintain a queue (i.e. a first-in-first-out linear data structure) of processes, put at the end the queue the current process when it must be paused, and run the first process in the queue.

However, it's easy to realize that this simple two-state model does not work. Given that the number of processes that an OS can manage is limited by the available resources, and that I/O events occur at much larger time scale that CPU events, it may well be the case that the first process of the queue must still wait for an I/O event before being able to restart; even worse, it may happen that most of the processes in the queue must wait for I/O. In this condition the scheduler would just waste its time sifting the queue in search of a runnable process.

A solution is to split the not-running process class according to two possible conditions: processes blocked in the wait for an I/O event to occur, and processes in pause, but nonetheless ready to run when given a chance. A process would then be put from running into blocked state on account of an event wait transition, would go running to ready state due to a timeout transition, and from blocked to ready due to event occurred transition.

This model would work fine if the OS had a very large amount of main memory available and none of the processes hogged too much of it, since in this case there would always be a fair number of ready processes. However, because the costs involved this scenario is hardly possible, and again the likely result is a list of blocked processes all waiting for I/O .

Modern architectures rely on a hierarchical organization of the available memory, only part of which exists in silicon on the machine board, while the rest is paged on a high speed storage peripheral like a hard disk. We will address this topic in detail later on; for now it will suffice to say that while the processor has the capability to address memory well above the amount of available RAM, it physically accedes the RAM only, and dedicated hardware is used to copy (or page) from the disk to RAM those pieces of memory which the processor refers to and are not available on RAM, and copy back areas of RAM onto disk when RAM space needs be made available. In particular, the act of copying entire processes back and forth from RAM to disk and vice versa is called swapping. As we'll see further on, swapping is often mandated by efficiency considerations even if paged memory is available, since the performance of a paged memory system can fall abruptly if too many active processes are maintained in main memory.

2.3.3 Process State with the Pending

Process state with the pending is shown in Figure 2.3.

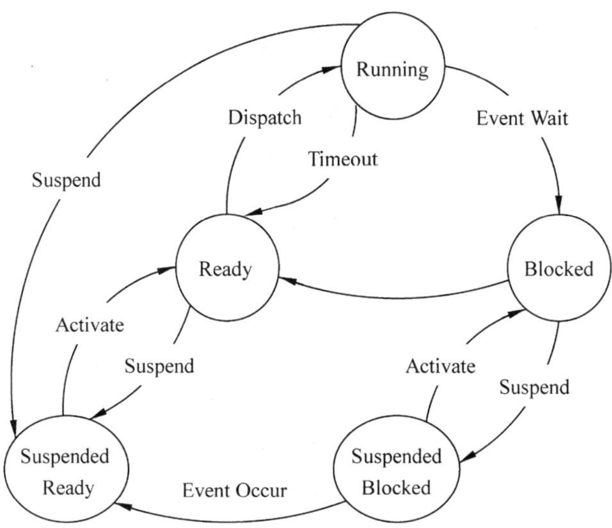

Figure 2.3 Process state with the pending

The OS could then perform a suspend transition on blocked processes, swapping them on disk and marking their state as suspended (after all, if they must wait for I/O, they might as well do it out of costly RAM), load into main memory a previously suspended process, activating into ready state and go on, However, swapping is an I/O operation in itself, and so at a first sight things might seem to get even worse doing this way. Again the solution is to carefully reconsider the reasons why processes are blocked and swapped, and recognize that if a process is blocked because it waits for I/O, and then suspended, the I/O event might occur while it swapped on the disk, We can thus classify suspended processes into two classes: ready-suspended for those suspended process whose restarting condition has occurred, and blocked-suspended for those who must still wait instead. This classification allows the OS to pick from the good pool of ready-suspended processes, when it wants to revive the queue in main memory. Provisions must be made for passing processes between the new states. In our model this means allowing for new transitions: activate and suspend between ready and ready-suspended, and between blocked-suspended and blocked as well, end event-occurred transitions from blocked to ready, and from blocks-suspended to ready-suspended as well.

2.4 Operations on Processes

The processes in most systems can execute concurrently, and they may be created and deleted dynamically. Thus, these systems must provide a mechanism for process creation and termination.

2.4.1 Process Creation

1. Process Tree

A process may create several new processes, via a create-process system call, during the course of execution. The creating process is called a parent process, and the new processes are called the children of that process. Each of these new processes may in turn create other processes, forming a tree of processes.

Most operating systems (including UNIX and the Windows family of operating systems) identify processes according to a unique process identifier (or pid), which is typically an integer number. Figure 2.4 illustrates a typical process tree for the Solaris operating system, showing the name of each process and its pid.

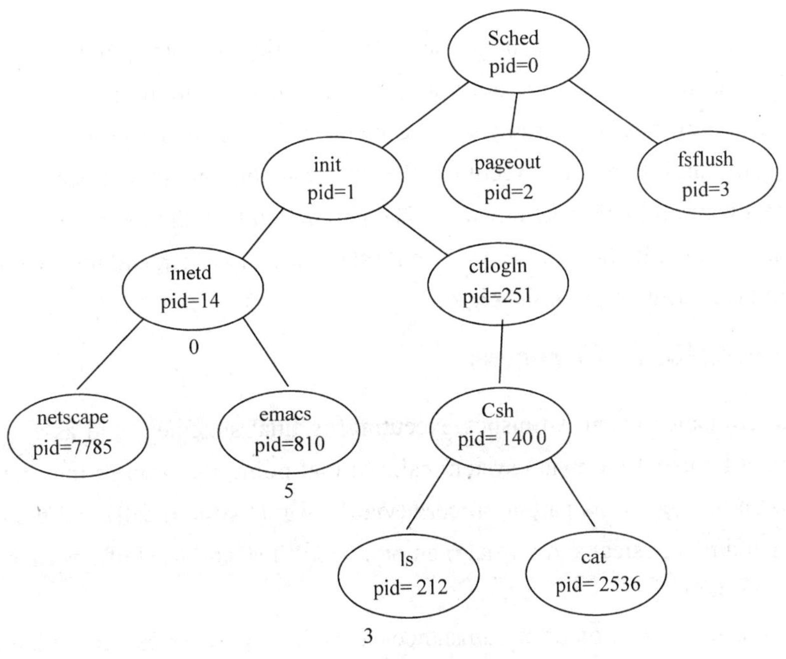

Figure 2.4 Process tree

2. Child Process

In addition to the various physical and logical resources that a process obtains when it is created, initialization data (input) may be passed along by the parent process to the child process. For example, consider a process whose function is to display the contents of a file—say, img.jpg— on the screen of a terminal. When it is created, it will get, as an input from its parent process, the name of the file img.jpg, and it will use that file name, open the file, and write the contents out. It may also get the name of the output device. Some operating systems pass resources to child processes. On such a system, the new process may get two open files, img.jpg and the terminal device, and may simply transfer the datum between the two.

When a process creates a new process, two possibilities exist in terms of execution:

(1) The parent continues to execute concurrently with its children;

(2) The parent waits until some or all of its children have terminated.

There are also two possibilities in terms of the address space of the new process:

(1) The child process is a duplicate of the parent process (it has the same program and data as the parent);

(2) The child process has a new program loaded into it.

3. Creation of Process

A new process is created by the fork() system call. The new process consists of a copy of the address space of the original process. This mechanism allows the parent process to communicate easily with its child process. Both processes (the parent and the child) continue execution at the instruction after the fork , with one difference: The return code for the fork() is zero for the new (child) process, whereas the (nonzero) process identifier of the child is returned to the parent.

Typically, the exec() system call is used after a fork() system call by one of the two processes to replace the process's memory space with a new program. The exec() system call loads a binary file into memory (destroying the memory image of the program containing the exec() system call) and starts its execution. In this manner, the two processes are able to communicate and then go their separate ways. The parent can then create more children; or, if it has nothing else to do while the child runs, it can issue a wait() system call to move itself off the ready queue until the termination of the child.

2.4.2 Termination of Process

A process terminates when it finishes executing its final statement and asks the operating system to delete it by using the exit() system call. At that point, the process may return a status value (typically an integer) to its parent process (via the wait() system call). All the resources of the process—including physical and virtual memory, open files, and I/O buffers—are deallocated by the operating system.

Termination can occur in other circumstances as well. A process can cause the termination of another process via an appropriate system call (for example, TerminateProcess() in Win32).

Usually, such a system call can be invoked only by the parent of the process that is to be terminated. Otherwise, users could arbitrarily kill each other's jobs. Note that a parent needs to know the identities of its children. Thus, when one process creates a new process, the identity of the newly created process is passed to the parent.

A parent may terminate the execution of one of its children for a variety of reasons, such as:

(1) The child has exceeded its usage of some of the resources that it has been allocated (To determine whether this has occurred, the parent must have a mechanism to inspect the state of its children);

(2) The task assigned to the child is no longer required;

(3) The parent is exiting, and the operating system does not allow a child to continue if its parent terminates.

Some systems, including VMS, do not allow a child to exist if its parent has terminated. In such systems, if a process terminates (either normally or abnormally), then all its children must also be terminated. This phenomenon, referred to as cascading termination, is normally initiated by the operating system.

To illustrate process execution and termination, consider that, in UNIX, we can terminate a process by using the exit() system call; its parent process may wait for the termination of a child process by using the wait() system call. The wait() system call returns the process identifier of a terminated child so that the parent can tell which of its possibly many children has terminated. If the parent terminates, however, all its children have assigned as their new parent the unit process. Thus, the children still have a parent to collect their status and execution statistics.

2.5 CPU Scheduling

The objective of multiprogramming is to have some process running at all times, to maximize CPU utilization. The objective of time sharing is to switch the CPU among processes so frequently that users can interact with each program while it is running. To meet these objectives, the process scheduler selects an available process (possibly from a set of several available processes) for program execution on the CPU. For a single-processor system, there will never be more than one running process. If there are more processes, the rest will have to wait until the CPU is free and can be rescheduled.

2.5.1 Queues and Schedulers

As processes enter the system, they are put into a job queue, which consists of all processes in the system. The processes that are residing in main memory and are ready and waiting to execute are kept on a list called the ready queue.

The system also includes other queues. When a process is allocated the CPU, it executes for a while and eventually quits, is interrupted, or waits for the occurrence of a particular event, such as the completion of an I/O request.

A common representation for a discussion of process scheduling is a queueing diagram, such as that in Figure 2.5. Each rectangular box represents a queue. Two types of queues are present: the ready queue and a set of device queues. The circles represent the resources that serve the queues, and the arrows indicate the flow of processes in the system.

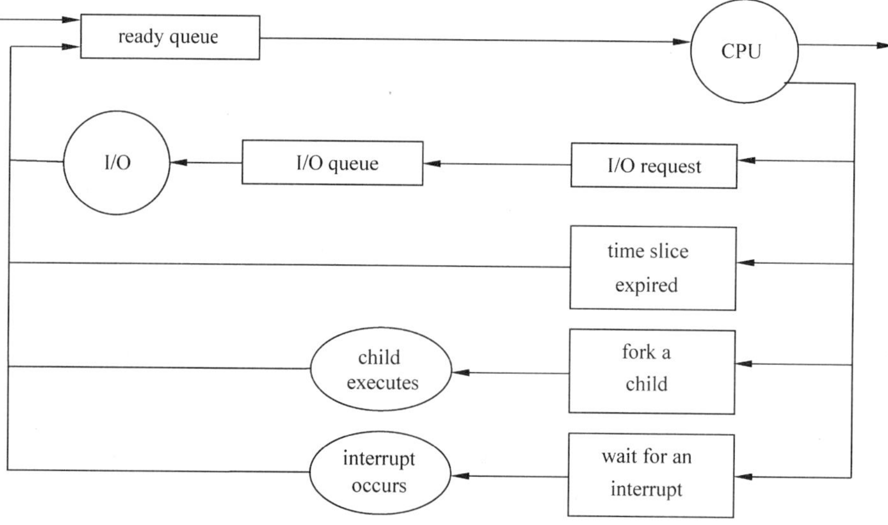

Figure 2.5 Queueing-diagram representation of process scheduling

A new process is initially put in the ready queue. It waits there until it is selected for execution, or is dispatched. Once the process is allocated the CPU and is executing, one of several events could occur:

(1) The process could issue an I/O request and then be placed in an I/O queue;

(2) The process could create a new subprocess and wait for the subprocess's termination;

(3) The process could be removed forcibly from the CPU, as a result of an interrupt, and be put back in the ready queue.

In the first two cases, the process eventually switches from the waiting state to the ready state and is then put back in the ready queue. A process continues this cycle until it terminates, at which time it is removed from all queues and has its PCB and resources deallocated.

A process migrates among the various scheduling queues throughout its lifetime. The operating system must select, for scheduling purposes, processes from these queues in some fashion. The selection process is carried out by the appropriate scheduler.

Often, in a batch system, more processes are submitted can be executed immediately. These processes are spooled to a mass-storage device (typically a disk), where they are kept for later execution.

The long-term scheduler, or job scheduler, selects processes from this pool and loads them into memory for execution. The short-term scheduler, or CPU scheduler, selects from among the processes that are ready to execute and allocates the CPU to one of them.

The primary distinction between these two schedulers lies in frequency of execution. The short-term scheduler must select a new process for the CPU frequently. A process may execute for only a few milliseconds before waiting for an I/O request. Often, the short-term scheduler executes at least once every 100 milliseconds. Because of the short time between executions, the short-term scheduler must be fast. If it takes 10 milliseconds to decide to execute a process for 100 milliseconds, then 10/(100 + 10) = 9 percent of the CPU is being used (wasted) simply for scheduling the work.

The long-term scheduler executes much less frequently; minutes may separate the creation of one new process and the next. The long-term scheduler controls the degree of multiprogramming (the number of processes in memory). If the degree of multiprogramming is stable, then the average rate of process creation must be equal to the average departure rate of processes leaving the system. Thus, the long-term scheduler may need to be invoked only when a process leaves the system. Because of the longer interval between executions, the long-term scheduler can afford to take more time to decide which process should be selected for execution.

It is important that the long-term scheduler makes a careful selection. In general, most processes can be described as either I/O bound or CPU bound. An I/O-bound process is one that spends more of its time doing I/O than it spends doing computations. A CPU-bound process, in contrast, generates I/O requests infrequently, using more of its time doing computations. It is important that the long-term scheduler selects a good process mix of I/O-bound and CPU-bound processes. If all processes are I/O bound, the ready queue will almost always be empty, and the short-term scheduler will have little to do. If all processes are CPU bound, the I/O waiting queue will almost always be empty, devices will go unused, and again the system will be unbalanced. The system with the best performance will thus have a combination of CPU-bound and I/O-bound processes.

Some operating systems, such as time-sharing systems, may introduce an additional, intermediate level of scheduling. This medium-term scheduler is diagrammed in Figure 2.6.

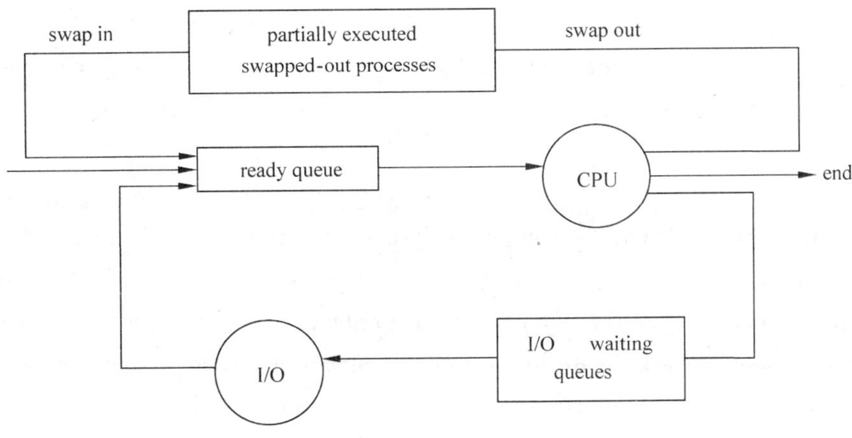

Figure 2.6 Addition of medium-term scheduling to the queueing diagram

The key idea behind a medium-term scheduler is that sometimes it can be advantageous to remove processes from memory (and from active contention for the CPU) and thus reduce the degree of multiprogramming. Later, the process can be reintroduced into memory, and its execution can be continued where it left off. This scheme is called swapping. The process is swapped out, and is later swapped in, by the medium-term scheduler. Swapping may be necessary to improve the process mix or because a change in memory requirements has overcommitted available memory, requiring memory to be freed up.

2.5.2 Schedule Criteria

Different CPU scheduling algorithms have different properties, and the choice of a particular algorithm may favor one class of processes over another. In choosing which algorithm to use in a particular situation, we must consider the properties of the various algorithms.

Many criteria have been suggested for comparing CPU scheduling algorithms. Which characteristics are used for comparison can make a substantial difference in which algorithm is judged to be best. The criteria include the following.

(1) CPU utilization. We want to keep the CPU as busy as possible. Conceptually, CPU utilization can range from 0 to 100 percents. In a real system, it should range from 40 percent (for a lightly loaded system) to 90 percents (for a heavily used system).

(2) Throughput. If the CPU is busy executing processes, then work is being done. One measure of work is the number of processes that are completed per time unit, called *throughput*. For long processes, this rate may be one process per hour; for short transactions, it may be 10 processes per second.

(3) Turnaround time. From the point of view of a particular process, the important criterion is how long it takes to execute that process. The interval from the time of submission of a process to the time of completion is the *turnaround time*. Turnaround time is the sum of the periods spent waiting to get into memory, waiting in the ready queue, executing on the CPU, and doing I/O.

(4) Waiting time. The CPU scheduling algorithm does not affect the amount of time during which a process executes or does I/O; it affects only the amount of time that a process spends waiting in the ready queue. *Waiting time* is the sum of the periods spent waiting in the ready queue.

(5) Response time. In an interactive system, turnaround time may not be the best criterion. Often, a process can produce some output fairly early and can continue computing new results while previous results are being output to the user. Thus, another measure is the time from the submission of a request until the first response is produced. This measure, called *response time*, is the time it takes to start responding, not the time it takes to output the response. The turnaround time is generally limited by the speed of the output device.

It is desirable to maximize CPU utilization and throughput and to minimize turnaround time, waiting time, and response time. In most cases, we optimize the average measure. However, under some circumstances, it is desirable to optimize the minimum or maximum values rather than the average. For example, to guarantee that all users get good service, we may want to minimize the maximum response time.

2.5.3 First-Come, First-Served Scheduling

By far the simplest CPU-scheduling algorithm is the first-come, first-served (FCFS) scheduling algorithm. With this scheme, the process that requests the CPU first is allocated the CPU first. The implementation of the FCFS policy is easily managed with a FIFO queue. When a process enters the ready queue, its PCB is linked onto the tail of the queue. When the CPU is free, it is allocated to the process at the head of the queue. The running process is then removed from the queue. The code for FCFS scheduling is simple to write and understand.

The average waiting time under the FCFS policy, however, is often quite long. Consider the following set of processes that arrive at time 0, with the length of the CPU burst given in milliseconds(Table 2.1).

Table 2.1

Process	Burst Time
P_1	24
P_2	3
P_3	3

If the processes arrive in the order P_1, P_2, P_3, and are served in FCFS order, we get the result shown in the following Gantt chart:

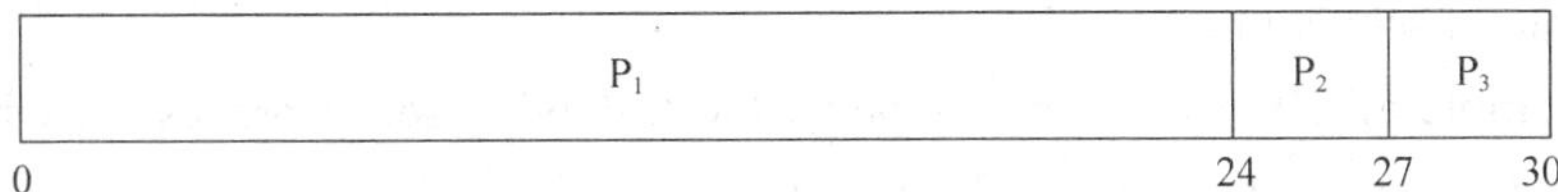

The waiting time is 0 milliseconds for process P_1, 24 milliseconds for process P_2, and 27 milliseconds for process P_3. Thus, the average waiting time is $(0 + 24 + 27)/3 = 17$ milliseconds. If the processes arrive in the order P_2, P_3, P_1, however, the results will be as shown in the following Gantt chart:

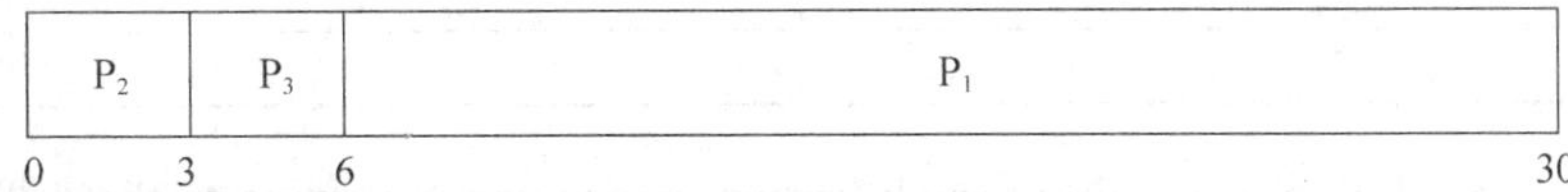

The average waiting time is now $(6 + 0 + 3)/3 = 3$ milliseconds. This reduction is substantial. Thus, the average waiting time under a FCFS policy is generally not minimal and may vary substantially if the process's CPU burst times vary greatly.

In addition, consider the performance of FCFS scheduling in a dynamic situation. Assume we have one CPU-bound process and many I/O-bound processes. As the processes flow around the system, the following scenario may result. The CPU-bound process will get and hold the CPU. During this time, all the other processes will finish their I/O and will move into the ready queue, waiting for the CPU. While the processes wait in the ready queue, the I/O devices are idle. Eventually, the CPU-bound process finishes its CPU bursts and moves to an I/O device. All the I/O-bound processes, which have short CPU bursts, execute quickly and move back to the I/O queues. At this point, the CPU sits idle. The CPU-bound process will then move back to the ready queue and be allocated the CPU. Again, all the I/O processes end up waiting in the ready queue until the CPU-bound process is done. There is a convoy effect as all the other processes wait for the one big process to get off the CPU. This effect results in lower CPU and device utilization than might be possible if the shorter processes were allowed to go first.

The FCFS scheduling algorithm is nonpreemptive. Once the CPU has been allocated to a process, that process keeps the CPU until it releases the CPU, either by terminating or by requesting I/O. The FCFS algorithm is thus particularly troublesome for time-sharing systems, where it is important that each user gets a share of the CPU at regular intervals. It would be disastrous to allow one process to keep the CPU for an extended period.

2.5.4 Shortest-Job-First Scheduling

A different approach to CPU scheduling is the shortest-job-first (SJF) scheduling algorithm. This algorithm associates with *each* process the length of the process's next CPU burst. When the CPU is available, it is assigned to the process that has the smallest next CPU burst. If the next CPU bursts of two processes are the same, FCFS scheduling is used to break the tie. Note that a more appropriate term for this scheduling method would be the shortest-next-CPU-burst algorithm, because scheduling depends on the length of the next CPU burst of a process, rather than its total length. We use the term SJF because most people and textbooks use this term to refer to this type of scheduling.

As an example of SJF scheduling, consider the following set of processes, with the length of the CPU burst given in milliseconds(Table 2.2).

Table 2.2

Process	Burst Time
P_1	6
P_2	8
P_3	7
P_4	3

Using SJF scheduling, we would schedule these processes according to the following Gantt chart:

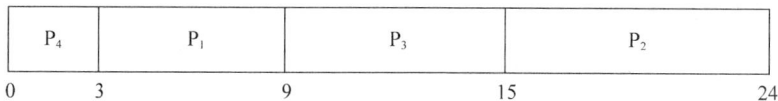

P_4	P_1	P_3	P_2

0 3 9 15 24

The waiting time is 3 milliseconds for process P_1, 16 milliseconds for process P_2, 9 milliseconds for process P_3, and 0 milliseconds for process P_4. Thus, the average waiting time is

(3 + 16 + 9 + 0)/4 = 7 milliseconds. By comparison, if we were using the FCFS scheduling scheme, the average waiting time would be 10.25 milliseconds.

The SJF scheduling algorithm is provably optimal, in that it gives the minimum average waiting time for a given set of processes. Moving a short process before a long one decreases the waiting time of the short process more than it increases the waiting time of the long process. Consequently, the average waiting time decreases.

The real difficulty with the SJF algorithm is knowing the length of the next CPU request. For long-term (job) scheduling in a batch system, we can use as the length the process time limit that a user specifies when he submits the job. Thus, users are motivated to estimate the process time limit accurately, since a lower value may mean faster response. (Too low a value will cause a time-limit-exceeded error and require resubmission.) SJF scheduling is used frequently in long-term scheduling.

Although the SJF algorithm is optimal, it cannot be implemented at the level of short-term CPU scheduling. There is no way to know the length of the next CPU burst. One approach is to try to approximate SJF scheduling. We may not know the length of the next CPU burst, but we may be able to predict its value. We expect that the next CPU burst will be similar in length to the previous ones. Thus, by computing an approximation of the length of the next CPU burst, we can pick the process with the shortest predicted CPU burst.

The next CPU burst is generally predicted as an exponential average of the measured lengths of previous CPU bursts. Let t_n be the length of the nth CPU burst, and let T_{n+1} be our predicted value for the next CPU burst. Then, for a, $0 \leqslant a \leqslant 1$, define

T_{n+1}=a t,, + (1–a)T,,.

This formula defines an exponential average. The value of t, contains our most recent information; T,, stores the past history. The parameter a controls the relative weight of recent and past history in our prediction. If $a = 0$, then $T_{n+1} = T_n$, and recent history has no effect (current conditions are assumed to be transient); if $a = 1$, then $T_{n+1} = t_n$, and only the most recent CPU bursts matters (history is assumed to be old and irrelevant). More commonly, $a = 1/2$, so recent history and past history are equally weighted. The initial T_0 can be defined as a constant or as an overall system average.

To understand the behavior of the exponential average, we can expand the formula for T_{n+1} by substituting for T_n to find

$$T_{n+1} = at_n + (1-a)at_{n-1} + \ldots + (1-a)^j at_{n-j} + \ldots + (1-a)^{n+1} T_0$$

Since both a and (1–a) are less than or equal to 1, each successive term has less weight than its predecessor.

The SJF algorithm can be either preemptive or nonpreemptive. The choice arises when a new process arrives at the ready queue while a previous process is still executing. The next CPU burst of the newly arrived process may be shorter than what is left of the currently executing process. A preemptive SJF algorithm will preempt the currently executing process, whereas a nonpreemptive SJF algorithm will allow the currently running process to finish its CPU burst.

Preemptive SJF scheduling is sometimes called shortest-remaining-time-first scheduling.

2.5.5 Priority Scheduling

The SJF algorithm is a special case of the general priority scheduling algorithm. A priority is associated with each process, and the CPU is allocated to the process with the highest priority. Equal-priority processes are scheduled in FCFS order. A SJF algorithm is simply a priority algorithm where the priority (p) is the inverse of the (predicted) next CPU burst. The larger the CPU bursts, the lower the priority, and vice versa.

Note that we discuss scheduling in terms of high priority and low priority. Priorities are generally indicated by some fixed range of numbers, such as 0 to 7 or 0 to 4,095. However, there is no general agreement on whether 0 is the highest or lowest priority. Some systems use low numbers to represent low priority; others use low numbers for high priority. This difference can lead to confusion. In this text, we assume that low numbers represent high priority.

As an example, consider the following set of processes, assumed to have arrived at time 0, in the order P_1, P_2,... P_5, with the length of the CPU burst given in milliseconds(Table 2.3).

Table 2.3

Process	Burst Time	Priority
P_1	10	3
P_2	1	1
P_3	2	4
P_4	1	5
P_5	5	2

Using priority scheduling, we would schedule these processes according to the following Gantt chart:

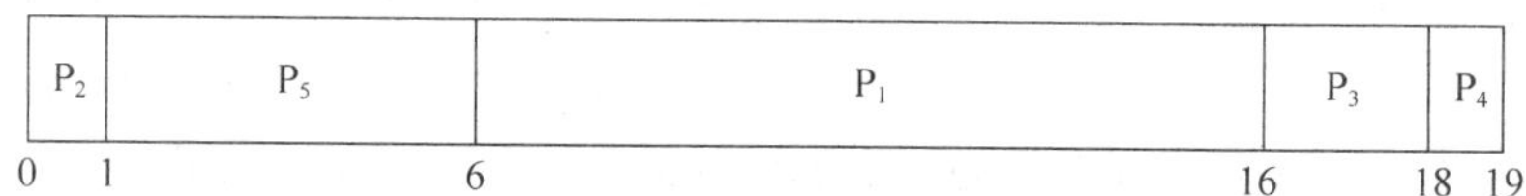

The average waiting time is 8.2 milliseconds.

Priorities can be defined either internally or externally. Internally defined priorities use some measurable quantity or quantities to compute the priority of a process. For example, time limits, memory requirements, the number of open files, and the ratio of average I/O burst to average CPU burst have been used in computing priorities. External priorities are set by criteria outside the operating system, such as the importance of the process, the type and amount of funds being paid for computer use, the department sponsoring the work, and other.

A major problem with priority scheduling algorithms is indefinite blocking, or starvation. A process that is ready to run but waiting for the CPU can be considered blocked. A priority scheduling algorithm can leave some low-priority processes waiting indefinitely. In a heavily loaded computer system, a steady stream of higher-priority processes can prevent a low-priority process from ever getting the CPU.

Generally, one of two things will happen. Either the process will eventually be run (at 2 A.M. Sunday, when the system is finally lightly loaded), or the computer system will eventually crash and lose all unfinished low-priority processes. (Rumor has it that, when they shut down the IBM 7094 at MIT in 1973, they found a low-priority process that had been submitted in 1967 and had not yet been run.)

A solution to the problem of indefinite blockage of low-priority processes is aging. Aging is a technique of gradually increasing the priority of processes that wait in the system for a long time. For example, if priorities range from 127 (low) to 0 (high), we could increase the priority of a waiting process by 1 every 15 minutes. Eventually, even a process with an initial priority of 127 would have the highest priority in the system and would be executed. In fact, it would take no more than 32 hours for a priority-127 process to age to a priority-0 process.

2.5.6 Round-Robin Scheduling

The round-robin (RR) scheduling algorithm is designed especially for timesharing systems. It is similar to FCFS scheduling, but preemption is added to switch between processes. A small unit of time, called a time quantum or time slice, is defined. A time quantum is generally from 10 to 100 milliseconds. The ready queue is treated as a circular queue. The CPU scheduler goes around the ready queue, allocating the CPU to each process for a time interval of up to 1 time quantum.

To implement RR scheduling, we keep the ready queue as a FIFO queue of processes. New processes are added to the tail of the ready queue. The CPU scheduler picks the first process from the ready queue, sets a timer to interrupt after 1 time quantum, and dispatches the process.

One of two things will then happen. The process may have a CPU burst of less than 1 time quantum. In this case, the process itself will release the CPU voluntarily. The scheduler will then proceed to the next process in the ready queue. Otherwise, if the CPU burst of the currently running process is longer than 1 time quantum, the timer will go off and will cause an interrupt to the operating system. A context switch will be executed, and the process will be put at the tail of the ready queue. The CPU scheduler will then select the next process in the ready queue.

The average waiting time under the RR policy is often long. Consider the following set of processes that arrive at time 0, with the length of the CPU burst given in milliseconds(Table 2.4).

Table 2.4

Process	Burst Time
P_1	24
P_2	3
P_3	3

If we use a time quantum of 4 milliseconds, then process P_1 gets the first 4 milliseconds. Since it requires another 20 milliseconds, it is preempted after the first time quantum, and the

CPU is given to the next process in the queue, process P_2. Since process P_2 does not need 4 milliseconds, it quits before its time quantum expires. The CPU is then given to the next process, process P_3. Once each process has received 1 time quantum, the CPU is returned to process P_1 for an additional time quantum. The resulting RR schedule is shown below:

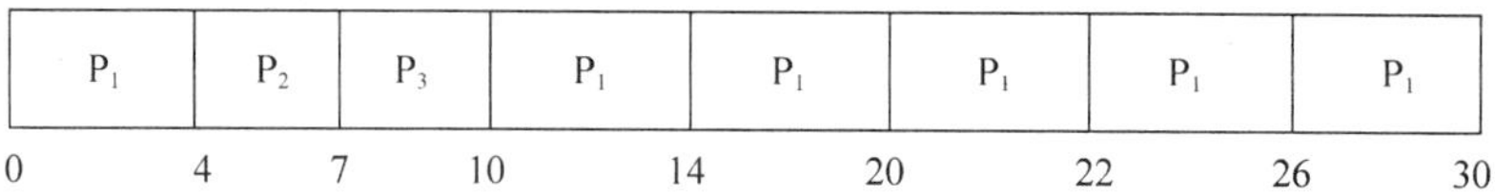

P_1	P_2	P_3	P_1	P_1	P_1	P_1	P_1

0 4 7 10 14 20 22 26 30

The average waiting time is 17/3 = 5.66 milliseconds.

In the RR scheduling algorithm, no process is allocated the CPU for more than 1 time quantum in a row (unless it is the only runnable process). If a process's CPU burst exceeds 1 time quantum, that process is preempted and is put back in the ready queue. The RR scheduling algorithm is thus preemptive.

If there are n processes in the ready queue and the time quantum is q, then each process gets $1/n$ of the CPU time in chunks of at most q time units. Each process must wait no longer than $(n-1) \times q$ time units until its next time quantum. For example, with five processes and a time quantum of 20 milliseconds, each process will get up to 20 milliseconds every 100 milliseconds.

The performance of the RR algorithm depends heavily on the size of the time quantum. At one extreme, if the time quantum is extremely large, the RR policy is the same as the FCFS policy. If the time quantum is extremely small (say, 1 millisecond), the RR approach is called processor sharing and (in theory) creates the appearance that each of n processes has its own processor running at $1/n$ the speed of the real processor. This approach was used in Control Data Corporation (CDC) hardware to implement ten peripheral processors with only one set of hardware and ten sets of registers. The hardware executes one instruction for one set of registers, then goes on to the next. This cycle continues, resulting in ten slow processors rather than one fast one. (Actually, since the processor was much faster than memory and each instruction referenced memory, the processors were not much slower than ten real processors would have been.)

习　题

一、选择题

1. 分配到必要的资源并获得处理机时的进程状态是________。

   A．就绪状态　　B．执行状态　　C．阻塞状态　　D．撤销状态

2. 进程被建立之后进入的第一个状态是________。

   A．挂起状态　　B．阻塞状态　　C．就绪状态　　D．执行状态

3. 一作业进入内存后，该作业的进程初始时处于________状态。

   A．就绪状态　　B．阻塞状态　　C．执行状态　　D．收容状态

4. 正在运行的进程由于时间片用完从运行状态进入________。

A．就绪状态　　B．阻塞状态
C．挂起状态　　D．静止状态

5．在进程管理中，当________时，进程从阻塞状态变为就绪状态。
A．进程被进程调度程序选中　　B．等待某一事件
C．等待的事件发生　　D．时间片用完

6．进程从运行状态进入就绪状态的原因可能是________。
A．被选中占有处理机　　B．等待某一事件
C．等待的事件已发生　　D．时间片用完

7．进程的并发性是指若干个进程执行时________。
A．在时间上是不能重叠的　　B．在时间上是可以重叠的
C．不能交替占用 CPU　　D．必须独占资源

8．进程的三个基本状态在一定条件下可以相互转化，进程由就绪状态变为运行状态的条件是___①___；由运行状态变为阻塞状态的条件是___②___。
A．时间片用完　　B．等待某事件发生
C．等待的某事件已发生　　D．被进程调度程序选中

9．下列的进程状态变化中，________变化是不可能发生的。
A．运行→就绪　　B．运行→阻塞
C．阻塞→运行　　D．阻塞→就绪

10．一个运行的进程用完了分配给它的时间片后，它的状态变为________。
A．就绪　　B．阻塞　　C．运行　　D．由用户自己确定

11．下面对进程的描述中，错误的是________。
A．进程是动态的概念　　B．进程执行需要处理机
C．进程是有生命期的　　D．进程是指令的集合

12．程序的顺序执行通常在___①___的工作环境中，具有___②___特征；程序的并发执行在___③___的工作环境中，具有___④___特征。
A．单道程序　　B．多道程序　　C．程序的可再现性　　D．资源共享

13．多道程序环境下，操作系统分配资源以________为基本单位。
A．程序　　B．指令　　C．进程　　D．作业

14．在一个单处理机系统中，存在 5 个进程，则最多有___①___个进程处于阻塞状态，最多有___②___个进程处于就绪状态。
A．5　　B．4　　C．1　　D．0

15．对进程的管理和控制使用________。
A．指令　　B．原语　　C．信号量　　D．信箱通信

16．进程控制就是对系统中的进程实施有效的管理，通过使用________，进程撤销，进程阻塞，进程唤醒等进程控制原语实现。
A．进程运行　　B．进程管理　　C．进程创建　　D．进程同步

17．操作系统通过________对进程进行管理。
A．进程　　B．进程控制快　　C．进程启动程序　　D．进程控制区

18．一个进程被唤醒意味着________。

A. 该进程重新占有了 CPU　　B. 它的优先权变为最大

C. 其 PCB 移至等待队列首　　D. 进程变为就绪状态

19. 在操作系统中，JCB 是指________。

A. 作业控制块　B. 进程控制块　C. 文件控制块　D. 程序控制块

20. 一种既有利于短小作业又兼顾到长作业的作业调度算法是________。

A. 先来先服务　　B. 轮转

C. 最高响应比优先　　D. 均衡调度

21. 一作业 5:00 到达系统，估计运行时间为 2 小时，若 7:00 开始执行该作业，其响应比是________。

A. 3　B. 1　C. 2　D. 0.5

22. 一作业 7:00 到达系统，估计运行时间为 2 小时，若 8:00 开始执行该作业，其响应比是________。

A. 3　B. 1　C. 2　D. 1.5

二、简答题

1. 简述进程的概念。

2. 简述程序与进程的区别。

3. 简述 PCB 的作用。

4. 画图说明带挂起的进程状态及其转换。

5. 简述处理机调度的三个层次。

三、计算题

1. 根据下表中作业到达时间和作业服务时间，计算表中按先来先服务（FCFS）和短作业优先服务（SJF）调度算法，各作业的开始执行时间和完成时间，并计算两种调度算法下的平均周转时间。

作业情况 / 调度算法	进程名	A	B	C	D	E	平均周转时间
	到达时间	0	1	2	3	4	
	服务时间	3	5	2	6	1	
FCFS	开始时间						
	完成时间						
	周转时间						
SJF	开始时间						
	完成时间						
	周转时间						

2. 若在后备作业队列中等待运行的同时有作业 J_1、J_2、J_3，已知它们各自的运行时间为 a、b、c，且满足关系 $a<b<c$，求采用短作业优先调度算法的各作业的开始时间、完成时间，并求出平均周转时间。

作业	开始时间	完成时间	周转时间
J_1			
J_2			
J_3			

3．有四道作业，它们的提交时间和运行时间如下表。

作业号	到达时间	运行时间/小时
1	8:00	2.0
2	8:50	0.5
3	9:00	0.2
4	9:50	0.3

如果系统采用单道程序设计技术，并分别采用先来先服务和短作业优先作业调度算法，请填下面的表格，并指出调度作业的顺序，计算作业的平均周转时间 T 和平均带权周转时间 W。

（1）先来先服务：

作业号	开始时间	结束时间	周转时间/分钟	带权周转时间
1				
2				
3				
4				
平均				

（2）短作业优先：

作业号	开始时间	结束时间	周转时间/分钟	带权周转时间
1				
2				
3				
4				
平均				

第3章 进 程 同 步

进程的并发执行，相互协同工作就是进程同步。进程同步可以用信号量控制，信号量是一个整数，代表可用资源数。在一组进程中，每个进程都无限地等待被该组进程中另外的进程所占有的资源，因而永远无法得到该资源，系统出现一种僵局，在无外力的作用下，无法推进的现象称为死锁，这一组进程就称为死锁进程。

3.1 计算机程序的并发执行

并发性是指两个或多个事件在同一时间间隔内发生。在多道程序环境下，允许多个进程并发执行，程序的并发性是指宏观上在一段时间内多道程序同时运行，但在单处理器系统中，每一时刻仅能执行一道程序，进程的执行并非“一气呵成”，而是“走走停停”，在微观上这些程序是在交替执行的。

3.1.1 程序的并发执行

1. 程序的顺序执行

根据程序逐条执行的基本原理，程序的执行一般是要顺序执行的。当前一操作（程序段）执行完后，才能执行后继操作。例如，在进行计算时，总须先输入用户的程序和数据，然后进行计算，最后才能打印计算结果，如图 3-1（a）所示。程序的语句也是按次序执行的，如图 3-1（b）所示。

S_1:a=x+y;

S_2:b=a–7;

S_3:c=b+3;

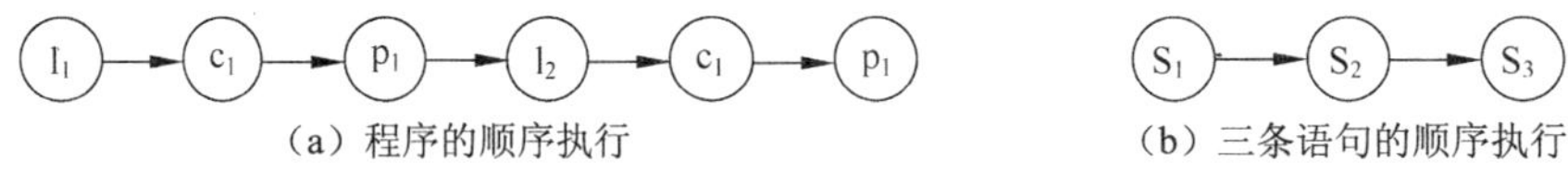

（a）程序的顺序执行　　（b）三条语句的顺序执行

图 3-1　程序的顺序执行

程序顺序执行时，程序的最终结果由给定的初始条件据顶，程序的结果是可再现的，同样的初始条件，会得到相同结果。总结起来有三个特征。

（1）顺序性。当顺序程序在 CPU 上执行时，CPU 严格地顺序执行程序规定的动作，每个动作都必须在前一动作结束后才能开始。除了人为干预造成机器暂时停顿外，前一动作的结束就意味着后一动作的开始，前一条指令的执行结束是后一条指令执行开始的充分必要条件。程序和计算机执行程序的活动严格一一对应。

（2）独占资源。程序在执行过程中，独占全部资源，除了初始状态外，只有程序本身规定的动作才能改变这些资源的状态。

（3）结果无关性。顺序执行的结果与其执行速度无关。也就是说，CPU 在执行程序的两个动作之间如有停顿不会影响程序的执行结果，如果程序的初始条件不变，当重复执行时，一定能得到相同的结果。

上述特点概括起来就是程序的顺序性、封闭性和可再现性。所谓顺序性指的是程序的各部分能够严格地按程序所确定的逻辑次序顺序地执行。所谓封闭性指的是程序一旦开始执行，其计算结果就只取决于程序本身，除了人为改变机器运行状态或机器故障外，不受外界因素的影响。所谓可再现性是指当该程序重复执行时，必将获得相同的结果。

2．程序的并发执行

在多道程序中，多道程序同时运行，系统的输入输出操作可以有通道控制，通过 CPU 的计算并行操作，可以大大地提高 CPU 的利用率。例如系统中有多个进程被提交运行，当进程 1 在做计算时，进程 2 可以同时做输入操作；当进程 1 在做输出操作时，进程 2 可以做计算操作，进程 3 可以做输入操作等。这样可以大大地提高系统的工作效率，使系统的各个设备尽量处于“忙”的状态，如图 3-2 所示。

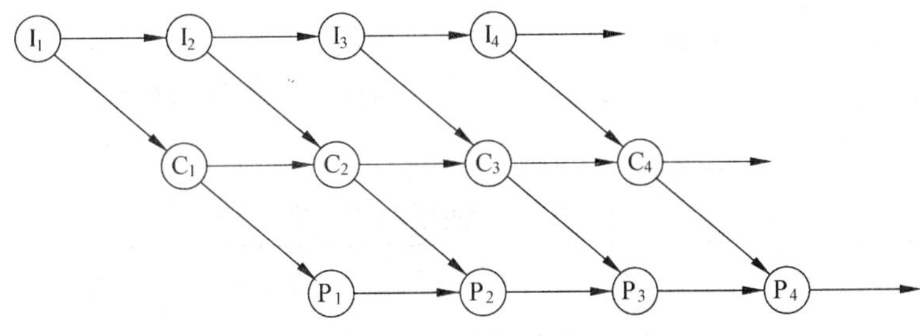

图 3-2　进程的并发执行

程序的并发执行，是否还能保持顺序执行时的特性呢？下面我们通过例子来说明程序并发执行时的特性。

设有两个进程，观察者进程和报告者进程，它们并行工作。在一条单向行驶的公路上经常有卡车通过。观察者不断观察并对通过的卡车计数（计数器加 1），报告者定时地将观察者的计数值打印出来，然后将计数器重新清 0。此时我们可以写出如下程序，其中 parbegin 表示多个程序段可以并发执行。

```
int n=0;
void observer(void)
{
    while (1)
    {
        n=n+1;
        remainder of observer;
    }
}
void reporter(void)
{
```

```
        while(1)
        {
            print(n);
            n=0;
            remainder of reporter;
        }
    }
    void main( )
    {
        parbegin(observer( ),reporter( ));
    }
```

由于观察者和报告者各自独立地并行工作；*n*=*n*+1 的操作，既可以在报告者的 print (*n*) 和 *n*=0 操作之前，也可能在 print(*n*)和 *n*=0 之间，还可能在两个语句之后。即可能出现以下三种执行序列：

（1）*n*=*n*+1；print(*n*)；*n*=0；

（2）print(*n*)；*n*=*n*+1；*n*=0；

（3）print(*n*)；*n*=0 ；*n*=*n*+1；

n 的值为 0；则在完成一个循环后，对上述三个执行序列打印机打印的 *n* 值和执行后的 *n* 值如表 3-1 所示。

表 3-1　三种执行序列所得结果

执行序列	（1）	（2）	（3）
打印的值	1	0	0
执行后的值	0	0	1

由表 3-1 可见，由于观察者和报告者的执行速度不同，导致了计算结果的不同，这就是说，程序并发执行已丧失了顺序执行所保持的封闭性和可再现性。

（1）失去封闭性。程序在并发执行时，多个程序共享系统中的资源，这些资源的使用状态将由多个进程改变，使程序的运行失去了封闭性。

（2）不可再现性。由于程序的并发执行，打破了由某一道程序独占系统资源时的封闭性，程序执行的结果变得不一定了，失去了可再现性。

这是不对的。我们必须设置一种机制，控制两个进程的推进过程，使两个进程能够协调地、无差错地运行。这就需要进程的同步。

3.1.2　前趋图

为了描述一个程序的各部分（程序段或语句）间的依赖关系，或者是一个大的计算的各个子任务间的因果关系，常常采用前趋图方式。

前趋图（Precedence Graph）是一个有向无循环图，记为 DAG（Directed Acyclic Graph），用于描述进程之间执行的前后关系。图中的每个节点可用于描述一个程序段或进程，乃至一条语句；节点间的有向边则用于表示两个节点之间存在的偏序（Partial Order）或前趋关系(Precedence Relation)，用“→”表示。

$\rightarrow$={(P_i,P_j)|P_i must complete before P_j may start}，如果(P_i,P_j)∈→,可写成 $P_i \rightarrow P_j$，称 P_i 是 P_j 的直接前趋，而称 P_j 是 P_i 的直接后继。在前趋图中，把没有前趋的节点称为初始节点（Initial Node），把没有后继的节点称为终止节点（Final Node）。

每个节点还具有一个重量（Weight），用于表示该节点所含有的程序量或节点的执行时间。

对于图 3-3（a）所示的前趋图， 存在下述前趋关系：

$P_1 \rightarrow P_2$,P1$\rightarrow P_3$,$P_1 \rightarrow P_4$,$P_2 \rightarrow P_5$,$P_2 \rightarrow P_9$,$P_3 \rightarrow P_5$,$P_3 \rightarrow P_6$,$P_4 \rightarrow P_7$,$P_5 \rightarrow P_8$,$P_6 \rightarrow P_8$,$P_7 \rightarrow P_8$,$P_8 \rightarrow P_9$

或表示为：

P={P_1, P_2, P_3, P_4, P_5, P_6, P_7, P_8, P_9}

$\rightarrow$={ (P_1, P_2), (P_1, P_3), (P_1, P_4), (P_2, P_5) , (P_2, P_9), (P_3, P_5), (P_3, P_6), (P_4, P_7), (P_5, P_8), (P_6, P_8), (P_7, P_8), (P_8, P_9)}

应当注意，前趋图中必须不存在循环，但在图 3-3（b）中却有着下述的前趋关系：

$S_2 \rightarrow S_3$,$S_3 \rightarrow S_2$，

是不正确的前趋图。

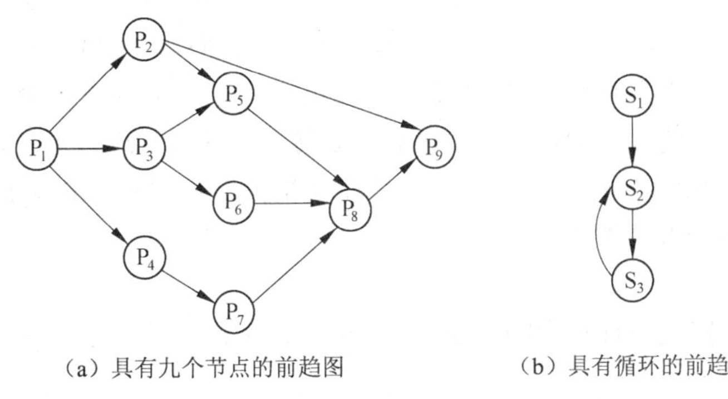

（a）具有九个节点的前趋图　　（b）具有循环的前趋图

图 3-3　前趋图

3.2 进 程 同 步

异步环境下的一组并发进程因直接制约而互相发送消息而进行互相合作、互相等待，使得各进程顺利执行的过程称为进程间的同步。具有同步关系的一组并发进程称为合作进程，合作进程间互相发送的信号称为消息或事件。

3.2.1 临界资源与临界区

系统中同时存在有许多进程，它们共享各种资源，然而有许多资源在某一时刻只能允许一个进程使用，这种每次只允许一个进程访问的资源叫临界资源。属于临界资源的硬件有打印机、磁带机等，软件有消息缓冲队列、变量、数组、缓冲区等。

这类资源必须被保护，避免两个或多个进程同时访问。几个进程若共享同一临界资源，必须控制它们，使它们以互相排斥的方式使用这个临界资源，即当一个进程正在使用某个临界资源且尚未使用完毕时，其他进程必须等待，只有当使用该资源的进程释放该资源时，其他进程才可使用该资源。这种以互相排斥等待方式，使用临界资源的方式称为互斥。互

斥其实是一种特殊的同步方式。

每个进程中访问临界资源的那段代码称为临界区（Critical Section）。显然，若能保证诸进程不能同时进入自己的临界区，便可实现诸进程对临界资源的互斥访问。为此，每个进程在进入临界区之前，应先对欲访问的临界资源进行检查，看它是否正被访问。如果此刻该临界资源未被访问，进程便可进入临界区对该资源进行访问，并设置它正被访问的标志；如果此刻该临界资源正被某进程访问，则本进程不能进入临界区。

3.2.2 信号量

怎样解决多进程的互斥问题呢？我们知道马路上的十字路口是不能允许两条路上的车辆同时驶入的，十字路口就是个临界资源，两条路上的车可以看成是进程。我们可以在十字路口设置一个信号灯，并控制两条路上的车交替通行，使车辆互斥地使用十字路口这个临界资源。

1．信号量

模仿信号灯，我们可以在系统中设置信号量（Semaphore），在多进程程环境下，进程在进入一个临界代码段之前，进程必须获取一个信号量；一旦临界代码段完成了，释放这个信号量。其他想进入临界代码段的进程程必须等待直到那个进程释放信号量。

信号量是一个整数（S），对于一个临界资源，设 S=1，进程进入临界段之前先判断信号量 S 是否大于 0，如果不大于 0，则等待；如果大于 0，则做 S=S−1 操作，进入临界区，退出临界区后做 S=S−1 操作。

2．互斥

利用信号量实现进程互斥的算法如下：

```
semaphore  S=1;
Parbegin
   Process1: Begin
          Repeate
             while S<=0 do   no-op;
             S = S-1;
             Critical section;
             S = S+1;
           Until false;
         End
   Process2: Begin
          Repeate
             while S<=0 do   no-op;
             S = S-1;
             Critical section;
             S = S+1;
           Until false;
         End
Parend.
```

Parbegin 和 Parend 之间的 Process1 和 Process2 为两个并发的进程。这两个进程中的临界区 Critical Section 是不能同时执行到的，它们是互斥的。这里信号量 S 起到了作用。

为了方便编程，我们可以把控制信号量的代码包装起来，由操作系统提供。

```
Wait (S):  while S<=0 do   no-op;
          S =S - 1;
Signal (S): S = S + 1;
```

有了上述的两个函数，进程的互斥算法实现起来就比较简单。只要在临界段的前后加上 Wait 和 Signal 函数即可。

```
Repeate
   Wait (S);
   Critical section;
   Signal (S);
Until false;
```

这里需要注意两个问题。① 信号量的含义。信号量是一个整数，它的值代表系统中某种资源的可用的数量。在互斥操作中，多个进程都想用一个资源，所以它的初值为 1。在 wait(S)函数中判断 S<=0，S<=0 的含义是系统中已经没有这种资源了。这时它就得等待。等待的时候执行 no-op，no-op 是空操作的意思。② Wait 函数是申请一个资源，如果有资源，申请成功，执行 S=S−1；表示占用一个资源；Signal 函数是释放一个资源，要执行信号量加 1 的操作。

3.2.3 进程同步

由于资源有限，进程的并发执行存在着资源的竞争，有时有些进程暂时停下来，等待别的进程把它需要的资源释放出来，在接着向前推进。进程之间的这种协同工作，就是进程同步。

举个例子，假设停车场只有三个车位，一开始三个车位都是空的。这时如果同时来了五辆车，看门人允许其中三辆直接进入，然后放下车拦，剩下的车则必须在入口等待，此后来的车也都不得不在入口处等待。这时，有一辆车离开停车场，看门人得知后，打开车拦，放入外面的一辆进去，如果又离开两辆，则又可以放入两辆，如此往复。

在这个停车场系统中，车位是公共资源，每辆车好比一个进程，看门人起的就是信号量的作用。

信号量的值代表可用资源的个数，通过 Wait（S）函数和 Signal（S）函数可以控制进程的同步。

例 1 设系统中有两个进程 P_1、P_2，P_1 进程负责计算数据，将结果放入缓冲区 Buf，P2 进程从缓冲区 Buf 中取数据输出。试写出两个进程的同步算法。

分析：这道题中 Buf 是一个临界资源，进程 P_1 需要一个空的 Buf，向里面放数据，进程 P_2 需要一个有数据的 Buf，从里面取数据。开始的时候 Buf 是空的。P_1 进程需要一个 Buf 而且是空的，P_2 进程也需要 Buf，但是它要有数据的。为此我们需要设置两个信号量，

be 和 bf，它们分别代表空缓冲区和有数据的缓冲区。be 初值为 1，bf 初值为 0。

```
semaphore be=1, bf=0;
Parbegin
   P1: Begin
          Repeate
            计算数据Data;
             Wait (be);
             Data=>Buf;
             Signal (bf);
           Until false;
       End
   P2: Begin
          Repeate
             Wait (bf);
             取Data;
             Signal(be);
           Until false;
       End
Parend.
```

例 2 假设系统中有 6 个进程 S_1、S_2、…、S_6，进程的前趋后继关系如图 3-4 所示。用信号量控制的方法，写出它们的同步算法。

分析：在图 3-4 中，可以看出除 S_1 外，每个进程都有一个有向边指向它，说明它需要等待上一个进程结束。共有 7 条有向边，设置 7 个信号量 a，b，c，d，e，f，g，它们的初值都是 0。

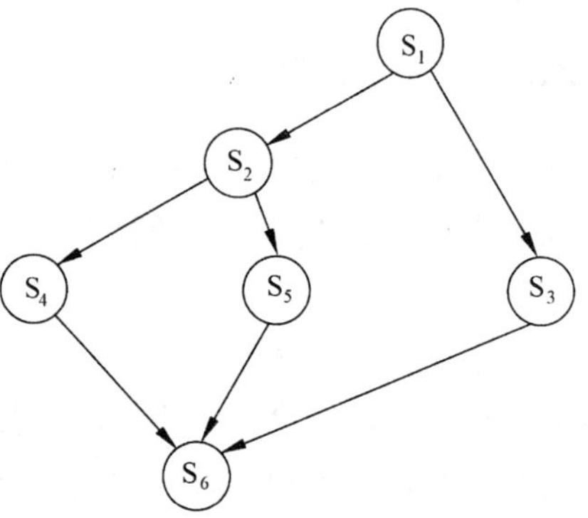

图 3-4 前趋图

```
semaphore a=0, b=0, c=0, d=0, e=0, f=0, g=0;
Parbegin
   Begin S1; Signal (a); Signal (b); End
   Begin Wait(a); S2; Signal (c); Signal (d);
   End
   Begin Wait(b); S3; Signal (e);  End
   Begin Wait(c); S4; Signal (f);  End
   Begin Wait(d); S5; Signal (g);  End
   Begin Wait(e); Wait(f); Wait(g); S6;  End
Parend.
```

3.2.4 改进的信号量机制

1. 记录型信号量

前面介绍了信号量的概念及应用，通过信号量控制进程的互斥与进程同步。这是由于资源有限，进程的并发执行存在着资源的竞争，有时有些进程暂时停下来，等待别的进程把它需要资源释放出来，它才能继续推进。

深入研究信号量的应用，主要是调用两个函数：

```
Wait (S):    while S<=0 do   no-op;
           S = S-1;
Signal (S):   S = S+1;
```

Wait 函数中的 while 是个循环，当信号量为 0 时，它会不停地执行空操作（no-op），直到把当前的时间片用完。这样很影响 CPU 的利用率。当信号量为 0 时，应该将进程阻塞起来，马上让出 CPU 给别的进程使用，这样它也不参与 CPU 的调度，提高 CPU 利用率；当有进程释放资源时，如果有进程在等待这个资源而阻塞，就唤醒它，使它得到 CPU 和相应的资源，继续推进。

这样需要对信号量机制加以改进，首先信号量的描述改成记录类型：

```
Type semaphore=record
     value: integer;
     L: list of process;
End
```

它的属性 value 为正整数时，代表的是可用资源个数；value 为负数时，value 的绝对值是系统中因等待这个资源而被阻塞的进程的个数。L 是一个集合，集合列表中的元素是因等待该信号量代表的资源而被阻塞起来的进程号。

对应的 Wait、Signal 函数需要改成：

```
semaphore  S;
Procedure Wait (S)
     begin
          S.value = S.value - 1;
          If(S.value<0) then block(S.L);
        end
Procedure Signal (S)
     begin
          S.value = S.value + 1;
          If(S.value<=0) then wakeup(S.L);
        end
```

Wait 函数申请信号量 S 所代表的资源，S.value=S.value-1；当 S.value 为 0 时表示资源没有了，再有资源申请时 S.value 就小于 0 了，这时要把申请资源的进程阻塞起来，让出 CPU，并把阻塞起来的进程的进程号存入 S.L 集合列表中记录下来。Signal 函数释放资源，S.value=S.value+1；如果 S.value 小于等于 0，说明有进程因为等待这个资源而阻塞，从 S.L 中选择一个进程唤醒。

这种改进的信号量机制避免了进程执行空操作的方式等待资源，提高了 CPU 的利用率。Wait、Signal 函数也被叫做 P、V 操作，可以用系统的原语操作实现。P、V 操作的流程图如图 3-5 所示。

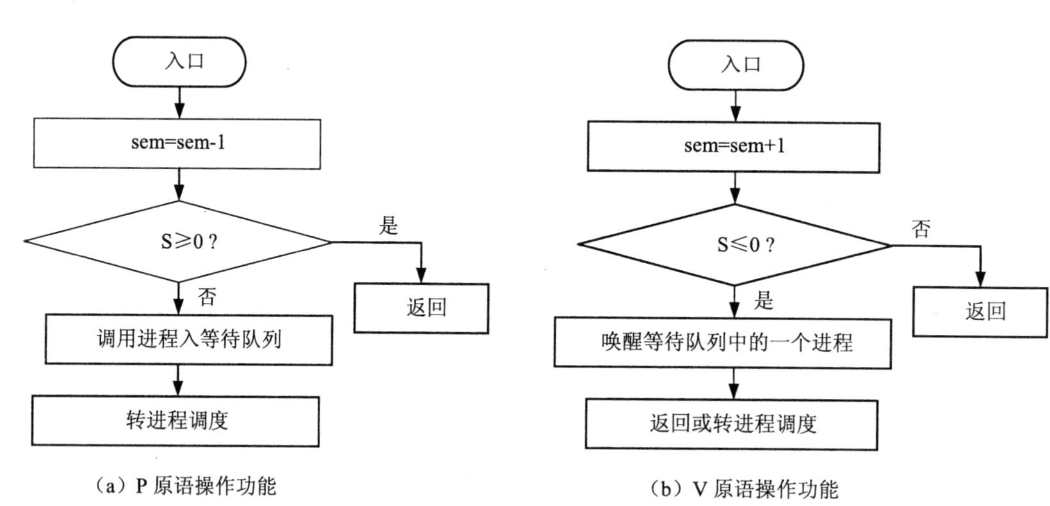

（a）P 原语操作功能

（b）V 原语操作功能

图 3-5　P、V 操作的流程图

2．集合型信号量

假设系统中有进程 P1、进程 P2，有资源 R1、资源 R2，进程 P1、P2 都需要使用资源 R1、资源 R2。进程 P1、P2 的同步算法如下：

```
semaphore SR1=1，SR2=1;
Parbegin
   P1: Begin
          Repeate
             Wait（SR1);      ...........a
             Wait（SR2);      ...........b
                P1进程的临界区;
             Signal（SR2);
             Signal（SR1);
           Until  false;
       End
   P2: Begin
          Repeate
             Wait（SR2);      ...........c
             Wait（SR1);      ...........d
                P2进程的临界区;
             Signal（SR1);
             Signal（SR2);
           Until  false;
       End
Parend.
```

P1 进程与 P2 进程是并发执行，假设 P1 进程执行完 a 行代码，申请了 R1 资源，时间片到，P1 进程回到就绪状态；轮到 P2 进程执行，P2 进程执行 c 行代码，申请了 R2 资源，再执行 d 行代码，由于信号量 SR1 已经被 P1 进程做了减 1 操作，执行 d 行代码的结果是

P2 进程被阻塞，等待进程 P1 释放资源 R1，并唤醒它。又轮到了 P1 进程执行，接下来执行 b 行代码，再申请 R2 资源，已经申请不到了，因为资源 R2 已经分配进程 P2 了，P1 也进入了阻塞状态，等待进程 P2 释放资源 R2，并唤醒它。

现在的情形是进程 P1、P2 都在等对方释放资源，唤醒自己，它们都不能向前推进了。这种现象叫做死锁，后面我们会专门讨论死锁的情况。

出现这种状况是我们不希望的，究其原因，是它们申请资源的次序出了问题。针对这种情况，如果我们把多个资源捆绑在一起，要申请就全部申请，用完之后全部释放，即引入集合类型的信号量机制，就可以解决这样的问题了。

集合类型的信号量机制是这样的，进程申请资源时，给出申请资源的列表，并说明每种资源需要多少个，列表（S1,d1,S2,d2,⋯,Sn,dn）中的 S1,⋯,Sn 是每种资源的信号量，d1,⋯,dn 是需要的对应资源的个数。对应 P、V 操作的函数 Swait、Ssignal 如下：

```
semaphore S1,S2,…,Sn;
Procedure Swait (S1,d1,S2,d2,…,Sn,dn)
     begin
        If  S1>=d1 and S2>=d2 and …and Sn>=dn  Then
            For  i = 1 to n  Do
                Si = Si-di;
            Next i
         Else
            block(L);  把进程送入系统的因等待资源而阻塞的进程队列
         Endif
         end
Procedure Ssignal (S1,d1,S2,d2,…,Sn,dn)
     begin
         For  i = 1 to n  Do
            Si = Si+di;
         Next i
         If (L不空)  Then
            Wakeup (l∈L) ;  唤醒L中满足条件的一个进程
         Endif
         end
```

采用集合型的信号量机制，可以解决前面进程 P1、P2 申请资源 A、B 的问题了。采用集合型信号量机制的算法描述如下：

```
semaphore Sa = 1, Sb = 1;
Parbegin
   P1: Begin
          Repeate
             Swait (Sa,1,Sb,1);
                P1进程的临界区;
             Ssignal (Sa,1,Sb,1);
           Until false;
```

```
        End
    P2: Begin
          Repeate
             Swait (Sa,1,Sb,1);
                P2进程的临界区;
             Ssignal (Sa,1,Sb,1);
           Until false;
        End
Parend.
```

3.3 经典同步问题

3.3.1 生产者-消费者问题

生产者-消费者（Producer-Consumer）问题，也称作有界缓冲区（Bounded-Buffer）问题，是一个经典的进程同步问题，该问题最早由 Dijkstra 提出，用以演示信号量机制。一组生产者进程向一组消费者进程提供数据，它们共享一个有界缓冲池，生产者进程向其中投放数据，消费者进程从中取出数据。生产者要不断将数据放入共享的缓冲，消费者要不断从缓冲取出数据。生产者必须等消费者取走数据后才能再放新数据，不允许覆盖数据，消费者必须等生产者放入新数据后才能去取，一组数据能且只能被一个消费者进程使用，数据不能被重复使用。

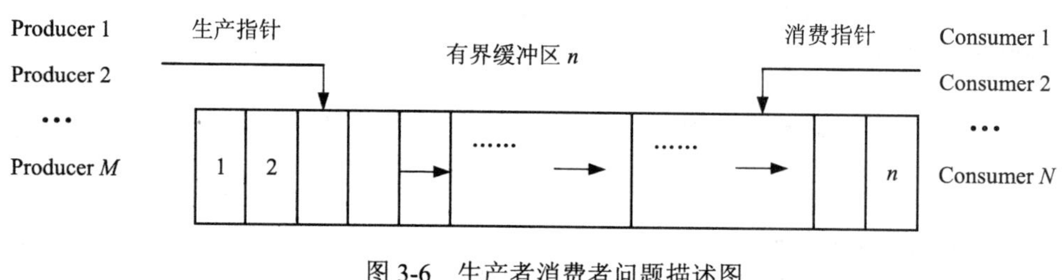

图 3-6　生产者消费者问题描述图

这是一个经典的同步问题，可以通过信号量来解决。生产者进程向缓冲中写数据，数据不能被覆盖，生产者需要的是空的缓冲区，所以可以信号量 empty，代表空的缓冲区；消费者进程从缓冲区中取数据，必须等到生产者进程把数据写到缓冲区中，也就是说，需要的是满的缓冲区，设一个信号量 full，代表满的缓冲区。

注意：在解决进程同步问题时，较难的问题是怎样设置信号量的问题，这里告诉大家一个技巧，就是根据进程的需要设置信号量，即，进程必须等待什么，就把什么设成信号量。这样进程的同步问题就好解决了。

生产者进程向缓冲区依次放数据，必须设置一个整数做指针，指示当前生产者应该放数据的位置；同样的道理，需要设置取数据的指针。

缓冲区被看作一个循环缓冲区，如图 3-7 所示，放数据的指针 in 和取数据的指针 out 的值必须按缓冲区的大小 n 取模，信号量 empty=n，full＝0。

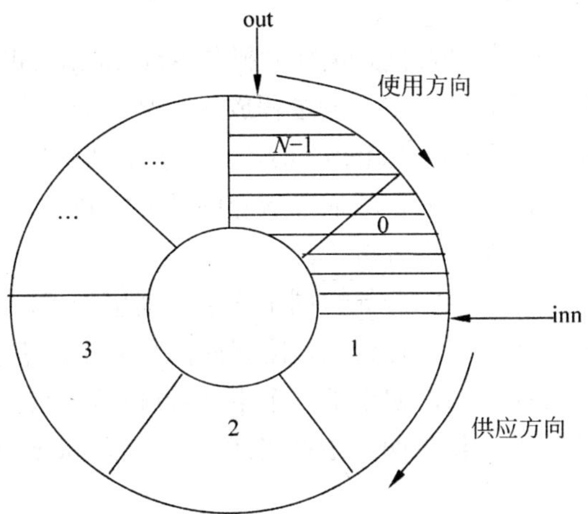

图 3-7　循环缓冲区

```
semaphore empty=n, full=0;
item  buffer [0, …, n-1] ;
integer  in=0, out=0;
Parbegin
  proceducer:begin
         repeat
           item = produce_item();
           Wait(empty);
           buffer(in) = item;........................a
           in = (in+1) mod n;
           Signal(full);
         until false;
       end
  consumer:begin
        repeat
           Wait(full);
           item = buffer(out);
           out =(out+1) mod n;
           Signal(empty);
           consumer the item;
        until false;
       end
Parend
```

生产者进程生产一个产品 item，申请一个空的缓冲区，执行 P 操作 wait(empty)，把这个产品放入缓冲区 buffer(in)=item，然后调整指针 in=(in+1) mod n，使下一个生产者的产品放入下一个缓冲区，这时，缓冲区中多放了一个产品，对于消费者等待的满缓冲区资源就增加了一个，所以要释放一个满缓冲区资源，执行 V 操作 signal(full)，如果有消费者进程被阻塞，则唤醒一个消费者进程。

消费者进程先申请一个满的缓冲区，执行 P 操作 wait(full)，full 的初值为 0，消费者进程刚启动时，都将进入阻塞状态，当生产者进程放入产品后会唤醒消费者进程。同样的道理，消费者进程取出一个产品 item = buffer(out)之后，也要调整指针 out = (out+1) mod n，下一个消费者从下一个缓冲区取产品，然后释放空的缓冲区资源，如果有生产者进程因等待缓冲区而被阻塞，则唤醒一个生产者进程。

这样生产者消费者进程似乎很完美的协同工作了，细想一下，看看会不会出问题呢？例如，某一个生产者进程，执行完 a 行代码，刚好时间片到了，需要停下来。别忘了，生产者进程和消费者进程是有很多的，如果下一个生产者进程再执行 a 行代码，想一想会出什么问题呢？由于指针没调整，结果两个生产者进程把产品 item 放到了同一个缓冲区中了，出现了数据覆盖。

因此，对上述的算法还需改进。放产品的代码与指针调整的代码，中间不应该被另一个生产者进程加进来，也就是说生产者进程之间放产品的代码与指针调整的代码应该互斥，不能让两个生产者同时执行这两行代码。同理，消费者进程也不允许两个进程从同一个缓冲区中取数据。

应该增加一个互斥信号量 mutex，使不能同时执行的代码段，不被同时执行到。

```
semaphore  mutex=1,empty=n, full=0 ;
item  buffer [0, …, n-1] ;
integer  in=0, out=0;
Parbegin
   proceducer:begin
            repeat
               item = produce_item();
               Wait(empty);
               Wait(mutex);
               buffer(in)  = item;
               in  =  (in+1)  mod n;
               Signal  (mutex);
               Signal(full);
            until false;
          end
   consumer:begin
           repeat
               Wait(full);
               Wait(mutex);
               item = buffer(out);
               out  =  (out+1)  mod n;
               Signal  (mutex);
               Signal(empty);
               consumer the item;
           until false;
```

```
        end
Parend
```

3.3.2 读者-写者问题

读者-写者问题的定义如下：有一个多进程共享的数据区，这个数据区可以是一个文件或者主存的一块空间。有一些进程只读取这个数据区的数据，我们叫它“读者”进程（Reader）；一些进程只往数据区写数据，我们叫它“写者”进程（Writer），如图 3-8 所示。读者进程和写者进程需要满足以下条件：

（1）任意多个读进程可以同时读这个文件；

（2）一次只有一个写进程可以往文件中写；

（3）如果一个写进程正在进行操作，禁止任何读进程度文件。

图 3-8　读者写者问题示意图

这是一个典型的互斥问题。不是两个进程的互斥关系，而是两组进程的互斥关系，首先写者进程同写者进程之间互斥，写者进程同读者进程之间互斥，读者进程同读者进程不互斥，如表 3-2 所示。

表 3-2　典型读者-写者互斥关系

	写者进程	读者进程
写者进程	互斥	互斥
读者进程	互斥	不互斥

解决读者-写者问题的关键，是把所有的读者进程当成是一个进程，那么就是写者进程同写者进程互斥，写者进程同读者进程互斥了。前面讲过互斥的控制方法，这样的互斥操作比较好实现。

怎样把所有的读者进程当成是一个进程呢？我们可以为读者进程设置一个计数器 readcount，判断 readcount，当第一读者进程来时申请同写者之间的互斥信号量，如果有写者就等待，无写者就占有这个资源；其他读者再来，就不用判断互斥信号量了，读者进程同读者进程不再互斥了；判断 readcount，当最后一位读者进程离开时释放申请同写者之间的互斥信号量。

```
semaphore  wrt=1;
```

```
integer  readcount=0;
Parbegin
    Reader:begin
          repeat
             if readcount=0 then wait(wrt);........................a
             readcount++;                   .........................b
             perform read operation;
             Readcount--;
             if readcount=0 then signal(wrt);
          until false;
       end
    Writer:begin
          repeat
             wait(wrt);
             perform write operation;
             signal(wrt);
          until false;
      end
Parend
```

读者进程先判断是不是第一个读者，通过读者计数器 readcount 判断，如果 readcount=0 说明是第一个读者，执行 wait(wrt)实现与写者的互斥；如果没有写者再写，执行读者计数器 readcount 加 1 操作，执行读操作。如果这时再来读者就可以直接进来，读者计数器也加 1，执行读操作，读者与读者之间不互斥。读者读操作做完后，读者计数器 readcount 减 1，再判断是不是最后一位读者，判断方法还是通过读者计数器 readcount 判断，如果 readcount=0 说明是最后一位读者，执行 V 操作，释放资源 signal(wrt)。

写者进程相对简单，只需要控制写操作的互斥。即在写操作的前后加上 P、V 操作。

我们在深入研究一下读者操作。如果在 a 行代码执行完，b 行代码没执行时，时间片到，调度其他进程，如果又来了一个读者进程，会出现什么情况呢？由于前面的读者进程，判断了 readcount 为 0，执行了 wait(wrt)操作，但 b 行代码没执行，所以 readcount 没有做加 1 操作，所以下一个读者来时，判断 readcount 还是为 0，又会执行 wait(wrt)操作，就出错了。解决这个问题的办法，还需要设置一个读者的互斥信号量 read，使 a、b 两行代码段互斥，避免两个读者都做 readcount 的判断，而没有加 1 的操作。修正后的读者-写者问题的算法描述如下：

```
semaphore  wrt = 1, read = 1;
integer  readcount=0;
Parbegin
    Reader:begin
          repeat
             wait(read);
             if readcount=0 then wait(wrt);
             readcount++;
             signal(read);
```

```
            perform read operation;
            wait(read);
            readcount--;
            if readcount=0 then signal(wrt);
            signal(read);
        until false;
      end
   Writer:begin
        repeat
            wait(wrt);
            perform write operation;
            signal(wrt);
        until false;
     end
Parend
```

3.3.3 哲学家问题

哲学家问题描述如下：有五个哲学家，他们的生活方式是交替地进行思考和进餐。哲学家们公用一张圆桌，圆桌中央有一盘通心粉，周围放有五把椅子，每人坐一把，每人面前有一只空盘子，每两人之间放一只筷子。哲学家的行为就是思考，饥饿时便试图取用其左、右最靠近他的筷子，但他可能拿不到。只有在他拿到两根筷子时，方能进餐，进餐完后，放下筷子又继续思考，如图 3-9 所示。

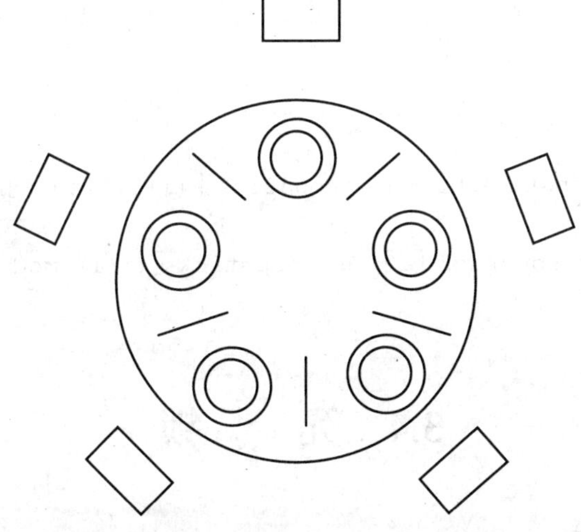

图 3-9 哲学家进餐问题示意图

每位哲学家，在吃饭时，需要得到左右两只筷子，所以筷子是它们需要的资源，我们为每支筷子设置一个信号量 chopstick［i］（i=0，…，4），它们的同步算法如下：

```
semaphore  chopstick[0…4];
begin
```

```
    for (i = 0 to 4)
      chopstick [i] = 1;
    next i
end
Parbegin
    repeat
        think;
        wait(chopstick [i] );
        wait(chopstick [(i+1) mod 5] );
        eat;
        signal(chopstick [i] );
        signal(chopstick [(i+1) mod 5] );
    until false;
Parend
```

这种解法可能发生死锁。例如说恰好五个哲学家均拿起了每个人左边的筷子，这样五个哲学家都争不到右边的筷子，都不能吃饭，发生死锁。

我们可以考虑使用集合型信号量的方法解决这个问题。每个哲学家要么把左右两边的筷子同时拿起，要么就一只都不拿，这样就不会死锁了。

```
semaphore  chopstick [0…4];
begin
    for (i = 0 to 4)
      chopstick [i] = 1;
    next i
end
Parbegin
    repeat
     think;
        Swait(chopstick [i], 1, chopstick [(i+1) mod 5], 1);
         eat;
        Ssignal(chopstick [i], 1, chopstick [(i+1) mod 5], 1);
    until false;
Parend
```

3.4 死　锁

3.4.1 死锁的产生

死锁是指两个或两个以上的进程在执行过程中，因争夺资源而造成的一种互相等待的现象，若无外力推动，它们都将无法推进下去，此时称系统处于死锁状态或者说系统产生了死锁。

前面讲过一个例子，系统中有 P_1、P_2 两个进程，有 R_1、R_2 两个资源，P_1、P_2 进程都是需要同时得到 R_1、R_2 两个资源才能运行结束。在某一时刻，P_1 进程申请并得到了 R_1 资源，P_2 进程申请并得到了 R_2 资源，接下来，P_1 进程等待 P_2 进程释放 R_2 资源，否则它不会释放 R_1 资源，而 P_2 进程在等待 P_1 进程释放 R_1 资源，否则它也不会释放 R_2 资源。这时，P_1、P_2 进程都将无限期地等待下去，都将无法运行结束，这就出现了死锁。

不难看出，死锁进程是针对两个或两个以上的进程而言的，所以，一个系统中一旦发生死锁，死锁的进程至少有两个，一个进程不存在死锁的问题；另外死锁与资源竞争有关，死锁的进程至少有两个进程占有资源；所有的死锁进程必须在等待资源。

1．产生死锁的原因

总结起来，产生死锁的原因有两个主要方面。

（1）资源不够，资源的数量不是足够多，不能同时满足所有进程提出的资源申请，这就造成了资源的竞争，而且资源的使用不允许剥夺。

（2）进程的推进不当，进程的推进次序影响系统对资源的使用。例如，上述的 P_1、P_2 进程，如果让 P_1 进程申请 R_1 资源，再申请 R_2 资源，然后 P_2 申请 R_2，可能这时 P_2 暂时因得不到资源而阻塞，但 P_1 进程需要的资源都已满足，P_1 进程会使用资源结束，释放资源并唤醒 P_2 进程。这样的推进方式就不会死锁了。

若 P_1 保持了资源 R_1,P_2 保持了资源 R_2，系统处于不安全状态，因为这两个进程再向前推进，便可能发生死锁。例如，当 P_1 运行到 P_1：Request（R_2）时，将因 R_2 已被 P_2 占用而阻塞。当 P_2 运行到 P_2：Request（R_1）时，也将因 R_1 已被 P_1 占用而阻塞，于是发生进程死锁。

当进程 P_1 和 P_2 并发执行时，如果按照下述顺序推进，P_1：Request（R_1）。P_1：Request（R_2）。P_1:Relese（R_1）。P_1:Relese（R_2）。P_2：Request（R_2）。P_2：Request（R_1）。P_2:Relese（R_2）。P_2:Relese（R_1）。这两个进程便可顺利完成，这种不会引起进程死锁的推进顺序是合法的。

如图 3-10 所示，①的推进路线，P_2 进程使用资源 R_1、R_2，全部释放后，P_1 进程申请和使用资源 R_1、R_2，不会发生死锁。同理③的推进路线，P_1 先使用资源，P_2 后使用资源，也不会发生死锁。但是②的推进路线，P_1 占有资源 R_1，P_2 占有资源 R_2，最终一定会发生死锁。

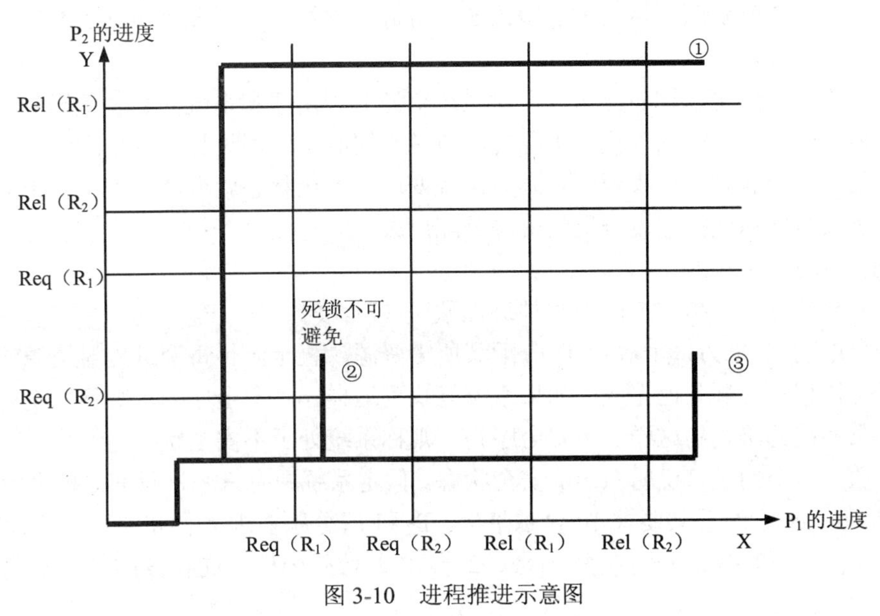

图 3-10　进程推进示意图

2．产生死锁的必要条件

虽然进程在运行过程中，可能发生死锁，但死锁的发生也必须具备一定的条件，死锁的发生必须具备以下四个必要条件。

（1）互斥条件：指进程对所分配到的资源进行排他性使用，即在一段时间内某资源只由一个进程占用。如果此时还有其他进程请求资源，则请求者只能等待，直至占有资源的进程用毕释放。

（2）请求和保持条件：指进程已经保持至少一个资源，但又提出了新的资源请求，而该资源已被其他进程占有，此时请求进程阻塞，但又对自己已获得的其他资源保持不放。如图 3-11 所示。

（3）不剥夺条件：指进程已获得的资源，在未使用完之前，不能被剥夺，只能在使用完时由自己释放。

（4）环路等待条件：指在发生死锁时，必然存在一个进程——资源的环形链，即进程集合{P_0，P_1，P_2，…，P_n}中的 P_0 正在等待一个 P_1 占用的资源；P_1 正在等待 P_2 占用的资源，……，P_n 正在等待已被 P_0 占用的资源。

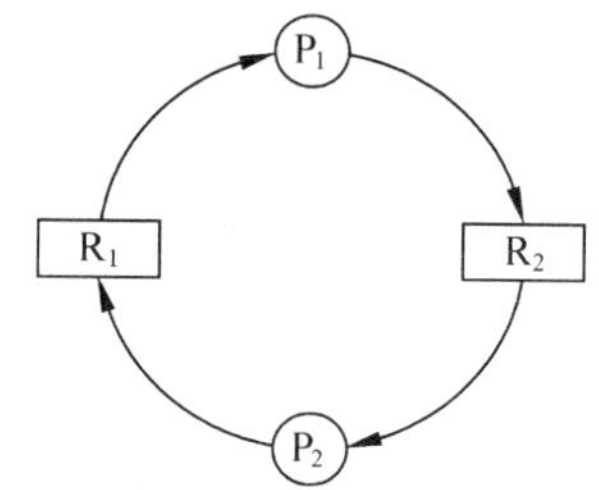

图 3-11　请求保持条件

3.4.2　预防死锁

预防死锁有许多办法，只要使产生死锁的必要条件不能成立，就不会发生死锁。例如禁止“请求保持条件”，进程需要的资源一次性提出，系统一次性分配，避免“占有资源请求其他资源”的情况发生，就不会发生死锁现象了。但是，这样做降低了进程并发的机会，降低了系统的吞吐量；进程获得了资源可能长时间不能用到，造成资源浪费，所以这种做法不可取。

将系统资源按类型进行线性排序，为系统中各种资源分配序号，分配资源时按照序号的次序去申请资源，例如按由小到大的次序申请资源，如果某进程需要 5 号资源和 2 耗资源，就要先申请 2 号资源，然后才能申请 5 号资源。这样可以避免“环路等待”的必要条件，系统不会死锁。

这几种方法虽然可以预防死锁，但对系统资源的利用率和系统的吞吐量影响较大。较好一点的做法是采用银行家算法，进程提出资源申请时，先假设把资源分配给它，再检查系统是否处于安全状态，如果分配资源后系统变成不安全状态就取消分配；如果分配后系统还是处于安全状态就把假设分配的资源分配出去。

1．安全状态

所谓安全状态，是指系统能按某种进程顺序（P_1，P_2，…，P_n）（称<P_1，P_2，…，P_n>序列为安全序列），来为每个进程 P_i 分配其所需资源，直至满足每个进程对资源的最大需求，使每个进程都可顺利地完成。如果系统能找到这样一个安全序列，则称系统处于安全状态。如果系统无法找到这样一个安全序列，则称系统处于不安全状态。

我们通过一个例子来说明系统的安全状态。假定系统中有三个进程 P_1、P_2 和 P_3，共有 12 台磁带机。进程 P_1 总共要求 10 台磁带机，P_2 和 P_3 分别要求 4 台和 9 台。假设在 T_0 时刻，进程 P_1、P_2 和 P_3 已分别获得 5 台、2 台和 2 台磁带机，尚有 3 台空闲未分配，如表 3-3 所示。

表 3-3 P_1、P_2、P_3状态与需求

进　程	最大需求	已分配	可　用
P_1	10	5	3
P_2	4	2	
P_3	9	2	

系统当前的状态是安全的。因为系统中还有 3 台磁带机，如果分配给进程 P_2 两台，满足了 P_2 的最大需求，P_2 可以运行结束，并最终释放它的全部资源，系统中将有 5 台磁带机；如果再把这 5 台磁带机分配给 P_1 进程，就满足了 P_1 进程的最大需求，P_1 进程也将能够运行结束，释放它的全部资源。系统中将会有 10 台磁带机，再分配 7 台给进程 P_3，P_3 也就满足最大需求，也会运行结束，释放全部资源。这样就存在着这样一个安全序列<P_2，P_1，P_3>，按照这个序列分配资源的话，系统中的进程会全部运行结束，所以说当前的状态是安全状态。

如果不按照安全序列分配资源，则系统可能会由安全状态进入不安全状态。例如，在 T_0 时刻以后，P_3 又请求 1 台磁带机，若此时系统把剩余 3 台中的 1 台分配给 P_3，则系统便进入不安全状态。因为，此时也无法再找到一个安全序列，例如，把其余的 2 台分配给 P_2，这样，在 P_2 完成后只能释放出 4 台，既不能满足 P_1 尚需 5 台的要求，也不能满足 P_3 尚需 6 台的要求，致使它们都无法推进到完成，彼此都在等待对方释放资源，即陷入僵局，结果导致死锁。

有了安全状态的概念，在进程提出资源申请后，先假设分配给它，然后检查系统的安全状态，看看是不是安全的，如果是安全的就分配资源，如果是不安全的就拒绝分配，这样能即能提高系统资源的利用率，又能避免死锁。这种做法也被叫做“银行家”算法。

2．银行家算法

通过检查系统的安全状态，决定是否将资源分配给进程的算法，就是银行家算法。先看一个例子，例如有一个银行家，手里有 100 万元的资金，他看中三个投资项目，这三个投资项目所需资金分别是 60 万、30 万、70 万。这三个项目的总投资 160 万元，超出了他的现有资金。但是这三个项目的投资回报率都很高，他都不想放弃，怎么办？他会根据每个项目在不同的阶段提出资金申请时，检查一下资金投入的状态是否安全。三个项目启动时的启动资金分别是 20 万、20 万、30 万，他认为现在的资金投入是安全的，如表 3-4 所示。

表 3-4 三个项目需求现状表

项目号	需求总资金	已获得资金	还需要资金	银行家剩余资金
1	60	20	40	
2	30	20	10	30
3	70	30	40	

这时银行家剩余资金 30 万元，可以给项目 2 投入 10 万，项目 2 能够运行结束，已投入资金能够收回来，他将有 50 万资金；再给项目 1 投入 40 万元，项目 1 也能运行结束，资金回笼，将有 70 万元；再投入 3 号项目 40 万，项目 3 结束，全部资金可以收回。也就

是说，现在存在这样一个投资序列（项目 2，项目 1，项目 3），可以使三个项目都能运行结束，所以说当前的状态是安全的。聪明的银行家用 100 万元可以运作总共需要 160 万元的三个项目。

在当前状态下如果项目 1 提出申请资金 20 万，银行家是否批准呢？他将假设同意为项目提供资金，然后再测算一下，资金的安全性，如表 3-5 所示。

表 3-5　项目 1 提出 20 万

项目号	需求总资金	已获得资金	还需要资金	银行家剩余资金
1	60	40	20	10
2	30	20	10	
3	70	30	40	

这时银行家剩余资金 10 万元，可以给项目 2 投入 10 万，项目 2 能够运行结束，已投入资金能够收回来，他将有 30 万资金；再给项目 1 投入 20 万元，项目 1 也能运行结束，资金回笼，将有 70 万元；再投入 3 号项目 40 万，项目 3 结束，全部资金可以收回。也就是说，现在还是存在这样一个投资序列（项目 2，项目 1，项目 3），可以使三个项目都能运行结束，所以说当前的状态也是安全的。那么项目 1 的这个资金申请就可以批准，为项目 1 提供 20 万的资金。

如果这时项目 3 提出 5 万元的资金要求，银行家是否同意呢？先假设他同意，那么系统的状态就变为如表 3-6 所示。

表 3-6　项目 3 提出 5 万

项目号	需求总资金	已获得资金	还需要资金	银行家剩余资金
1	60	40	20	5
2	30	20	10	
3	70	35	35	

这时，银行家剩余的资金分配给哪一个项目，都不能使项目运行结束，他已投入的资金也收不回来，现在的系统状态就是不安全状态，所以项目 3 提出的 5 万元资金的申请，不能同意。

上述例子中银行家的做法就是判断系统的安全性，动态地决定资源的分配。用到的是安全状态的概念。

银行家算法是一种最有代表性的避免死锁的算法。在避免死锁方法中允许进程动态地申请资源，但系统在进行资源分配之前，应先计算此次分配资源的安全性，若分配不会导致系统进入不安全状态，则分配，否则不分配。

3．银行家算法的实现

1）银行家算法中的数据结构

（1）可利用资源向量 Available。这是一个含有 m 个元素的数组，其中的每一个元素代表一类可利用的资源数目，其初始值是系统中所配置的该类全部可用资源的数目，其数值随该类资源的分配和回收而动态地改变。如果 Available［j］=K，则表示系统中现有 R_j 类资源 K 个。

（2）最大需求矩阵 Max。这是一个 $n \times m$ 的矩阵，它定义了系统中 n 个进程中的每一个进程对 m 类资源的最大需求。如果 Max［i, j］=K，则表示进程 i 需要 R_j 类资源的最大数目为 K。

（3）分配矩阵 Allocation。这也是一个 $n \times m$ 的矩阵，它定义了系统中每一类资源当前已分配给每一进程的资源数。如果 Allocation［i, j］=K，则表示进程 i 当前已分得 R_j 类资源的数目为 K。

（4）需求矩阵 Need。这也是一个 $n \times m$ 的矩阵，用以表示每一个进程尚需的各类资源数。如果 Need［i,j］=K，则表示进程 i 还需要 R_j 类资源 K 个，方能完成其任务。

Need［i,j］=Max［i,j］-Allocation［i,j］

2）银行家算法

设 Requesti 是进程 P_i 的请求向量，如果 Requesti［j］=K，表示进程 P_i 需要 K 个 R_j 类型的资源。当 P_i 发出资源请求后，系统按下述步骤进行检查：

（1）如果 Request i［j］≤Need［i,j］，便转向步骤（2）；否则认为出错，因为它所需要的资源数已超过它所宣布的最大值。

（2）如果 Request i［j］≤Available［j］，便转向步骤（3）；否则，表示尚无足够资源，P_i 须等待。

（3）系统试探着把资源分配给进程 P_i，并修改下面数据结构中的数值：

Available［j］：=Available［j］-Request i［j］；

Allocation［i,j］：=Allocation［i,j］+Request i［j］；

Need［i,j］：=Need［i,j］-Request i［j］；

（4）系统执行安全性算法，检查此次资源分配后，系统是否处于安全状态。若安全，才正式将资源分配给进程 P_i，以完成本次分配；否则， 将本次的试探分配作废，恢复原来的资源分配状态，让进程 P_i 等待。

3）安全性算法

（1）设置两个向量。① 工作向量 Work: 它表示系统可提供给进程继续运行所需的各类资源数目，它含有 m 个元素，在执行安全算法开始时，Work：=Available。② Finish:它表示系统是否有足够的资源分配给进程，使之运行完成。开始时先做 Finish［i］：=false，当有足够资源分配给进程时，再令 Finish［i］：=true。

（2）从进程集合中找到一个能满足下述条件的进程：

① Finish［i］=false; ② Need［i, j］≤Work［j］；若找到，执行步骤（3），否则，执行步骤（4）。

（3）当进程 P_i 获得资源后，可顺利执行，直至完成，并释放出分配给它的资源，故应执行：

Work［j］：=Work［i］+Allocation［i,j］；

Finish［i］：=true;

go to step 2;

（4）如果所有进程的 Finish［i］=true 都满足， 则表示系统处于安全状态；否则，系统处于不安全状态。

4）银行家算法举例

假定系统中有五个进程｛P_0, P_1 ,P_2, P_3, P_4｝和三类资源｛A, B, C｝，各种资源的数量分别为 10、5、7，在 T_0 时刻的资源分配情况如表 3-7 所示。

表 3-7　T_0 时刻资源分配情况

资源情况 / 进程	Max			Allocation			Need			Available		
	A	B	C	A	B	C	A	B	C	A	B	C
P_0	8	5	2	1	1	0	7	4	2	2 (1	3 3	3 1)
P_1	3	2	2	2 (3	0 0	0 2)	1 (0	2 2	2 0)			
P_2	8	0	2	3	0	1	5	0	1			
P_3	2	2	3	2	1	1	0	1	2			
P_4	4	3	3	0	0	2	4	3	1			

（1）T_0 时刻的安全性：

如表 3-8 所示，系统中存在一个安全序列（P_1，P_3，P_4，P_2，P_0），如果按照安全序列的次序运行，每个进程都能够运行结束，所以说，T_0 时刻系统是安全的，不会发生死锁。

表 3-8　T_0 时刻安全序列

资源情况 / 进程	Work			Need			Allocation			Work+Allocation			Finish
	A	B	C	A	B	C	A	B	C	A	B	C	
P_1	2	3	3	1	2	2	2	0	0	4	3	3	true
P_3	4	3	3	0	1	2	2	1	1	6	4	4	true
P_4	6	4	4	4	3	1	0	0	2	6	4	6	true
P_2	6	4	6	5	0	1	3	0	1	9	4	7	true
P_0	9	4	7	7	2	3	1	1	0	10	5	7	true

（2）T_1 时刻 P_1 发出请求向量 Request1(1，0，2)，系统按银行家算法进行检查。

① Request1(1, 0, 2)≤Need1(1, 2, 2)

② Request1(1, 0, 2)≤Available1(2, 3, 3)

③ 系统先假定可为 P_1 分配资源，并修改 Available, Allocation1 和 Need1 向量，由此形成的资源变化情况如表 3-7 中的圆括号所示。

④ 再利用安全性算法检查此时系统是否安全。

如表 3-9 所示，系统中仍然存在着安全序列（P_1，P_3，P_4，P_2，P_0），如果按照这个安全序列的次序运行，每个进程都能够运行结束，所以说，T_1 时刻系统也是安全的，不会发生死锁。T_1 时刻 P_1 发出的请求向量 Request1(1，0，2)可以分配。

表 3-9　T_1 时刻安全序列

资源情况 / 进程	Work			Need			Allocation			Work+Allocation			Finish
	A	B	C	A	B	C	A	B	C	A	B	C	
P_1	1	3	1	0	2	0	3	0	2	4	3	3	true
P_3	4	3	3	0	1	2	2	1	1	6	4	4	true
P_4	6	4	4	4	3	1	0	0	2	6	4	6	true
P_2	6	4	6	5	0	1	3	0	1	9	4	7	true
P_0	9	4	7	7	2	3	1	1	0	10	5	7	true

（3）T_2 时刻，P_4 发出请求向量 Request4(1，2，0)，系统按银行家算法进行检查：

① Request4(1, 2, 0)≤Need4(4, 3, 1)；

② Request4(1, 2, 0) <Available(1, 3, 1)；

③ 系统先假定可为 P_4 分配资源，并修改 Available, Allocation4 和 Need4 向量，由此形成的资源变化情况如表 3-10 所示的资源分配表。

表 3-10　资源分配表

资源情况 / 进程	Max			Allocation			Need			Available		
	A	B	C	A	B	C	A	B	C	A	B	C
P_0	8	5	2	1	1	0	7	4	2	0	1	1
P_1	3	2	2	3	0	2	0	2	0			
P_2	8	0	2	3	0	1	5	0	1			
P_3	2	2	3	2	1	1	0	1	2			
P_4	4	3	3	1	2	2	3	1	1			

如表 3-10 和图 3-12 所示，系统不存任何的安全序列能够使所有进程运行结束，系统将进入不安全状态，将会发生死锁。所以，T_2 时刻 P_4 发出请求向量 Request4(1，2，0)，不予分配。

3.4.3　死锁的检测与解除

死锁的预防和避免在某些时候可能花费较大的代价。另一种方法是检测到系统出现死锁时，通过采用一定的技术来恢复系统。因此，首先需要确定以下两个问题。

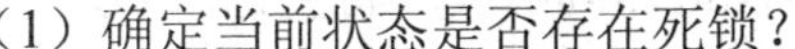

（1）确定当前状态是否存在死锁？

（2）存在死锁时包含哪些进程？

图 3-12　资源分配图

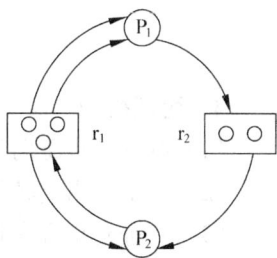

1．资源分配图(Resource Allocation Graph)

（1）把 N 分为两个互斥的子集，即一组进程节点 $P=\{p_1,p_2,\ldots,p_n\}$ 和一组资源节点 $R=\{r_1,r_2,\ldots,r_n\}$，$N=P\cup R$。$P=\{p_1,p_2,\ldots,p_n\}$，$R=\{r_1,r_2,\ldots,r_n\}$，$N=\{p_1,p_2,\ldots,p_n\}\cup\{r_1,r_2,\ldots,r_n\}$。

（2）凡属于 E 中的一个边 $e\in E$，都连接着 P 中的一个节点和 R 中的一个节点，$e=\{p_i, r_j\}$ 是资源请求边，由进程 p_i 指向资源 r_j，它表示进程 p_i 请求一个单位的 r_j 资源。$e=\{r_j, p_i\}$ 是资源分配边，由资源 r_j 指向进程 p_i,它表示把一个单位的资源 r_j 分配给进程 p_i，如图 3-12 所示。

当进程 p_i 请求资源类 r_i 的一个实例时，将一条请求边加入资源分配图，如果这个请求是可以满足的，则该请求边立即转换成分配边；当进程随后释放了某个资源时，则删除分配边。

2．资源分配图的化简

化简的方法如下。

（1）在资源分配图中，找出一个既非等待又非孤立的进程节点 P_i，如果 P_i 能获得它所需要的全部资源，那么它会运行完成，然后会释放它所占有的全部资源，故可在资源分配图中消去 P_i 所有的申请边和分配边，使之成为既无申请边又无分配边的孤立节点。

（2）将 P_i 所释放的资源分配给其他申请资源的进程，即在资源分配图中将这些进程对资源的申请边改为分配边。

（3）重复（1）、（2）两步骤，直到找不到符合条件的进程节点。

经过化简后，若能消去资源分配图中的所有边，使所有进程都成为孤立节点，则该图是可完全化简的，否则不可化简的。

3．死锁定理

基于资源分配图的定义，可给出判定死锁的法则，又称为死锁定理。如果资源分配图是可完全化简的，则系统是安全的，如果资源分配图是不可化简的，则系统处于不安全状态，会发生死锁。例如针对图 3-12 的资源分配图的化简，如图 3-13 所示。

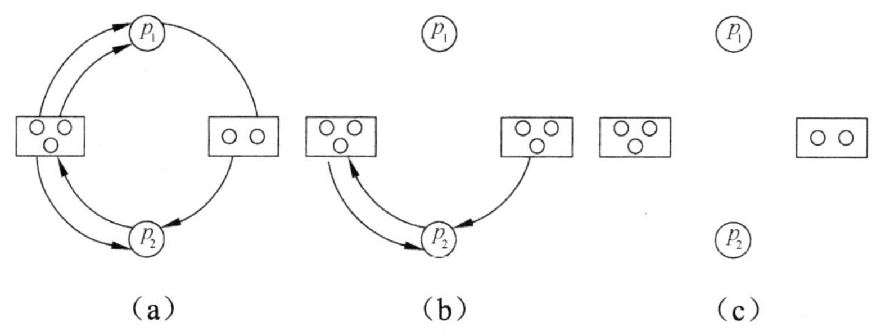

图 3-13　资源分配图的简化

4．死锁检测

（1）可利用资源向量 Available，它表示了 m 类资源中每一类资源的可用数目。

（2）把不占用资源的进程(向量 Allocation：=0)记入 L 表中，即 $L_i \cup L$。

（3）从进程集合中找到一个 Requesti≤Work 的进程，做如下处理：① 将其资源分配图简化，释放出资源，增加工作向量 Work：=Work+Allocationi。② 将它记入 L 表中。

（4）若不能把所有进程都记入 L 表中，便表明系统状态 S 的资源分配图是不可完全简化的。因此，该系统状态将发生死锁。算法描述如下：

```
Work: =Available;
L: ={Li|Allocationi=0∩Requesti=0}
for  all  Li ∈ L do
      begin
           for all Requesti≤Work do
               begin
                  Work: =Work+Allocationi;
                  Li∪L;
               end
      end
deadlock：  = ¬ (L={p1, p2, …, pn});
```

5．死锁的解除

当检测到系统中已发生死锁时，须将进程从死锁状态中解脱出来。常用的实施方法是撤销或挂起一些进程，以便回收一些资源，再将这些资源分配给已处于阻塞状态的进程，使之转为就绪状态，以继续运行。解除死锁的常用办法有剥夺资源法、撤销进程法和回退法等。

1）剥夺法恢复

将某些资源从其他进程强占过来分配给另一些进程。要求强占不影响原进程恢复后的执行。与资源的属性有关，难实现。

2）撤销进程

这是常用的解除死锁的方法，从系统中撤销某些进程，释放资源以解除死锁。要求保证系统的数据等的一致性，难于判断。

3）回退法恢复

系统执行过程中设置若干断点，并保存现场。采用回滚方式释放资源以解除死锁。要求保护的现场不能频繁覆盖。

3.5 管　　程

系统中的各种硬件资源和软件资源，均可用数据结构抽象地描述其资源特性，即用少量信息和对资源所执行的操作来表征该资源，而忽略了它们的内部结构和实现细节。

利用共享数据结构抽象地表示系统中的共享资源，而把对该共享数据结构实施的操作定义为一组过程。

代表共享资源的数据结构，以及由对该共享数据结构实施操作的一组过程所组成的资源管理程序，共同构成了一个操作系统的资源管理模块，我们称之为管程。

Hansan 为管程所下的定义：一个管程定义了一个数据结构和能为并发进程所执行（在该数据结构上）的一组操作，这组操作能同步进程和改变管程中的数据。

有上述定义可知管程由四部分组成，如图 3-14 所示：

（1）管程名称；

（2）管程内部的共享变量；

（3）管程内部并行执行的进程；

（4）对局部于管程内部的共享数据设置初始值的语句。

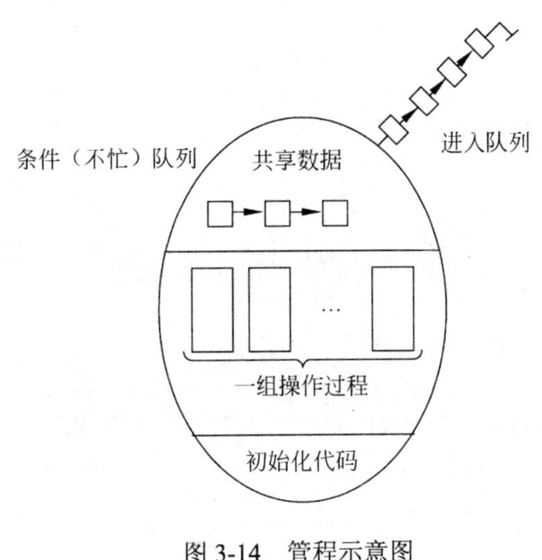

图 3-14　管程示意图

利用管程实现生产者-消费者进程的同步。

管程：buffer=MODULE;

（假设已实现一基本管程 monitor，提供 enter,leave,signal,wait 等操作）

```
notfull,notempty:condition;
   - notfull控制缓冲区不满,notempty控制缓冲区不空;
count,in,out: integer;
   - count记录共有几件物品，in记录第一个空缓冲区，out记录第一个不空的缓冲区
buf:array [0…k-1] of item_type;
define deposit,fetch;
use monitor.enter,monitor.leave,monitor.wait,monitor.signal;
procedure deposit(item);
{
   if(count=k) monitor.wait(notfull);
   buf[in]=item;
   in:=(in+1) mod k;
   count++;
   monitor.signal(notempty);
 }
procedure fetch:Item_type;
 {
   if(count=0) monitor.wait(notempty);
   item=buf[out];
   in:=(in+1) mod k;
   count--;
   monitor.signal(notfull);
   return(item);
 }
{
    count=0;
    in=0;
    out=0;
}
进程: producer,consumer;
producer (生产者进程):
Item_Type item;
{
    while (true)
    {
        produce(&item);
        buffer.enter();
        buffer.deposit(item);
        buffer.leave();
    }
```

```
}
consumer（消费者进程）:
Item_Type  item;
{
    while (true)
    {
        buffer.enter();
        item=buffer.fetch();
        buffer.leave();
        consume(&item);
    }
}
```

Chapter 3 Process Synchronization

A cooperating process is one that can affect or be affected by other processes executing in the system. Cooperating processes can either directly share a logical address space (that is, both code and data) or be allowed to share data only through files or messages. The former case is achieved through the use of lightweight processes or threads. Concurrent access to shared data may result in data inconsistency. In this chapter, we discuss various mechanisms to ensure the orderly execution of cooperating processes that share a logical address space, so that data consistency is maintained.

3.1 Process Synchronization

3.1.1 Critical Section

In concurrent programming a critical section is a piece of code that accesses a shared resource (data structure or device) that must not be concurrently accessed by more than one thread of execution. A critical section will usually terminate in fixed time, and a process will have to wait a fixed time to enter it (also known as bounded waiting). Some synchronization mechanism is required at the entry and exit of the critical section to ensure exclusive use, for example a semaphore.

By carefully controlling which variables are modified inside and outside the critical section (usually, by accessing important state only from within), concurrent access to that state is prevented. A critical section is typically used when a multithreaded program must update multiple related variables without a separate thread making conflicting changes to that data. In a related situation, a critical section may be used to ensure a shared resource, for example a printer, can only be accessed by one process at a time.

3.1.2 Semaphore

A semaphore S is an integer variable that, apart from initialization, is accessed only through two standard atomic operations: wait() and signal(). The wait() operation was originally termed P (from the Dutch proberen, "to test"); signal() was originally called V (from verhogen, "to increment"). The definition of wait() is as follows:

```
wait(s) {
    while S <= 0; // no-op
```

```
        S--;
}
```

The definition of signal() is as follows:

```
signal(S) {
    S++;
}
```

All the modifications to the integer value of the semaphore in the wait() and signal() operations must be executed indivisibly. That is, when one process modifies the semaphore value, no other process can simultaneously modify that same semaphore value. In addition, in the case of wait(S), the testing of the integer value of S (S <= 0), and its possible modification (S--), must also be executed without interruption.

Operating systems often distinguish between counting and binary semaphores. The value of a counting semaphore can range over an unrestricted domain. The value of a binary semaphore can range only between 0 and 1. On some systems, binary semaphores are known as mutex locks, as they are locks that provide mutual exclusion.

We can use binary semaphores to deal with the critical-section problem for multiple processes. The n processes share a semaphore, mutex, initialized to 1. Each process P, is organized as shown as follows:

```
do {
    waiting(mutex);
    // critical section
    Signal(mutex);
    // remainder section
}while(TRUE);
```

Counting semaphores can be used to control access to a given resource consisting of a finite number of instances. The semaphore is initialized to the number of resources available. Each process that wishes to use a resource performs a wait() operation on the semaphore (thereby decrementing the count). When a process releases a resource, it performs a signal() operation (incrementing the count). When the count for the semaphore goes to 0, all resources are being used. After that, processes that wish to use a resource will block until the count becomes greater than 0.

We can also use semaphores to solve various synchronization problems. For example, consider two concurrently running processes: P1 with a statement S1 and P2 with a statement S2. Suppose we require that S2 be executed only after S1has completed. We can implement this scheme readily by letting P1 and P2 share a common semaphore synch, initialized to 0, and by inserting the statements

```
S1;
```

```
Signal(synch);
```

in process P1, and the statements

```
wait(synch);
S2;
```

in process P2. Because synch is initialized to 0, P2 will execute S2 only after P1 has invoked signal(synch), which is after statement S1 has been executed.

The main disadvantage of the semaphore definition given here is that it requires busy waiting. While a process is in its critical section, any other process that tries to enter its critical section must loop continuously in the entry code. This continual looping is clearly a problem in a real multiprogramming system, where a single CPU is shared among many processes. Busy waiting wastes CPU cycles that some other process might be able to use productively. This type of semaphore is also called a spinlock because the process "spins" while waiting for the lock (Spinlocks do have an advantage in that no context switch is required when a process must wait on a lock, and a context switch may take considerable time. Thus, when locks are expected to be held for short times, spinlocks are useful; they are often employed on multiprocessor systems where one thread can "spin" on one processor while another thread performs its critical section on another processor).

To overcome the need for busy waiting, we can modify the definition of the wait() and signal() semaphore operations. When a process executes the wait() operation and finds that the semaphore value is not positive, it must wait. However, rather than engaging in busy waiting, the process can block itself. The block operation places a process into a waiting queue associated with the semaphore, and the state of the process is switched to the waiting state. Then control is transferred to the CPU scheduler, which selects another process to execute.

A process that is blocked, waiting on a semaphore S , should be restarted when some other process executes a signal() operation. The process is restarted by a wakeup() operation, which changes the process from the waiting state to the ready state. The process is then placed in the ready queue (The CPU may or may not be switched from the running process to the newly ready process, depending on the CPU-scheduling algorithm.)

To implement semaphores under this definition, we define a semaphore as a "C" struct:

```
typedef struct {
    int value;
    struct process *list;
} semaphore;
```

Each semaphore has an integer value and a list of processes list. When a process must wait on a semaphore, it is added to the list of processes. A signal() operation removes one process from the list of waiting processes and awakens that process.

The wait() semaphore operation can now be defined as

```
wait(semaphore *S) {
    S->value--;
    if (S->value < 0) {
        add this process to S->list;
        block();
    }
}
```

The signal() semaphore operation can now be defined as

```
signal(semaphore *S) {
    S->value++;
    if (S->value <= 0) {
        remove a process from P from S->list;
        wakeup(P);
    }
}
```

The block() operation suspends the process that invokes it. The wakeup(P) operation resumes the execution of a blocked process P. These two operations are provided by the operating system as basic system calls.

Note that, although under the classical definition of semaphores with busy waiting the semaphore value is never negative, this implementation may have negative semaphore values. If the semaphore value is negative, its magnitude is the number of processes waiting on that semaphore. This fact results from switching the order of the decrement and the test in the implementation of the wait() operation.

The list of waiting processes can be easily implemented by a link field in each process control block (PCB). Each semaphore contains an integer value and a pointer to a list of PCBs. One way to add and remove processes from the list in a way that ensures bounded waiting is to use a FIFO queue, where the semaphore contains both head and tail pointers to the queue. In general, however, the list can use any queueing strategy. Correct usage of semaphores does not depend on a particular queueing strategy for the semaphore lists.

The critical aspect of semaphores is that they be executed atomically. We must guarantee that no two processes can execute wait() and signal() operations on the same semaphore at the same time. This is a critical-section problem; and in a single-processor environment (that is, where only one CPU exists), we can solve it by simply inhibiting interrupts during the time the wait() and signal() operations are executing. This scheme works in a single-processor environment because, once interrupts are inhibited, instructions from different processes cannot be interleaved. Only the currently running process executes until interrupts are reenabled and the scheduler can regain control.

In a multiprocessor environment, interrupts must be disabled on every processor; otherwise, instructions from different processes (running on different processors) may be interleaved in some arbitrary way. Disabling interrupts on every processor can be a difficult task and furthermore can seriously diminish performance. Therefore, SMP systems must provide alternative locking techniques—such as spinlocks — to ensure that wait() and signal() are performed atomically.

It is important to admit that we have not completely eliminated busy waiting with this definition of the wait() and signal() operations. Rather, we have removed busy waiting from the entry section to the critical sections of application programs. Furthermore, we have limited busy waiting to the critical sections of the wait() and signal() operations, and these sections are short (if properly coded, they should be no more than about ten instructions).

Thus, the critical section is almost never occupied, and busy waiting occurs rarely, and then for only a short time. An entirely different situation exists with application programs whose critical sections may be long (minutes or even hours) or may almost always be occupied. In such cases, busy waiting is extremely inefficient.

3.2 Classic Problems of Synchronization

3.2.1 Producer-Consumer (Bounded-Buffer) Problem

A producer process produces information that is consumed by a consumer process. For example, a compiler may produce assembly code, which is consumed by an assembler. The assembler, in turn, may produce object modules, which are consumed by the loader. The producer—consumer problem also provides a useful metaphor for the client—server paradigm. We generally think of a server as a producer and a client as a consumer. For example, a web server produces (that is, provides) HTML files and images, which are consumed (that is, read) by the client web browser requesting the resource.

One solution to the producer—consumer problem uses shared memory. To allow producer and consumer processes to run concurrently, we must have available a buffer of items that can be filled by the producer and emptied by the consumer. This buffer will reside in a region of memory that is shared by the producer and consumer processes. A producer can produce one item while the consumer is consuming another item. The producer and consumer must be synchronized, so that the consumer does not try to consume an item that has not yet been produced.

Two types of buffers can be used. The unbounded buffer places no practical limit on the size of the buffer. The consumer may have to wait for new items, but the producer can always produce new items. The bounded buffer assumes a fixed buffer size. In this case, the consumer must wait if the buffer is empty, and the producer must wait if the buffer is full.

We assume that the pool consists of n buffers, each capable of holding one item. The mutex semaphore provides mutual exclusion for accesses to the buffer pool and is initialized to the value 1. The empty and full semaphores count the number of empty and full buffers. The semaphore empty is initialized to the value n; the semaphore full is initialized to the value 0.

The code for the producer process :

```
do {
     ...
     // produce an item in next p
     ...
     wait(empty);
     wait(mutex);
     ...
     // add next p to buffer
     ...
     signal(mutex);
     signal(full);
}while(TRUE);
```

The code for the consumer process :

```
do {
     wait(full);
     wait(mutex);
     ...
     // remove an item from buffer to next c
     ...
     signal(mutex);
     signal(empty);
     ...
     // consume the item in next c
     ...
}while(TRUE);
```

Note the symmetry between the producer and the consumer. We can interpret this code as the producer producing full buffers for the consumer or as the consumer producing empty buffers for the producer.

3.2.2 The Readers-Writers Problem

A database is to be shared among several concurrent processes. Some of these processes may want only to read the database, whereas others may want to update (that is, to read and write) the database. We distinguish between these two types of processes by referring to the former as readers and to the latter as writers. Obviously, if two readers access the shared data simultaneously, no adverse affects will result. However, if a writer and some other thread (either a reader or a writer) access the database simultaneously, chaos may ensue.

To ensure that these difficulties do not arise, we require that the writers have exclusive access to the shared database. This synchronization problem is referred to as the readers—writers problem. Since it was originally stated, it has been used to test nearly every new synchronization primitive. The readers—writers problem has several variations, all involving priorities. The simplest one, referred to as the first readers—writers problem, requires that no reader will be kept waiting unless a writer has already obtained permission to use the shared object. In other words, no reader should wait for other readers to finish simply because a writer is waiting. The second readers—writers problem requires that, once a writer is ready, that writer performs its write as soon as possible. In other words, if a writer is waiting to access the object, no new readers may start reading.

A solution to either problem may result in starvation. In the first case, writers may starve; in the second case, readers may starve. For this reason, other variants of the problem have been proposed. In this section, we present a solution to the first readers—writers problem.

In the solution to the first readers—writers problem, the reader processes share the following data structures:

```
semaphore mutex, wrt;
int readcount;
```

The semaphores mutex and wrt are initialized to 1; readcount is initialized to 0. The semaphore wrt is common to both reader and writer processes. The mutex semaphore is used to ensure mutual exclusion when the variable readcount is updated. The readcount variable keeps track of how many processes are currently reading the object. The semaphore wrt functions as a mutual-exclusion semaphore for the writers. It is also used by the first or lastreader that enters or exits the critical section. It is not used by readers who enter or exit while other readers are in their critical sections.

The code for a writer process:

```
do {
    wait(wrt);
    ...
    // writing is performed
    ...
    signal(wrt);
}while(TRUE);
```

The code for a reader process:

```
do {
    wait(mutex);
    readcount++;
    if (readcount == 1)
```

```
        wait(wrt);
    signal(mutex);
    ...
    //reading is performed
    ...
    wait(mutex);
    readcount--;
    if (readcount == 0)
        signal(wrt);
    signal(mutex);
}while (TRUE);
```

Note that, if a writer is in the critical section and n readers are waiting, then one reader is queued on wrt, and n −1 readers are queued on mutex. Also observe that, when a writer executes signal (wrt), we may resume the execution of either the waiting readers or a single waiting writer. The selection is made by the scheduler.

The readers—writers problem and its solutions have been generalized to provide reader—writer locks on some systems. Acquiring a reader—writer lock requires specifying the mode of the lock: either read or write access. When a process only wishes to read shared data, it requests the reader—writer lock in read mode; a process wishing to modify the shared data must request the lock in write mode. Multiple processes are permitted to concurrently acquire a reader—writer lock in read mode; only one process may acquire the lock for writing as exclusive access is required for writers.

Reader—writer locks are most useful in the following situations.

(1) In applications where it is easy to identify which processes only read shared data and which threads only write shared data.

(2) In applications that have more readers than writers. This is because reader—writer locks generally require more overhead to establish than semaphores or mutual exclusion locks, and the overhead for setting up a reader—writer lock is compensated by the increased concurrency of allowing multiple readers.

3.3 Deadlocks

3.3.1 Causes of Deadlocks

A system consists of a finite number of resources to be distributed among a number of competing processes. The resources are partitioned into several types, each consisting of some number of identical instances. Memory space, CPU cycles, files, and I/O devices (such as printers and DVI drives) are examples of resource types. if a system has two CPUs, then the resource type CPU has two instances. Similarly, the resource type printer may have five instances.

If a process requests an instance of a resource type, the allocation of any instance of the type will satisfy the request. If it will not, then the instances are not identical, and the resource type classes have not been defined properly. For example, a system may have two printers. These two printers may be defined to be in the same resource class if no one cares which printer prints which outputs. However, if one printer is on the ninth floor and the other is in the basement, then people on the ninth floor may not see both printers as equivalent, and separate resource classes may need to be defined for each printer.

A process must request a resource before using it and must release the resource after using it. A process may request as many resources as it requires to carry out its designated task. Obviously, the number of resources requested may not exceed the total number of resources available in the system. In other words, a process cannot request three printers if the system has only two.

Under the normal mode of operation, a process may utilize a resource in only the following sequence.

(1) Request. If the request cannot be granted immediately (for example, if the resource is being used by another process), then the requesting process must wait until it can acquire the resource.

(2) Use. The process can operate on the resource (for example, if the resource is a printer, the process can print on the printer).

(3) Release. The process releases the resource.

The request and release of resources are system calls. Examples are the request() and release() device, open() and close() file, and allocate() and free() memory system calls. Request and release of resources that are not managed by the operating system can be accomplished through the wait()and signal() operations on semaphores or through acquisition and release of a mutex lock. For each use of a kernel-managed resource by a process or thread, the operating system checks to make sure that the process has requested and has been allocated the resource. A system table records whether each resource is free or allocated; for each resource that is allocated, the table also records the process to which it is allocated. If a process requests a resource that is currently allocated to another process, it can be added to a queue of processes waiting for this resource.

A set of processes is in a deadlock state when every process in the set is waiting for an event that can be caused only by another process in the set. The events with which we are mainly concerned here are resource acquisition and release. The resources may be either physical resources (for example, printers, tape drives, memory space, and CPU cycles) or logical resources (for example, files, semaphores, and monitors). However, other types of events may result in deadlocks .

To illustrate a deadlock state, consider a system with three CD RW drives. Suppose each of three processes holds one of these CD RW drives. If each process now requests another drive, the three processes will be in a deadlock state. Each is waiting for the event “CD RW is released,” which can be caused only by one of the other waiting processes. This example illustrates a deadlock involving the same resource type.

Deadlocks may also involve different resource types. For example, consider a system with one printer and one DVD drive. Suppose that process P_1 is holding the DVD and process P_2 is holding the printer. If P_1 requests the printer and requests the DVD drive, a deadlock occurs.

3.3.2 Deadlock Avoidance

1. Safe State

A state is safe if the system can allocate resources to each process (up to its maximum) in some order and still avoid a deadlock. More formally, a system is in a safe state only if there exists a safe sequence. A sequence of processes $\langle P_1, P_2, \cdots, P_n\rangle$ is a safe sequence for the current allocation state if, for each P_i, the resource requests that P_i can still make can be satisfied by the currently available resources plus the resources held by all P_i, with $j < i$. In this situation, if the resources that P_i needs are not immediately available, then P_i can wait until all P_j have finished. When they have finished, P_i can obtain all of its needed resources, complete its designated task, return its allocated resources, and terminate. When P_i terminates, P_i+1 can obtain its needed resources, and so on. If no such sequence exists, then the system state is said to be unsafe.

A safe state is not a deadlocked state. Conversely, a deadlocked state is an unsafe state. Not all unsafe states are deadlocks, however. An unsafe state may lead to a deadlock. As long as the state is safe, the operating system can avoid unsafe (and deadlocked) states. In an unsafe state, the operating system cannot prevent processes from requesting resources such that a deadlock occurs: The behavior of the processes controls unsafe states.

To illustrate, we consider a system with 12 magnetic tape drives and three processes: P_0, P_1, and P_2. Process P_0 requires 10 tape drives, process P_1 may need as many as 4 tape drives, and process P_2 may need up to 9 tape drives. Suppose that, at time t0, process P_0 is holding 5 tape drives, process P_1 is holding 2 tape drives, and process P_2 is holding 2 tape drives (Thus, there are 3 free tape drives)As is shoun in Table 3.1.

Table 3.1

	Maximum Needs	Current Needs
P_0	10	5
P_1	4	2
P_2	9	2

At time t_0, the system is in a safe state. The sequence <P_1, P_0, P_2> satisfies the safety condition. Process P_1 can immediately be allocated all its tape drives and then return them (the system will then have 5 available tape drives); then process P_0 can get all its tape drives and return them (the system will then have 10 available tape drives); and finally process P_2 can get all its tape drives and return them (the system will then have all 12 tape drives available).

A system can go from a safe state to an unsafe state. Suppose that, at time t_1, process P_2

requests and is allocated one more tape drive. The system is no longer in a safe state. At this point, only process P_1 can be allocated all its tape drives. When it returns them, the system will have only 4 available tape drives. Since process P_0 is allocated 5 tape drives but has a maximum of 10, it may request 5 more tape drives. Since they are unavailable, process P0 must wait. Similarly, process P_2 may request an additional 6 tape drives and have to wait, resulting in a deadlock. Our mistake was in granting the request from process P_2 for one more tape drive. If we had made P_2 wait until either of the other processes had finished and released its resources, then we could have avoided the deadlock.

Given the concept of a safe state, we can define avoidance algorithms that ensure that the system will never deadlock. The idea is simply to ensure that the system will always remain in a safe state. Initially, the system is in a safe state. Whenever a process requests a resource that is currently available, the system must decide whether the resource can be allocated immediately or whether the process must wait. The request is granted only if the allocation leaves the system in a safe state.

In this scheme, if a process requests a resource that is currently available, it may still have to wait. Thus, resource utilization may be lower than it would otherwise be.

2. Banker's Algorithm

The name "Banker's Algorithm" was chosen because the algorithm could be used in a banking system to ensure that the bank never allocated its available cash in such a way that it could no longer satisfy the needs of all its customers.

When a new process enters the system, it must declare the maximum number of instances of each resource type that it may need. This number may not exceed the total number of resources in the system. When a user requests a set of resources, the system must determine whether the allocation of these resources will leave the system in a safe state. If it will, the resources are allocated; otherwise, the process must wait until some other process releases enough resources.

3. Banker's Algorithm Implementation

1) Banker's Algorithm Structure

Several data structures must be maintained to implement the banker's algorithm. These data structures encode the state of the resource-allocation system. Let n be the number of processes in the system and m be the number of resource types. We need the following data structures.

(1) Available. A vector of length in indicates the number of available resources of each type. If Available[j] equals k, there are k instances of resource type R_j available.

(2) Max. A $n \times m$ matrix defines the maximum demand of each process. If Max[i][j] equals k, then process P_i may request at most k instances of resource type R_j.

(3) Allocation. A$n \times m$ matrix defines the number of resources of each type currently allocated to each process. If Allocation[i][j] equals k, then process P_i is currently allocated k instances of resource type R_j.

(4) Need. A$n \times m$ matrix indicates the remaining resource need of each process. If Need[i][j] equals k, then process P_i may need k more instances of resource type R_j to complete its task. Note

that Need[i][j] equals Max[i][j] – Allocation [i][j].

These data structures vary over time in both size and value.

To simplify the presentation of the 'banker's algorithm, we next establish some notation. Let X and Y be vectors of length n. We say that $X <= Y$ if and only if $X[i] <= Y[i]$ for all $1 = 1, 2, ..., n$. For example, if $X = (1,7,3,2)$ and Y $(0,3,2,1)$, then $Y <= X$. $Y < X$ if $Y <= X$ and $Y != X$.

We can treat each row in the matrices Allocation and Need as vectors and refer to them as Allocation$_i$ and Need$_i$. The vector Allocation$_i$ specifies the resources currently allocated to process P_i; the vector Need$_i$ specifies the additional resources that process P_i may still request to complete its task.

2) Banker's Algorithm

We now describe the algorithm which determines if requests can be safely granted.

Let Request$_i$ be the request vector for process P_i. If Request$_i$ [j] == k, then process P_i wants k instances of resource type R_j. When a request for resources is made by process P_i, the following actions are taken.

(1) If Request$_i$ < Need$_i$, go to step 2. Otherwise, raise an error condition, since the process has exceeded its maximum claim.

(2) If Request$_i$ < Available, go to step 3. Otherwise, P_i must wait, since the resources are not available.

(3) Have the system pretend to have allocated the requested resources to process P_i by modifying the state as follows:

Available =Available – Request$_\mathrm{i}$;

Allocation$_i$ = Allocation$_i$ + Request$_i$;

Need$_i$ = Need$_i$ – Request$_i$;

If the resulting resource-allocation state is safe, the transaction is completed, and process P_i is allocated its resources. However, if the new state is unsafe, then P_i must wait for Request$_i$, and the old resource-allocation state is restored.

3) Safety Algorithm

We can now present the algorithm for finding out whether or not a system is in a safe state. This algorithm can be described. as follows.

(1) Let Work and Finish be vectors of length m and n, respectively. Initialize Work = Available and Finish[i] = false for $i = 0, 1, ..., n – 1$.

(2) Find an i such that both

a. Finish[i] == false

b. Need$_\mathrm{i}$ <= Work

If no such i exists, go to step 4.

(3) Work = Work + Allocationi

Finish[i] = true

Go to step 2.

(4) If Finish[i] = true for all i, then the system is in a safe state.

This algorithm may require an order of $m \times n^2$ operations to determine whether a state is safe.

4) An Illustrative Example

Finally, to illustrate the use of the banker's algorithm, consider a system with five processes P0 through P_4 and three resource types A, B, and C. Resource type A has 10 instances, resource type B has 5 instances, and resource type C has 7 instances. Suppose that, at time T_0, the following snapshot of the system. has been taken(shown in Table3.2):

Table 3.2

	Allocation			Max			Available		
	A	B	C	A	B	C	A	B	C
P_0	0	1	0	7	5	3	3	3	2
P_1	2	0	0	3	2	2			
P_2	3	0	2	9	0	2			
P_3	2	1	1	2	2	2			
P_4	0	0	2	4	3	3			

The content of the matrix Need is defined to be Max - Allocation and is as follows(Table 3.3):

Table 3.3

	Need		
	A	B	C
P_0	7	4	3
P_1	1	2	2
P_2	6	0	0
P_3	0	1	1
P_4	4	3	1

We claim that the system is currently in a safe state. Indeed, the sequence <P_1, P_3, P_4, P_2, P_0> satisfies the safety criteria. Suppose now that Process P_i requests one additional instance of resource type A and two instances of resource type C, so Request1 = (1,0,2). To decide whether this request can be immediately granted, we first check that Request1 <= Available — that is, that (1,0,2) <= (3,3,2), which is true. We then pretend that this request has been fulfilled, and we arrive at the following new state (Table 3.4):

Table 3.4

	Allocation			Max			Available		
	A	B	C	A	B	C	A	B	C
P_0	0	1	0	7	4	3	2	3	0
P_1	3	0	2	0	2	0			
P_2	3	0	2	6	0	0			
P_3	2	1	1	0	1	1			
P_4	0	0	2	4	3	1			

We must determine whether this new system state is safe. To do so, we execute our safety algorithm and find that the sequence < P_1, P_3, P_4, P_0, P_2 > satisfies the safety requirement. Hence, we can immediately grant the request of process P_1.

You should be able to see, however, that when the system is in this state, a request for (3,3,0) by P_4 cannot be granted, since the resources are not available. Furthermore, a request for (0,2,0) by P_0 cannot be granted, even though the resources are available, since the resulting state is unsafe.

习　　题

一、选择题

1．P、V（wait、signal）操作是________。

A．两条低级进程通信原语　　B．两组不同的机器指令

C．两条系统调用命令　　D．两条高级进程通信原语

2．若 P、V（wait、signal）操作的信号量 S 初值为 2，当前值为−1，则表示有________等待进程。

A．0 个　　B．1 个　　C．2 个　　D．3 个

3．用 P、V（wait、signal）操作管理临界区时，信号量的初值应定义为________。

A．−1　　B．0　　C．1　　D．任意值

4．用 P、V（wait、signal）操作唤醒一个等待进程时，被唤醒进程的状态变为________。

A．阻塞　　B．就绪　　C．执行　　D．完成

5．进程间的同步是指进程在逻辑上的相互________关系。

A．联接　　B．制约　　C．继续　　D．调用

6．________是一种只能进行 P 操作和 V 操作的特殊变量。

A．调度　　B．进程　　C．同步　　D．信号量

7．________是解决进程间同步和互斥的一对低级通信原语。

A．lock 和 unlock　　B．P 和 V

C．W 和 S　　D．Send 和 Receive

8．下面叙述中正确的是________。

A．操作系统的一个重要概念是进程，因此不同进程所执行的代码也一定不同

B．为了避免发生进程死锁，各进程只能逐个申请资源

C．操作系统用 PCB 管理进程，用户进程可以从 PCB 中读出与本身运行状况相关的信息

D．进程同步是指某些进程之间在逻辑上的相互制约关系

9．在操作系统中，解决进程的____①____和____②____问题的一种方法是使用____③____。

A．调度　　B．互斥　　C．通信　　D．同步　　E．分派　　F．信号量

10．对于两个并发进程，设互斥信号量为 mutex，若 mutex=0，则________。

A．表示没有进程进入临界区

B．表示有一个进程进入临界区

C．表示有一个进程进入临界区，另一个进程等待进入

D．表示有两个进程进入临界区

11．两个进程合作完成一个任务。在并发执行中，一个进程要等待其合作伙伴发来消息，或者建立某个条件后再向前执行，这种制约性合作关系被称为进程的________。

A．同步　　B．互斥　　C．调度　　D．执行

12．为了进行进程协调，进程之间应当具有一定的联系，这种联系通常采用进程间交换数据的方式进行，这种方式称为________。

A．进程互斥　　B．进程同步　　C．进程制约　　D．进程通信

13．为了解决进程间的同步和互斥问题，通常采用一种称为________机制的方法。

A．调度　　B．信号量　　C．分派　　D．通信

14．若系统中有 5 个进程，每个进程都需要 4 个资源 R，那么使系统不发生死锁的资源 R 的最少数目是________。

A．20　　B．18　　C．16　　D．15

15．某系统中有 5 个并发进程，都需要同类资源 2 个，试问该系统不会发生死锁的最少资源数是________。

A．5　　B．6　　C．7　　D．10

16．用记录型信号量 S 实现对系统中 3 台打印机的互斥使用，S.value 的初值应设置为________。

A．−3　　B．0　　C．1　　D．3

17．临界区是指并发进程中访问临界资源的________段。

A．管理信息　　B．信息存储　　C．数据　　D．程序

18．通常不采用________方法来解除死锁。

A．终止一个死锁进程　　B．终止所有死锁进程

C．从死锁进程处抢夺资源　　D．从非死锁进程处抢夺资源

19．要求进程一次性申请所需的全部资源，是破坏了死锁必要条件中的________。

A．互斥　　B．请求与保持　　C．不剥夺　　D．循环等待

20．银行家算法是一种________算法。

A．死锁解除　　B．死锁避免　　C．产生死锁　　D．死锁检测

二、填空题

1. 信号量的物理意义是当信号量值大于 0 时表示可用资源的数目，当信号量值小于 0 时，其________为因请求该资源而被阻塞的进程的数目。

2. 对信号量 S 的 P 原语操作定义中，使进程进入相应等待队列等待的条件是________。

3. 在一个单处理机系统中，若有 5 个用户进程，且假设当前时刻为用户态，则处于就绪状态的用户进程最多有________个，最少有________个。

4. 有 M 个进程共享同一个临界资源，若使用信号量机制实现对临界资源的互斥访问，则信号量值的变化范围是 1 至________。

5. 用信号量 S 实现对系统中 4 台打印机的互斥使用，S 的初值应设置为________，若 S 的当前值为−1，则表示等待队列有________个等待进程。

三、简答题

1．简述临界资源与临界区。

2．简述死锁的概念。

3．简述死锁产生的原因。

4．简述死锁产生的必要条件。

四、综合题

1．有一个发送者进程和一个接收者进程，其流程图如图 3-15 所示，S 是用于实现进程同步的信号量，Mutex 是用于实现进程互斥的信号量。试问流程图中的 A,B,C,D 框中应填写什么？假定消息链长度没有限制，S 和 Mutex 的初值应为多少？

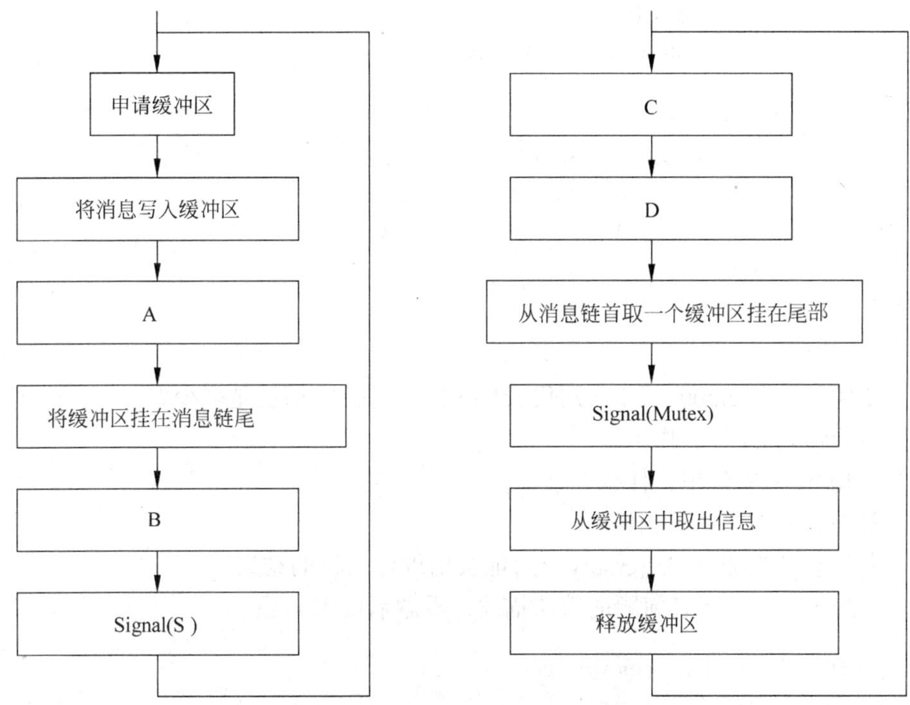

图 3-15　发送者和接收者进程的流程图

发送者进程接收者进程

A__________ B_________ C_________ D________ S____ Mutex_____

2．这是一个生产者-消费者问题的算法描述，看题后回答问题。

```
Var mutex, empty, full: semaphore :=1, n, 0;
buffer: array[0, …, n-1] of item;
in, out: integer := 0,0;
begin
  parbegin
    proceducer: begin
                  repeat
                     producer an item nextp;
```

```
                  wait(empty);-----------------------------------a
                  wait(mutex);-----------------------------------b
                  buffer(in)  := nextp;
                  in := ( in + 1 ) mod n;
                  signal(mutex);---------------------------------c
                  signal(full);------------------------------------d
               until false;
             end
    consumer: begin
               repeat
                  wait(full);
                  wait(mutex);
                  nextc := buffer(out);
                  out :=( out + 1 ) mod n;
                  ? ? ?
                  signal(empty);
                  consumer the item in nextc;
               until false;
             end
 parend
end
```

（1）说明信号量 empty, full 分别代表什么值，它们的初值是多少？

（2）解释语句 a、d 的作用。

（3）语句 b、c 的作用是什么？

（4）补充？？？的位置的语句。

（5）若语句 d 换成 signal(empty); 可能会出现什么样的结果？

3．这是一个读者-写者问题的算法描述，看题后回答问题。

```
Var Rmutex, Wmutex: semaphore :=1, 1;
Readcount: integer := 0;
begin
  parbegin
    Reader: begin
             repeat
               wait(Rmutex);
               if Readcount=0 then wait(Wmutex); -------------------------a
               Readcount := Readcount + 1; -------------------------------b
               signal(Rmutex);
               …
Perform read operation;
               …
               wait(Rmutex);
               Readcount := Readcount - 1;
               if Readcount=0 then signal(Wmutex);
```

```
                signal(Rmutex);
            until false;
          end
    Writer: begin
            repeat
                wait(Wmutex); -------------------------------------------c
                perform write operation;
                signal(Wmutex); -----------------------------------------d
            until false;
         end
  parend
end
```

（1）说明信号量 Rmutex, Wmutex 的作用，它们的初值是多少？

（2）解释变量 Readcount 的作用。

（3）解释语句 a、b。

（4）解释语句 c、d 的作用。

（5）如果读者进程改成下列语句，请描述一下会出现什么情况？

```
repeat
  wait(Rmutex);
  Perform read operation;
  signal(Rmutex);
until false;
```

4．请用信号量的方法实现如图 3-16 所示的前趋图的同步关系。写出同步算法。

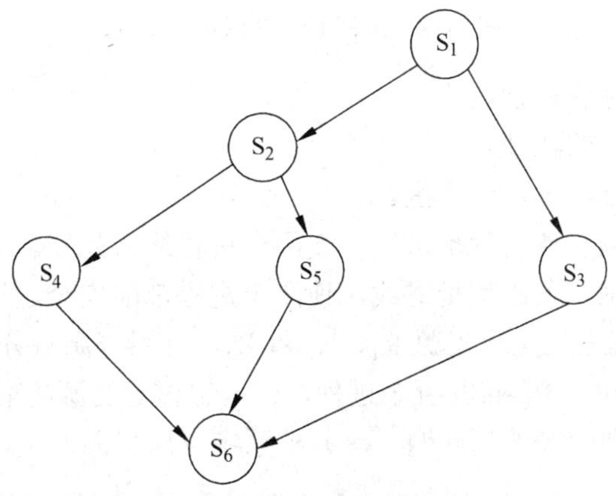

图 3-16　前趋图

5．桌上有一只盘子，每次只能放入一只水果。爸爸专向盘子中放削好的苹果（apple），妈妈专向盘子中放剥好的桔子（orange），一个儿子专等吃盘子中的桔子，一个女儿专等吃盘子中的苹果。试用 P、V 操作实现爸爸、妈妈、儿子和女儿同步活动的算法。

6．有一个文件 P 可供多个进程共享，进程平分成 A、B 两组，规定同组进程可以同时读文件 P，不同组进程不能同时读文件 P，试写出实现两组进程同步的算法。

7．有三个并发进程，通过 Buf1、Buf2 协作完成。如图 3-17 所示，试写出三个进程的同步算法。

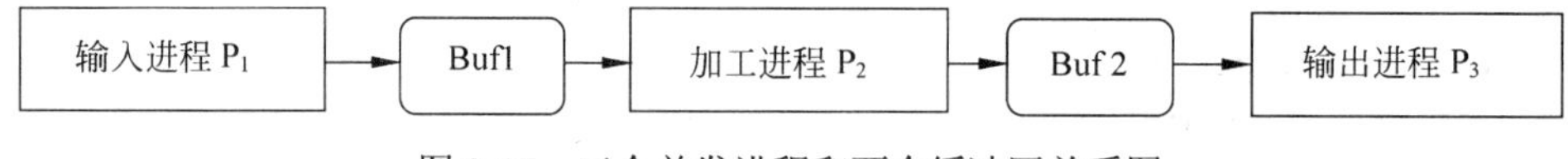

图 3-17　三个并发进程和两个缓冲区关系图

8．嗜睡的理发师，一个理发店由一个 N 张沙发的等候室和一个放有一张理发椅的理发室组成。没有顾客时，理发师便去睡觉。当一个顾客走进理发店时，如果所有的沙发都已占用，他便离开理发店，否则，如果理发师正在为其他顾客理发时，则该顾客就找一张空沙发坐下来等待，如果理发师因无顾客正在睡觉，则由新到的顾客唤醒并为其理发，在理发完成后，顾客必须付费后才能离开理发店，试用信号量实现这一同步问题。

9．图 3-18 的横坐标表示进程 P_1 的推进过程，纵坐标表示进程 P_2 的推进过程，R_1、R_2 代表系统的两个资源，Req(R)表示申请资源，Rel(R)表示释放资源。回答问题。

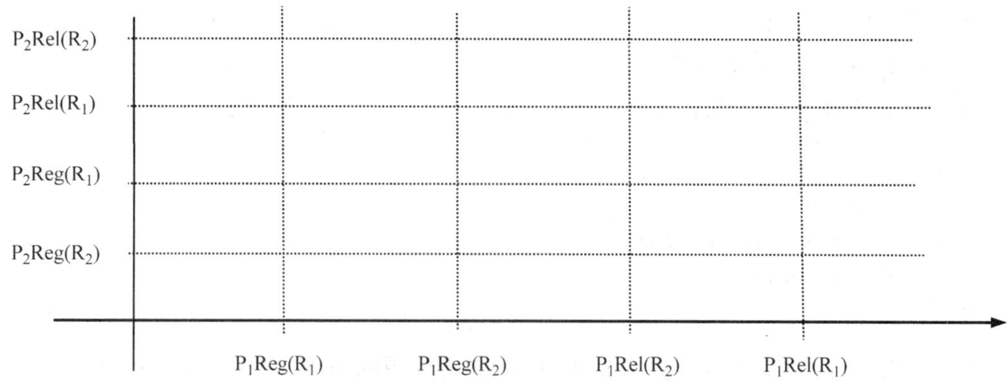

图 3-18　进程 P1、P2 推进示意图

（1）在图 3-18 中标注死锁区。

（2）表明图 3-18 中的死锁点。

（3）在图 3-18 中标注不可到达区域。

（4）在图 3-18 中画一条进程推进线，使两个进程终将死锁。

（5）在图 3-18 中画一条进程推进线，使两个进程都能安全运行结束。

10．假设系统中有三类互斥资源 R_1、R_2 和 R_3，可用资源数分别为 10、8 和 5。在 T_0 时刻系统中有 P_1、P_2、P_3、P_4 和 P_5 五个进程，这些进程对资源的最大需求量和已分配资源数如表 3-11 所示。分别回答下列问题。（1）求系统可用资源向量。（2）需求矩阵。

表 3-11　进程资源分配表

资源进程	最大需求量 R_1 R_2 R_3	已分配资源数 R_1 R_2 R_3
P_1	6　5　2	0　2　1
P_2	2　2　1	2　1　0
P_3	8　0　1	2　0　0
P_4	1　2　1	1　2　0
P_5	3　4　4	1　1　3

（3）问 T_0 时刻系统是否安全？如果安全，给出一个进程执行的安全序列。

（4）T_1 时刻 P_1 进程提出资源请求 Request（2，2，0），系统能否将资源分配给它？为什么？写出推导过程。

11．假设某进程中有三个资源（R_1，R_2，R_3），资源数分别为 9、3 和 6，在某时刻系统中有四个进程。进程 P_1、P_2、P_3、P_4 的最大资源需求向量和此时已分配资源向量分别为：

进程	已分配资源矩阵	最大资源需求矩阵
P_1	（1，0，0）	（3，2，2）
P_2	（5，1，1）	（6，1，3）
P_3	（2，1，1）	（3，1，4）
P_4	（0，0，2）	（4，2，2）

（1）求可用资源向量。

（2）写出四个进程的需求矩阵。

（3）当前的状态是否是安全的？如果是安全的，写出进程安全序列。

（4）如果进程 P_2 发出资源请求向量（1，0，1），能否将资源分配给它？为什么？

（5）如果进程 P_1 发出资源请求向量（1，0，1），能否将资源分配给它？为什么？

12．系统中有同类资源 M 个，供 N 个进程共享使用，如果每个进程对资源的最大需求量为 K，问表 3-12 所示的情况下是否会发生死锁。

表 3-12　进程资源数量关系表

序号	M	N	K	是否死锁
1	6	3	2	
2	6	3	3	
3	7	3	3	
4	9	4	3	
5	13	6	3	

第4章　进程通信与多线程

进程通信，是指进程之间的信息交换，是操作系统内核层中比较重要的部分。并发进程之间的相互通信是实现多进程间协作和同步的常用工具。多核处理器、多线程的应用在适当的环境中可以大大提高程序的性能。有两种典型的进程内多线程实现技术，即多线库实现技术和核心线程实现技术。许多操作系统都已经采用多线程思想。

4.1　进 程 通 信

并发进程之间的相互通信是实现多进程间协作和同步的常用工具。具有很强的实用性，进程通信是操作系统内核层极为重要的部分。

前面讲过的进程之间的互斥与同步也可以看做是进程之间的一种通信，但它们交换的信息量较少，也常被叫做低级通信。这里所说的进程之间的通信，指的是进程之间交换较多信息（数据）这样一种情况，也叫高级通信。

4.1.1　共享存储区通信

共享存储区通信可使若干进程共享主存中的某一个区域，且使该区域出现在多个进程的虚地址空间中。进程之间通过共享变量或数据结构进行通信，这种通信要处理好互斥进入的问题，如图 4.1 所示。

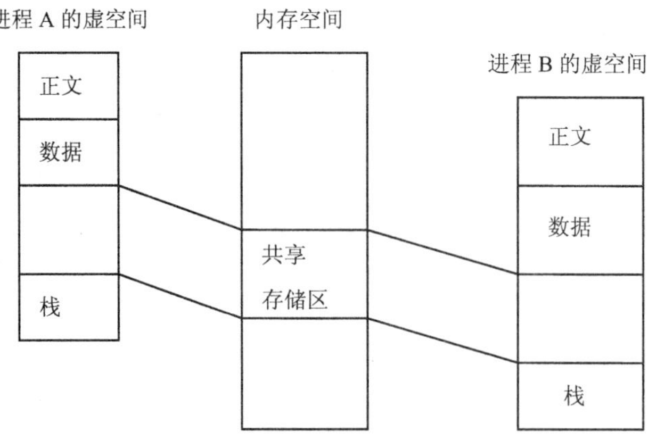

图 4-1　共享存储区通信

在这种通信方式中，要求各进程公用某个数据结构，进程通过它们交换信息。例如在生产者-消费者问题中，就是把缓冲池（有界缓冲区）这种数据结构用来作通信的。这时需要对公用数据设置进程间的同步问题。操作系统提供共享存储区，这种方式只适用传送少量的数据。

为了传送大量数据，在存储区中划出一块存储区，供多个进程共享，共享进程通过对这一共享存储区中的数据进行读或写来实现通信。

共享存储区是 UNIX 系统中通信速度最高的一种通信机制，它包含在进程通信软件包中。该方法当进程 A 要利用共享存储区进行通信时，用系统调用 shmget 建立一个共享存储区。执行该系统调用时，核心首先检查共享存储区表，该表是系统范围的数据结构，每个共享存储区在该表中占一表项。如未找到，则分配一系统空闲区作为共享区的页表区，分配相应的内存块，有关参数填入页表中。返回共享存储器的描述符 shmid。进程 B 要利用共享存储区与进程 A 通信时，也要先利用系统调用 shmget，此时核心检查共享存储区表找到了指定的表项，则返回该表项的描述符 shmid。

进程 A 和 B 在建立了一个共享存储区并获得了其描述符后，还需分别利用系统调用 shmat 将共享存储区附接到各自进程的虚地址空间上。此后共享存储区便成为进程虚地址空间的一部分，进程可采取与其他虚地址空间一样的存取方法来存取它。

4.1.2 消息传递系统

在消息传递系统中，进程间的数据交换以消息为单位，在计算机网络中，消息又称为报文。程序员直接利用系统提供的一组通信命令（原语）来实现通信。操作系统隐藏了通信的实现细节，大大简化了程序编制的复杂性，因而获得广泛的应用。

消息传递系统因其实现方式不同，又可分为直接通信方式和间接通信方式两种。

1．直接通信方式

这是指发送进程利用操作系统提供的发送命令直接把消息发送给接收进程，并将它挂在接收进程的消息缓冲队列上。接收进程利用操作系统提供的接收命令直接从消息缓冲队列中取得消息。此时要求发送进程和接收进程都以显示的方式提供对方的标识符，通常系统提供下述两条通信原语：

```
Send(Receiver, message);
Receive(Sender, message);
```

直接通信的实例-消息缓冲队列通信机制由美国 Hansan 提出，并在 RC4000 系统上实现，后来广泛应用于本地进程的通信中。

1）消息缓冲队列通信原理

消息缓冲队列通信机制的原理是：由系统管理一组缓冲区，其中每个缓冲区可以存放一个消息。当进程要发送消息时先要向系统申请一个缓冲区，然后把消息写进去，接着把该缓冲区连接到接收进程的消息缓冲队列中。接收可以在适当的时候从消息缓冲队列中摘下消息缓冲区，读取消息，并释放该缓冲区。消息缓冲队列通信的发送和接收过程如图 4-2 所示。

2）消息缓冲队列通信的数据结构

消息缓冲队列通信所用的数据结构有消息缓冲区和进程 PCB 中有关通信的扩充数据项，它们描述如下：

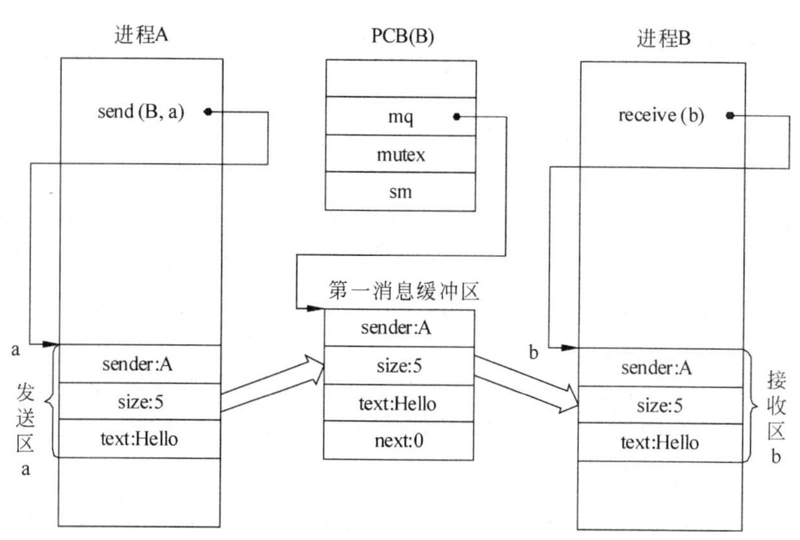

图 4-2　消息缓冲通信

① 消息缓冲区。

```
  type message buffer=record
  sender         ; 发送进程的标识符
  size           ; 消息长度
   text          ; 消息正文
   next          ; 指向下一个消息缓冲区的指针
end
```

② PCB 有关通信的扩充数据项。

```
type PCB = record
            ...
          mutex    ; 消息缓冲队列互斥信息量 ;
          Sm       ; 消息缓冲队列资源信息量 ;
          mq       ; 消息缓冲队列首指针    ;
            ...
end
```

3）发送原语

发送原语在利用发送原语发送消息之前应先在自己的内存空间设置发送区 a，把发送进程标识符、待发送消息的长度和消息的正文等信息填入其中。然后调用发送原语，把消息发送给接收进程。发送原语的工作分三步。首先根据发送区 a 中所设置的消息长度 a.size 来申请一块缓冲区 i；接着把发送区 a 中的信息复制到消息缓冲区 i 中；最后将消息缓冲区 i，挂在接收进程的消息缓冲队列末尾。由于该队列是临界资源，故在执行 insert 操作的前后，都要对 j.mutex 执行 P、V 操作，然后对 j.Sm 施加 V 操作通知接收进程。发送原语描述如下：

```
      Proceduce send (receiver,a)
      begin
            getbuf(a.size, i) ;
            i.sender = a.sender ;
   i.size = a.size ;
   i.text = a.text ;
   i.next = 0 ;
   getid (PCB set , receiver , j) ;
   P(j.mutex) ;
   insert(j.mq, i) ;
   V(j.mutex) ;
   V(j.Sm) ;
End
```

4）接收原语

接收进程在调用接收原语接收消息前，也应先在自己内存空间设置一个接收区 b，然后调用接收原语 receive。接收原语工作也分三步。首先它从自己的消息缓冲队列 mq 上摘取队首的一个消息缓冲区，然后将其中的消息复制到自己的接收区 b，最后释放该缓冲区。接收原语描述如下：

```
proceduce  receive(b)
begin
   j := internal name ;
   P (j.Sm) ;
   P (j.mutex) ;
   Remove(j.mq, i ) ;
   V(j.mutex) ;
   b.sender :=i.sender ;
   b.size :=i.size ;
   b.text :=i.text ;
   Releasebuf(i) ;
end
```

2. 间接通信方式

间接通信是消息传递的另一种方式。在这种情况，消息不直接从发送者发送到接收者，而是发送到暂存消息的共享数据结构组成的队列，这个实体称为信箱（Mailbox）。因此两个进程通信情况，一个进程发送一个消息到某个信箱，而另一个进程从信箱中摘取消息。

间接通信的使用好处是增加了使用消息的灵活性。发送者和接收者的关系可能是一对一、多对一、一对多或多对多。一对一的关系允许一个专用通信链路用于两个进程间的交互，它能使俩进程间交互不受其他进程错误干预的影响，多对一的关系对客户 / 服务器交互特别有用，一个进程对多个其他进程（用户）提供服务。在这种情况信箱经常称作为端口。一对多关系允许一个发送进程和多个接收进程交互，这可用来将消息广播给一组进程。

（1）信箱的创建和撤销。进程可利用信箱创建原语来建立一个新信箱。创建者进程应给出信箱名字、信箱属性(公用、私用或共享)；对于共享信箱，还应给出共享者的名字。当进程不再需要读信箱时，可用信箱插销原语将之插销。

（2）消息的发送和接收。当进程之间要利用信箱进行通信时，必须使用共享信箱，并利用系统提供的下述通信原语进行通信。

```
Send(mailbox, message)      ; 将一个消息发送到指定信箱
Receive(mailbox, message)   ; 从指定信箱中接收一个消息
```

信箱可由操作系统创建，也可由用户进程创建，创建者是信箱的拥有者。据此，可把信箱分为以下三类。

1）私用信箱

用户进程可为自己建立一个新信箱，并作为该进程的一部分。信箱的拥有者有权从信箱中读取消息，其他用户则只能将自己构成的消息发送到该信箱中。这种私用信箱可采用单向通信链路的信箱来实现。当拥有该信箱的进程结束时，信箱也随之消失。

2）公用信箱

它由操作系统创建，并提供给系统中的所有核准进程使用。核准进程既可把消息发送到该信箱中，也可从信箱中读取发送给自己的消息。显然，公用信箱应采用双向通信链路的信箱来实现。通常，公用信箱在系统运行期间始终存在。

3）共享信箱

由某进程创建，在创建时或创建后，指明它是可共享的，同时须指出共享进程（用户）的名字。信箱的拥有者和共享者，都有权从信箱中取走发送给自己的消息。

进程与信箱的联系有二种：静态和动态。端口经常与特定的进程保持静态的联系，即端口创建后永久地分配给某个进程。同样一对一关系是典型地规定为静态的和永久的。当有许多发送者时，一个发送者与信箱的联系可以动态发生，例如联接（Connect）和去联接原语可以应用于此用途。

4.1.3 管道通信

管道（pipe）是指用于连接一个读进程和一个写进程以实现他们之间通信的一个共享文件，又名 pipe 文件。向管道（共享文件）提供输入的发送进程（写进程），以字符流形式将大量的数据送入管道；而接受管道输出的接收进程（读进程），则从管道中接收（读）数据。由于发送进程和接收进程是利用管道进行通信的，故又称为管道通信。如图 4-3 所示。

管道分为无名管道和有名管道两种类型。无名管道是早期 UNIX 版本已提供的，它是一个只存在于打开文件机构中的一个临时文件，从结构上没有文件路径名，不占用文件目录项。无名管道利用系统调用 pipe(filedes)创建，只用该系统调用所返回的文件描述符 filedes[2]来标识该文件。只有调用 pipe 的进程及调用后该进程创建的子孙进程，才能识别此文件描述符，从而才能利用该文件（管道）进行父子或子、子之间通信。

UNIX 系统 V 的管道文件最大长度为 10 个盘块，核心为它设置了一个读指针和一个写指针。为了确保利用管道使进程间传送数据的正确性。系统在管道读写时设置互斥和同步机制，它首先保证写进程和读进程互斥地进行写和读，同时又要保证写和读的同步。即写进程将数据写入管道后，读进程才能来读取数据，如写进程未将数据写入管道，则读进程

必须阻塞等待。如写进程将数据写满管道后读进程未来读，则发出写溢出，此时写进程必须阻塞等待。与一般文件不同是管道数据的先进先出（FIFO）处理方式和管道文件数据不可再现性，即被读取的数据在管道中不能再利用。无名管道在关机后自动消失。

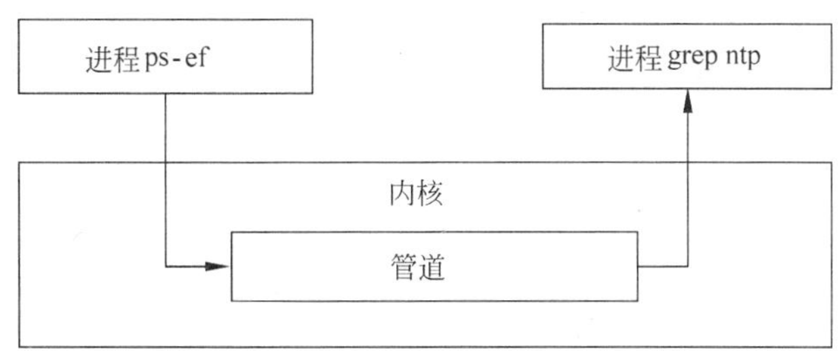

图 4-3　管道通信

为了克服无名管道使用上的局限性，以便让更多的进程也能利用管道进行通信，在以后版本 UNIX 系统中又增加了有名管道。有名管道是利用 mknod 系统调用建立的，它是可以在文件系统中长期存在的、具有路径名的文件，因而其他进程可以知道它的存在，并能利用该路径名来访问该文件。对有名管道的访问方式同访问其他文件一样，都需要先用 open 系统调用打开它。对有名管道读写方式同无名管道读写。直接用 read（filedes[0]，buf，size）和 write（filedes[1]，buf，size）系统调用对管道进行读、写。

在 UNIX、MS-DOS 系统中，用户可以在命令级使用管道，这时管道是一个程序的标准输出和另一个程序的标准输入之间的连接，管道命令用符号“|”表示。

UNIX 列出的文件命令 ls 能列出要求子目录下的所有文件和子目录，使用管道可以增强 ls 功能。例如 $ls －l /bin | more 复合命令可以分屏显示列出 / bin 子目录下的文件和子目录。$ls －l /bin | wc -l 复合命令可以统计 / bin 子目录下文件和子目录总个数（行数）。

Windows 2000 也提供无名管道和命名管道两种管道机制，它是利用操作系统核心的缓冲区（通常几十个 KB）来实现的一种单向通信。Windows 2000 的无名管道类似于 UNIX 系统的管道，但提供的安全机制比 UNIX 管道完善。利用 CreatePipe 可创建无名管道，并得到两个读写句柄；然后利用 ReadFile 和 WriteFile 可进行无名管道的读写。

Windows 2000 的命名管道是服务器进程与一个客户进程间的一条通信通道，可实现不同机器上的进程通信。它采用客户/服务器模式连接本机或网络中的两个进程。在建立命名管道时，存在一定的限制。即服务器方（创建命名管道的一方）只能在本机上创建命名管道，不能在其他机器上创建管道；但客户方（连接到一个命名管道实例的一方）可以连接到其他机器上的命名管道。服务器进程为每个管道实例建立单独的线程或进程。

4.1.4　Socket 通信

20 世纪 70 年代中，美国国防部高研署（DARPA）将 TCP/IP 的软件提供给加利福尼亚大学 Berkeley 分校后，TCP/IP 很快被集成到 UNIX 中，同时出现了许多成熟的 TCP/IP 应用程序接口（API）。这个 API 称为 Socket 接口。Socket 接口是 TCP/IP 网络最为通用的 API，也是在 Internet 上进行应用开发最为通用的 API。通过 Socket 可以使两台计算机上的进程交换数据，也可以使同一台计算机上的进程之间进行通信。

20 世纪 90 年代初，由 Microsoft 联合了其他几家公司共同制定了一套 Windows 下的网络编程接口，即 Windows Sockets 规范。它是 Berkeley Sockets 的重要扩充，主要是增加了一些异步函数，并增加了符合 Windows 消息驱动特性的网络事件异步选择机制。Windows Sockets 规范是一套开放的、支持多种协议的 Windows 下的网络编程接口。Windows 下的 Internet 软件都是基于 WinSock 开发的。

1. Socket 通信

Socket 实际在计算机中提供了一个通信端口，可以通过这个端口与任何一个具有 Socket 接口的计算机通信。应用程序在网络上传输，接收的信息都通过这个 Socket 接口来实现。在应用开发中就像使用文件句柄一样，可以对 Socket 句柄进行读、写操作。套接字是网络的基本构件。它是可以被命名和寻址的通信端点，使用中的每一个套接字都有其类型和一个与之相连进程。

在 TCP/IP 网络应用中，通信的两个进程间相互作用的主要模式是客户/服务器模式（Client/Server Model），即客户向服务器发出服务请求，服务器接收到请求后，提供相应的服务。通信机制为希望通信的进程间建立联系，为二者的数据交换提供同步，这就是基于客户/服务器模式的 TCP/IP。

客户/服务器模式在操作过程中采取的是主动请求方式，如图 4-4 所示。

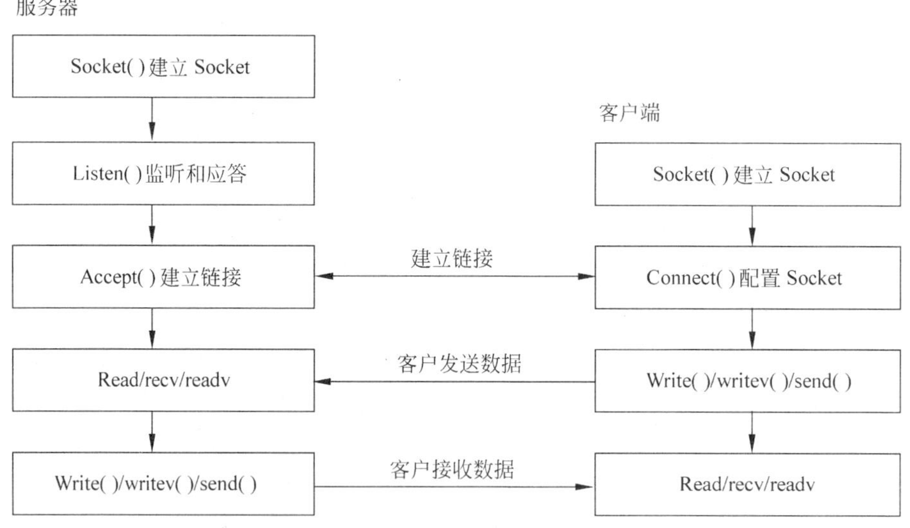

图 4-4　Socket 通信

服务器：

（1）首先服务器方要先启动，并根据请求提供相应服务；

（2）打开一通信通道并告知本地主机，它愿意在某一 IP 地址上接收客户请求；

（3）处于监听状态，等待客户请求到达该端口；

（4）接收到服务请求，处理该请求并发送应答信号，接收到并发服务请求，要激活一新进（线）程来处理这个客户请求。新进（线）程处理此客户请求，并不需要对其他请求做出应答。服务完成后，关闭此新进程与客户的通信链路，并终止；

（5）返回第二步，等待另一客户请求；

（6）关闭服务器。

客户端：

（1）打开一通信通道，并连接到服务器所在主机的特定端口；

（2）向服务器发服务请求报文，等待并接收应答；继续提出请求并重复；

（3）请求结束后关闭通信通道并终止。

2．Socket 通信示例

在实际工作中，常常有这样的应用，在服务器端接受可客户端发来的请求，为客户端程序做某些处理，把结果发给客户端。这样的程序就是我们常说的 C/S 结构的程序。这个实例就是实现这样结构程序的一个框架。

在服务器端，创建一个 Socket 接口，并使这个 Socket 接口处于监听状态，如果有客户端连接申请，就响应连接，读取客户端通过 Socket 发来请求命令，并创建一个新的线程，去处理这个客户端的请求，主线程继续处于监听状态，等待下一个客户端的请求。所以主线程的主要任务就是监听客户端请求，收到请求后，创建新的线程去处理，主线程可以设置一个死循环，创建新线程后又回到监听状态。

客户端，需要服务器提供服务时，就建立一个 Socket 连接，连接成功后，发出服务请求，再接收服务请求的结果。

本例中服务器端的程序采用 C#编程，客户端的程序采用 Visual Basic 编程。编程语言不一样，但它们都遵守 TCP/IP 协议，遵守 Socket 编程标准，它们之间就可以顺畅的通信，交换信息。

1）服务器端 C#程序

```
using System;
using System.Net;
using System.Net.Sockets;
using System.Collections;
using System.IO;
using System.Threading;
namespace sut.af.commonfunction{
    public class ListenCls
    {
        internal int ListenNo_=1;
        IPAddress localAddr=IPAddress.Any;  //取服务器的IP
        public void Listening(){
            try{
                System.Net.Sockets.TcpListener s;
                s = new System.Net.Sockets.TcpListener(localAddr, PORT );
                s.Start();   // 启动服务器端监听
                CommandProcCls cp;
                for (;;){
                    System.Net.Sockets.TcpClient incoming = s.AcceptTcp-
                    Client();
```

```
                    //等待客户端连接，收到客户端连接请求程序才继续执行
                    cp=new CommandProcCls(incoming);
                    Thread newthread = new Thread(new ThreadStart(cp.
                    CommandProc));
                    newthread.Start();    //创建线程处理客户端请求
                }
            }
            catch (System.Exception e3){
                ...
            }
            return ;
        }
    }
    public class CommandProcCls {  //处理客户端请求
        System.Net.Sockets.TcpClient tc;
        public CommandProcCls(System.Net.Sockets.TcpClient incoming){
            tc=incoming;   //获得客户端的请求信息
        }
        public void CommandProc(){
            if (tc!=null){
                ... //处理客户端的请求
            }
            return;
        }
    }
```

2）客户端 VB 程序

```
Public Function connect_toserver(ByVal SeverName As String, ByVal PORT As
Long) As Integer
  Winsock1.RemoteHost = SeverName
  Winsock1.RemotePort = PORT
  Winsock1.Connect    //请求连接到服务器的Socket接口
  if Winsock1.State = sckConnected then
     Winsock1.SendData Msg //发送客户端的请求信息
  End If
  Winsock1.Close
End Function
```

4.2 多核技术

多核技术的出现是计算机发展和应用需求的必然产物。当处理器时钟频率的提升遇到了不可逾越的瓶颈，芯片制造厂商便不约而同将目光瞄向了双核乃至多核架构。多核是计算机体系结构发展史上的划时代、革命性的里程碑。可以说，多核处理器实现了真正意义

上的并行，极大地提升了 CPU 的性能。

4.2.1 并行计算机

可以把现代计算机的发展历程分为两个时代：串行计算时代和并行计算时代。串行计算机只有单个处理单元，顺序执行计算程序，因此也称为顺序计算机。并行计算机是由一组处理单元组成，这组处理单元通过相互之间的通信与协作，以更快的速度共同完成一项大规模的计算任务。

并行计算机的发展可以追溯到 20 世纪 60 年代初期，由于晶体管及磁芯存储器的出现，使得处理单元变得越来越小，存储器也变得更加小巧和廉价，这些技术发展的结果导致了并行计算机的出现，这个时期的并行计算机多是规模不大的共享多处理器系统。到了 20 世纪 60 年代末期，同一处理器开始设置多个功能相同的功能单元并引入了流水线技术，与单纯提高时钟频率相比，这些并行特性在处理器内部的应用大大提高了并行计算机系统的性能。20 世纪 80 年代中期，微处理器的性能每年以 50%速度提升，而大型计算机和超级计算机的性能每年只提升 25%，微处理器的性能逐渐可以与那些集成度低且十分昂贵的其他处理器芯片相媲美，人们开始采用小型的、便宜的、低功耗的和批量生产的微处理器作为基本模块来构筑计算机系统。到了 20 世纪 90 年代，并行计算机已成为计算机技术中的关键技术。可以预见并行技术将会对未来计算机的发展产生更大的影响。

并行计算机有多种分类方法，其中最著名的是 Flynn 分类法。按照计算机的运行时指令流和数据流的不同组织方式，把计算机系统的结构分为：

（1）单指令流单数据流（Single Instruction stream Single Data stream, SISD）；

（2）单指令流多数据流（Single Instruction stream Multiple Data stream, SIMD）；

（3）多指令流多数据流（Multiple Instruction stream Multiple Data stream, MIMD）；

（4）多指令流单数据流（Multiple Instruction stream Single Data stream, MISD）；

其中指令流是指机器执行的指令序列；数据流是指令流调用的数据序列，包括输入数据和中间结果。

在这四种结构的计算机系统中，SIMD 和 MIMD 是典型的并行计算机。而 SISD 实际就是普通的、顺序处理的串行计算机。对于 MISD 代表何种计算机，实际存在争议，有的学者认为没有这种结构的计算机，而有的文献则把流水线结构的计算机看成 MISD 结构。

SIMD 计算机只有一个控制单元，每次只能执行一条指令，但有许多处理单元，通常将大量的处理单元构成阵列，因此 SIMD 计算机也称为阵列处理机。所有的处理单元在控制单元的统一控制下工作，同时执行一条指令，不过每个处理单元处理的数据有所不同。SIMD 计算机主要用于一些使用向量和阵列这样比较规整的数据结构来解决复杂的科学、工程计算的问题。

在 MIMD 计算机中，没有统一的控制部件，每个处理器都有独立的控制部件，各处理器可以分别执行不同的指令，处理器之间通过互连网络进行通信。它是目前主流的并行技术。MIMD 计算机又分成多处理器系统和多计算机系统两大类。其中多处理器系统的特点是：每个 CPU 不带或少量带有自己的内存，所有 CPU 共享同一个物理内存，整个内存空间由许多内存模块组成，由统一的操作系统管理。多计算机系统的特点是：每个 CPU 都有自己的内存，即自己独立的物理地址空间，执行自己的操作系统，CPU 通过通信处理器对外通信。

按照系统中所用的处理器是否相同，可将多处理器系统分为如下两类。

(1) 对称多处理器（SMP）。系统中所包含的各处理器单元，在功能和结构上都是相同的。大多数多处理器系统都属于对称结构。

(2) 非对称多处理器。系统中有多种类型的处理单元，它们的功能和结构各不相同，其中只有一个主处理器，有多个协处理器。

4.2.2 多核处理器

1．多核处理器的引入

根据摩尔定律，CPU 的速度每过 18 个月就翻一番。在过去的十几年内，芯片制造厂商不断地推出拥有更高时钟频率的处理器，一次次验证了摩尔定律的准确性。但在单核产品中，主要是通过提高频率和增大缓存这两种方式来提高性能，前者会导致芯片功耗的提升，后者则会让芯片内的晶体管规模激增，造成芯片成本大幅度上扬。随着晶体管的集成度越来越高，采用这两种措施所付出的代价也越来越高昂，而且这也只能带来性能小幅度的提升。当晶体管的集成度超过上亿个时，传统的处理器体系结构将面临瓶颈。

随着人们对计算机应用需求的不断提高，继续提升 CPU 的时钟频率已经不能解决广大用户对多线程操作的需求。有人提出在一个物理核心上整合两个或多个计算单元，生产出双核、多核处理器的设计思路。从技术上看，引入多核技术，可以在较低频率、较小缓存的条件下达到大幅提高系统性能的目的。相比大缓存的单核产品，耗费同样数量晶体管的多核心处理器拥有更出色的效能，同样在每瓦性能方面，多核设计也有明显的优势，多核心设计可谓是提高晶体管效能的最佳手段。

最早推出多核处理器的是 IBM 公司。IBM 在 2001 年发布了双核 RISC 处理器 POWER4，它将两个 64 位 PowerPC 处理器内核集成在同一颗芯片上，成为首款采用多核设计的服务器处理器。在 UNIX 阵营当中，两大巨头 HP 和 Sun 也相继在 2004 年 2 月和 3 月发布了名为 PA-RISC8800 和 UltraSPARC IV 的双内核处理器。

2006 年 5 月，英特尔发布了其服务器芯片 Xeon 系列的新成员——双核芯片 Dempsey。该产品使用了 65nm 制造工艺，其 5030 和 5080 型号的主频在 2.67GHz 和 3.73GHz 之间。不久英特尔又推出了另一款双核芯片 Woodcrest（Xeon 5100 系列），英特尔声称与奔腾 D 系列产品相比，其计算性能提高了 80%，能耗降低了 20%。

继双核之后，英特尔已经在 2006 年 11 月抢先推出了四核产品，AMD 也推出代号为巴塞罗那的四核处理器。

2．多核处理器的结构和优势

多核处理器也称 CMP，是指在一个芯片内含有多个处理核心而构成的处理器。所谓“核心”，通常指包含指令部件、算术/逻辑部件、寄存器堆和一级或者二级缓存的处理单元。在芯片内，多个核心通过某种方式互连，使它们之间能够交换数据，而对外则表现为一个统一的多核处理器。因为多核处理器能够通过划分任务，分配给多个内核并行执行线程，从而使处理器并发执行多个线程，大大提高了处理速度。

按照计算内核是否对等，可将多核处理器分为同构多核和异构多核。计算内核相同，地位对等的称为同构多核，现在 Intel 和 AMD 主推的双核处理器都是同构结构。计算内核不同、地位不对等的称为异构多核，异构多核多采用主处理核＋协处理核的设计，IBM 等联合设计的 Cell 处理器就是典型的异构多核处理器。从理论上看，异构多核结构似乎具有更好的性能。

在多核处理器的不同内核上执行的程序之间需要进行数据的共享和同步，因此多核处理器的硬件结构必须支持核间通信，高效的通信机制是多核处理器高性能的重要保障。目前比较主流的片上的通信机制有两种，一种是基于总线共享的 Cache 结构，一种是基于片上的互连结构。

（1）基于总线共享的 Cache 结构。这种结构是指每个 CPU 内核拥有共享的二级或三级 Cache，用于保存比较常用的数据，并通过连接核心的总线进行通信。这种结构的优点是结构简单，通信速度快，缺点是可扩展性差。

（2）基于片上的互连结构。它是指每个 CPU 内核具有独立的处理单元和 Cache，各个 CPU 内核通过交叉开关或片上网络等方式连接在一起，各个 CPU 内核之间通过消息通信。这种结构的优点是可扩展性好，数据带宽有保证，缺点是硬件结构复杂，且软件改动大。

多核处理器实现了真正意义上的多线程并行。在单核时代下，并发的效果是由软件模拟出来的。操作系统的进程管理部分负责调度各个进程和线程对 CPU 的控制权，把时间片按照设定的优先级别轮流分配给各个进程和线程。从时间段的角度来看，这些任务仿佛是被同时完成的，而事实上，只是 CPU 不停地切换服务对象，从而达到了并发的效果。这种并发机制耗费不少代价，因为 CPU 在管理和调度进程或线程时，需要一定的开销。实际上，单核时代下的并行，是以牺牲一定的 CPU 效率为前提的。而在多核的架构下，每个单一的内核都拥有自己独立的 Cache 以及一组相关的硬件资源。当并行的线程在两个或多个内核上执行时，彼此没有干扰和影响，这种做法从硬件的源头支持了并行，使处理器彻底、完全地并发执行程序的多个线程，大大提高了处理速度。

多核处理器全面提升了并行的效果，极大地提高了计算的性能，其前所未有的计算能力增强了多任务环境下的计算体验。如今的桌面操作系统几乎都是多任务的环境，特别是在数字娱乐、多媒体技术高度发展的今天，并行计算、并行处理的能力已经是衡量计算机性能的重要指标。

4.2.3 操作系统对多核处理器的支持方法

随着多核处理器的发展，操作系统对多核处理器的支持应当提升。如何管理线程的同步，如何控制内核间的通信，如何提供多核系统对单线程程序的兼容性，如何在多核架构下以更加充分地发挥 CPU 性能的方式来调度进程和线程，都是操作系统的设计人员面对多核时代来临时应该思考的问题。

1．分配与调度

多核操作系统的关键是任务的分配和调度。任务调度还会涉及到比较广泛的问题，例如负载均衡、实时性能等。

任务分配是多核时代提出的新概念。在单核时代，只有一个核的资源可以使用，没有核的任务分配问题。而在多核体系下，有多个核可以被使用，如果系统中有几个进程需要分配，任务分配要实现将进程分配到合理的处理器核上。由于不同的核在历史运行情况和共享性等方面有所不同。在任务分配时，需要考虑是将它们均匀地分配到各个核，还是一起分配到一个处理器核，或是按照一定的算法进行分配。还要考虑底层的系统结构是 SMP 构架还是 CMP 构架和在 CMP 构架中共享二级缓存的核的数量，如果将有数据共享的进程分配给有共享二级 Cache 的核上，将大大提升性能；反之，就有可能影响性能。

任务分配结束后，需要进行任务调度。对于不同的核，每个处理器核可以有自己独立的调度算法来执行不同的任务（实时任务或者交互性任务），也可以使用一致的调度算法。进行任务调度时，还可以考虑是否进行线程迁移，线程迁移是指一个进程上一个时间片运行在一个核上，下一个时间片运行在另外一个核上；如何直接调度实时任务和普通任务；系统的核资源是否要进行负载均衡等问题。

在单核处理器中，常见的调度策略有先到先服务（FCFS），最短作业调度（SJF），优先级调度，轮转法调度（RR）和多级队列调度等。例如在 Linux 操作系统中对实时任务采取 FCFS 和 RR 两种调度，普通任务调度采取优先级调度。

对于多核处理器系统的调度，目前还没有明确的标准和规范。由于系统中有多个处理器核可以使用，如何处理好负载均衡便成为调度策略的关键。一般任务调度算法有全局队列调度和局部队列调度。

（1）全局队列调度。它是指操作系统维护一个全局的任务等待队列，一旦系统中有一个 CPU 核心空闲时，操作系统就从全局任务等待队列中选取一个合适的就绪任务在此核心上执行。这种算法的优点是 CPU 核心利用率较高，但需要对进程进行上下文切换、锁的转换等操作，增加了执行时间，从而降低了性能。目前多数多核 CPU 操作系统采用的是基于全局队列的任务调度算法。

（2）局部队列调度。操作系统为每个 CPU 内核维护一个局部的任务等待队列，当系统中有一个 CPU 内核空闲时，便从该核心的任务等待队列中选取恰当的任务执行，这种算法的优点是任务无须在多个 CPU 核心之间切换，有利于提高 CPU 核心局部 Cache 的命中率。

与传统的对称多处理器并行架构相比较，多处理器 CMP 架构最大的特点在于片上的多个处理器核之间具有紧密的耦合关系。多核处理器的各个 CPU 核心会共享一些部件，如二级 Cache、I/O 端口等，当它们同时竞争访问这些资源的时候，就会发生冲突。

下面是几个具有代表性的多核调度算法。

（1）对任务的分配进行优化。使同一应用程序的任务尽量在一个核上执行。通过这种方式，保证了有共享数据的任务在一个核上进行，而没有共享数据量或者共享数据量很少的任务在不同核上进行。可以显著地提高缓存的命中率，极大地提升了系统的整体性能。

（2）对任务的共享数据优化。由于 CMP 体系结构拥有共享二级缓存，可以考虑改变任务在内存中的数据分布，使任务在执行时尽量增加二级缓存的命中率。

（3）对任务的负载均衡优化。当任务在调度时，出现了负载不均衡，考虑将较忙处理器中与其他任务最不相关的任务迁移出来，以尽量减少数据的冲突。

2．中断

多核处理器系统的中断处理和单核的中断处理有很大的不同。因为多核的各处理器之间需要通过中断方式进行通信，所以多个处理器之间的本地中断控制器和负责仲裁各核之间中断分配的全局中断控制器都封装在芯片内部。另外，多核处理器是一个多任务系统，由于不同任务会竞争共享资源，因此需要系统提供同步与互斥机制，而传统的、用于单核的解决机制并不能满足多核，需要利用硬件提供的“读－修改－写”的原子操作或其他同步互斥机制来保证。

3．存储管理

多核环境下，存储管理相对变化较小。其主要的一些改进包括以下两个方面。

（1）事务内存管理机制。为了充分使用多核的运算能力，很多的库函数都做成 non blocking 的，这会引起数据冲突或不同步，所以需要有支持数据同步的机制。事务内存管理就是这样的机制，能够协作程序，在并行运行的同时，保证数据的同步。

（2）多线程内存分配。为了提高内存分配的效率，可以使用多线程内存分配，这样可以提高效率，降低 Cache 冲突，特别适合空间和时间关联性强的内存操作。

4.3 线程与线程管理

随着多处理机的发展，为了方便用户开发多任务并行程序，进程内多线程概念被提了出来，使得用户可以开发共享地址空间的并发或并行程序。线程是操作系统分配处理器的基本单元，每个线程都有唯一的标识。

4.3.1 线程

1．线程概念的引入

由于进程既是程序执行时系统独立调度和分派的基本单位，同时又是系统资源分配的基本单位。每个进程都拥有自己的数据段、代码段和堆栈段，这就造成了进程在进行创建、切换和撤销等操作时，需要付出较大的系统开销。因此，在系统中设置的进程数目不宜太多，进程切换的频率也不宜过高，这一定程度上限制了并发程度的进一步提高。

为了能使多个程序更好地并发执行，同时又尽可能减少系统开销，人们提出将进程的两个属性分开，即作为系统独立调度和分派的基本单位，不能同时作为系统资源分配的基本单位。于是进程在演化中出现了另一个概念——线程。

线程是进程的一个实体，是被系统独立调度和分派的基本单位，每个线程都有一个唯一的标识符。线程自己基本不拥有系统资源，只拥有一点在运行中必不可少的资源，但它可与同属一个进程的其他线程共享所属进程拥有的全部资源。

一个进程可以有一个或多个线程，至少有一个线程。一个线程可以创建和插销另一个线程，同一进程中的多个线程之间可以并发执行。

2．线程与进程的关系

线程与进程有很多类似的性质，因此人们习惯上也称线程为轻量级进程（Light Weight Process，LWP)，而传统意义上的进程则被称为重量级进程（Heavy Weight Process，HWP)，从现代的角度来看，它就是只拥有一个线程的进程。线程与进程的根本区别是把进程作为资源分配单位，而线程是调度和执行单位。

下面从调度、并发性、拥有资源和系统开销等方面对线程和进程进一步进行比较。

1）调度

在传统的操作系统中，CPU 调度和分派的基本单位是进程。而在引入线程的操作系统中，则把线程作为 CPU 调度和分派的基本单位，进程则作为资源拥有的基本单位，从而使传统进程的两个属性分开，线程便能轻装运行，这样可以显著地提高系统的并发性。同一进程中线程的切换不会引起进程切换，从而避免了昂贵的系统调用。但是在由一个进程中的线程切换到另一个进程中的线程时，依然会引起进程切换。

2）并发性

在引入线程的操作系统中，不仅进程之间可以并发执行，而且在一个进程中的多个线程之间也可以并发执行，因而使操作系统具有更好的并发性，从而能更有效地使用系统资源和提高系统的吞吐量。例如，在一个未引入线程的单 CPU 操作系统中，若仅设置一个文件服务进程，当它由于某种原因被封锁时，便没有其他的文件服务进程来提供服务。

3）拥有资源

无论是传统的操作系统，还是引入线程的操作系统，进程都是拥有资源的独立单位，它可以拥有自己的资源。但线程自己不拥有系统资源，它可以访问其隶属进程的资源。一个进程的代码段、数据段及其他系统资源，可供该进程中的所有线程共享。

4）系统开销

由于在创建或撤销进程时，系统都要为之分配或回收资源，如内存空间、I/O 设备等。因此，操作系统所付出的开销将显著地大于在创建或撤销线程时的开销。类似地，在进行进程切换时，涉及到整个当前进程 CPU 环境的保存环境的设置，以及新被调度运行的进程的 CPU 环境的设置。而线程切换只需保存和设置少量寄存器的内容，并不涉及存储器管理方面的操作。可见，进程切换的开销也远大于线程切换的开销。此外，由于同一进程中的多个线程具有相同的地址空间，使得它们之间的同步和通信的实现也变得比较容易。在有的系统中，线程的切换、同步和通信都无须操作系统内核的干预。

3. 多线程

传统的操作系统中，资源分配和 CPU 调度的单位是进程，即一个程序的一次执行。进程在任何时候只有一个执行现场，即称为单线程结构。这种单线程结构的进程已不能很好地适应计算机的发展。首先，计算机硬件向多处理机、网络方向发展，这就要求操作系统能适应这种发展，合理地使用各处理机和网络上的其他处理机，许多工作可以分配到不同的处理机上同时运行。这一点上，传统的单线程进程不能有效地实现。其次，应用程序要求并发执行。如数据库中，可以同时有多个用户在交互执行，同时对几个文件或网络操作。又如窗口系统中，同时处理多个子窗口的请求等，这就要求操作系统提供一些机制，使得用户能按需求，在一个程序中设计出多个线程同时运行，而这又不是多个进程组合在一起能完成的，因为他们之间共享大部分资源。

基于上述原因，操作系统在系统结构上有了新的发展，与传统操作系统中的单线程结构相对应，提出了多线程结构的概念。多线程的好处在于：①提高应用程序响应；②使多处理器效率更高；③改善程序结构；④占用较少的系统资源；⑤把线程和远程过程调用 RPC 结合起来；⑥提高了系统性能等。

以 Microsoft Word 为例说明多线程的应用：在进行文件打印时，就默认选择了后台打印方式，打印处理会在继续编辑文档的同时异步进行。此时 Word 创建了独立的线程来进行打印处理，并将打印线程的优先级设定低于处理用户输入的线程，这在 Windows 2000 后的版本与 Windows NT 4. 0 中能够得到，它们真正采用抢占式多任务，支持多线程应用。此外，IIS、组件服务和 SQL Server 全部都是多线程的。在实际应用中包含多个独立任务的问题适合用多线程方法解决。

4.3.2 线程管理

1. 线程的描述

在 OS 中的每一个线程都可以利用线程标识符和一组状态参数进行描述。状态参数通常有这样几项：① 寄存器状态，它包括程序计数器 PC 和堆栈指针中的内容；②堆栈，在堆栈中通常保存有局部变量和返回地址；③线程运行状态，用于描述线程正处于何种运行状态；④优先级，描述线程执行的优先程度；⑤线程专有存储器，用于保存线程自己的局部变量拷贝；⑥信号屏蔽，即对某些信号加以屏蔽。

2. 线程的基本状态

与进程一样，各线程之间也存在着共享资源和相互使用合作的制约关系，使得线程在其执行过程中出现间断性。相应地，线程在运行时也具有以下三种基本状态。

（1）执行状态：线程正获得处理机运行。

（2）就绪状态：线程除处理机以外的其他资源已经全部获得。

（3）阻塞状态：线程在其执行过程中因某事件受阻而暂停执行。

但是，线程没有进程中的挂起状态，即线程是一个只与内存和寄存器相关的概念，它的内容不会因交换而进入外存。

针对线程的三种基本状态，存在以下五种基本操作来转换线程的状态。

（1）派生。线程在进程内派生出来，它既可由进程派生，也可由线程派生。用户一般用系统调用（或相应的库函数）派生自己的线程。一个新派生出来的线程具有相应的数据结构指针和变量，这些指针和变量作为寄存器上下文放在相应的寄存器和堆栈中。

（2）阻塞。若一个线程在执行过程中需要等待某个事件发生，则被阻塞。阻塞时寄存器上下文、程序计数器及堆栈指针都会得到保存。

（3）激活。若阻塞线程的事件发生，则该线程被激活并进入就绪队列。

（4）调度。选择一个就绪线程进入执行状态。

（5）结束。若线程执行结束，它的寄存器上下文及堆栈内容等将释放。

在某些情况下，某个线程被阻塞也可导致该线程所属的进程被阻塞。

线程的状态转换如图 4-5 所示。

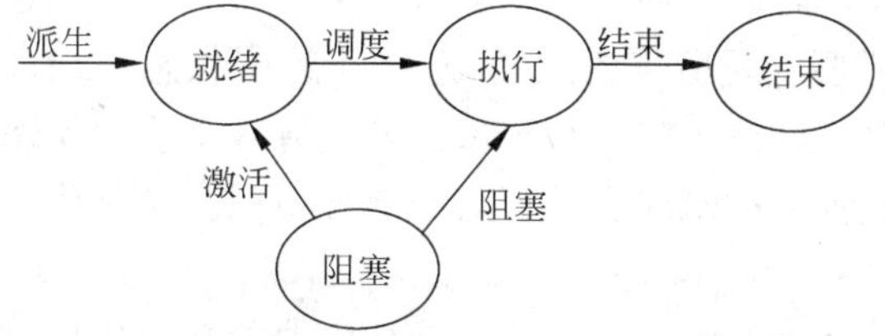

图 4-5　线程的状态与操作

3. 线程的创建、终止和切换

1）创建线程

应用程序在启动时，通常仅有一个线程在执行，该线程被人们称为“初始化线程”，它可以根据需要再去创建若干个线程。

2）线程的终止

终止线程的方式有两种：一种是在线程完成了自己的工作后自愿退出；另一种是线程在运行中出现错误或由于某种原因而被其他线程强行终止。

在大多数操作系统中，线程被终止后并不立即释放它所占有的资源，只有当进程中的其他线程执行分离函数后，被终止的线程才与资源分离，此时的资源才能被其他线程利用。

3）线程的切换

线程由相关堆栈（系统栈或用户栈）寄存器和线程控制表 TCB 组成。寄存器可被用来存储线程内的局部变量，但不能存储其他线程的相关变量。进程切换时将涉及到有关资源指针的保存及地址空间的变化等问题。线程切换时由于同一进程内的线程共享资源和地址空间，将不涉及资源信息的保存和地址变化问题，从而减少了操作系统的开销时间。而且，进程的调度与切换都是由操作系统内核完成，而线程即可由操作系统内核完成，也可由用户程序完成。

4．线程间的通信与同步

线程间通过共享空间进行通信。由于同一进程中的所有线程共享该进程的所有资源和地址空间，任何线程对资源的操作都会对其他相关线程带来影响。例如，如果一个线程想要对一个对象进行读写操作的同时，另一个线程也要对该变量进行写操作，这样前一个线程读入的变量值可能就是一个不稳定的值。因此，系统必须为线程的执行提供同步控制机制，以防止因线程的执行而破坏其他的数据结构和给其他线程带来不利的影响。

常用的线程同步机制主要有以下三种。

（1）互斥锁。互斥锁是一种比较简单的、用于实现进程间对资源互斥访问的机制。由于操作互斥锁的时间和空间开锁都较低，因而较适合于高频度使用的关键共享数据和程序段。互斥锁可以有两种状态， 即开锁（Unlock）和关锁（Lock）状态。相应地，可用两条命令（函数）对互斥锁进行操作。其中的关锁操作用于将 mutex 关上，开锁操作则用于打开 mutex。

（2）条件变量。每一个条件变量通常都与一个互斥锁一起使用，亦即，在创建一个互斥锁时便联系着一个条件变量。单纯的互斥锁用于短期锁定，主要是用来保证对临界区的互斥进入。而条件变量则用于线程的长期等待，直至所等待的资源成为可用的。线程首先对 mutex 执行关锁操作，若成功便进入临界区，然后查找用于描述资源状态的数据结构，以了解资源的情况。只要发现所需资源 R 正处于忙碌状态，线程便转为等待状态，并对 mutex 执行开锁操作后，等待该资源被释放；若资源处于空闲状态，表明线程可以使用该资源，于是将该资源设置为忙碌状态，再对 mutex 执行开锁操作。

（3）信号量机制。

① 私用信号量（Private Semaphore）。当某线程需利用信号量来实现同一进程中各线程之间的同步时，可调用创建信号量的命令来创建一个私用信号量，其数据结构是存放在应用程序的地址空间中。私用信号量属于特定的进程所有，OS 并不知道私用信号量的存在，因此，一旦发生私用信号量的占用者异常结束或正常结束，但并未释放该信号量所占有空间的情况时，系统将无法使它恢复为 0（空），也不能将它传送给下一个请求它的线程。

② 公用信号量（Public Semaphore）。公用信号量是为实现不同进程间或不同进程中各线程

之间的同步而设置的。由于它有着一个公开的名字供所有的进程使用，故而把它称为公用信号量。其数据结构是存放在受保护的系统存储区中，由 OS 为它分配空间并进行管理，故也称为系统信号量。如果信号量的占有者在结束时未释放该公用信号量，则 OS 会自动将该信号量空间回收，并通知下一进程。可见，公用信号量是一种比较安全的同步机制。

5. 线程调度

而线程调度即可由操作系统内核完成，也可由用户程序完成。有两种线程调度模型：分时调度模型和抢占式调度模型。

线程调度采用的调度算法有抢占式的动态优先级调度算法。线程调度程序按线程的优先级进行调度，高优先级的线程先被调度。线程在执行过程中优先级可以变化，调度程序调度时所依据的主要数据结构是多优先级就绪队列。

4.4 多线程的实现

线程的实现，需要直接或间接的取得内核的支持。系统有两种方式支持线程：内核支持线程可以直接利用系统调用为他服务，所以控制线程简单；用户级线程必须借助于某种形式的中间系统的帮助取得内核的服务，所以对线程的控制相对复杂一些。

4.4.1 典型的实现方式

线程已在许多操作系统中实现， 但实现的方式并不完全相同。有两种典型的实现线程的方法，一种是在用户层多实现的多线库支持线程方法，另一种是核心层实现的核心支持多线程方法。在现代操作系统中，往往结合这两种实现方法，以满足用户方便地进行多任务程序设计且使多任务高效地占用多处理机运行。为了区别，处于不同层次中的线程分别叫做用户级线程（User-Level Threads）和核心级线程（Kernel-Supported Threads）。

多线库支持线程不依赖于系统的内核。用户可以按需要编写拥有多个线程的程序，从理论上说，用户进程可以创建任意多个用户线程，这些线程可以在不同的处理机上同时执行。但从资源分配、处理机调度、线程切换、效率等诸多问题出发，并不是每创建一个线程， 就作为一个独立的单位，交给核心管理。不改变操作系统内核而开发一个多线程函数库，由函数库提供线程创建、结束、同步等函数，而进程主程序作为主线程运行，以后可调用多线程库的创建线程函数创建用户级线程。操作系统只看到进程，只做进程调度，而由多线库调度用户级线程分时地在进程中运行。也就是说，用户级线程仅存在于用户空间，只能在所属进程内进行资源竞争，内核并不能看到用户线程。这是一种非操作系统的实现方法，是一种过渡方法。

实际上，用户线程运行在一个中间系统上面，目前中间系统的实现有两种，即运行时系统（Runtime System）和内核控制线程。运行时系统实质上是用于管理和控制线程的函数(过程)的集合，其中包括用于创建和插销线程的函数、线程同步和通信的函数以及实现线程调度的函数等。正因为有这些函数，才能使用户级线程与内核无关。运行时系统中的所有函数都驻留在用户空间，并作为用户级线程与内核之间的接口。内核控制线程又称为轻型进程 LWP。 每一个进程都可拥有多个 LWP，同用户级线程一样，每个 LWP 都有自己的数据结构（如 TCB），其中包括线程标识符、优先级、状态，另外还有栈和局部存储区等，它们也可以共享进程所拥有的资源。LWP 可通过系统调用来获得内核提供的服务，这样，当一个用户级线程运行时，只要将它连接到一个 LWP 上，此时它便具有了内核支持线程的所有属性。

用户级线程的创建、结束、调度、现场保护与切换开销非常少，图 4-6 给出了在传统操作系统进程基础上实现用户级线程示意图，P 是传统操作系统进程。线程的创建、撤销和切换等都需要内核直接实现，即内核了解每一个作为可调度实体的线程，这些线程可以在整个系统内进行资源的竞争。内核空间为每一个内核支持线程设置了一个线程控制块（TCB），内核根据该控制块感知线程的存在并进行控制。在一定程度上核心级线程类似于进程，只是创建、调度的开销比进程小。

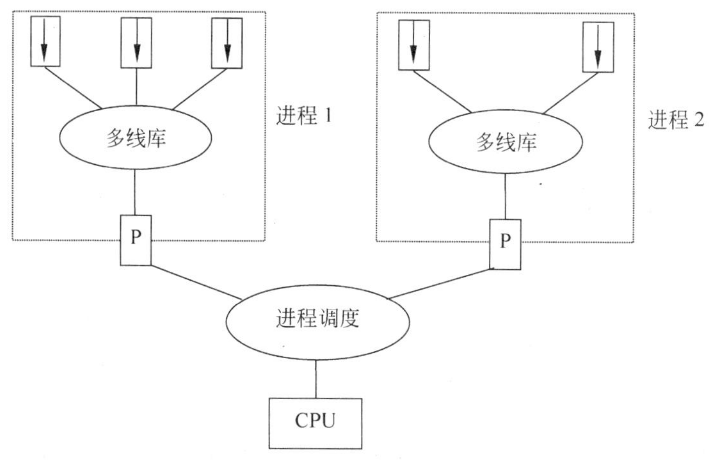

图 4-6　由传统进程支持实现用户级多线程示意图

核心级线程是由操作系统支持实现的线程，操作系统维护核心级线程的各种管理表格， 负责线程在处理机上的调度和切换，用户层上无须核心级线程的管理代码，操作系统提供了一系列系统调用界面让用户程序请求操作系统作线程创建，结束等操作。图 4-7 给出了核心级线程实现示意图，图 4-7 中 L 是 LWP 的缩写，代表了操作系统表示、管理的核心级线程的数据结构。每个进程可以有多个核心级线程，核心级线程用于运行并行执行的用户任务，当它被外部中断打断或自陷进入操作系统时，也运行操作系统内核程序。

当一个进程被创建时，系统同时创建一个核心级线程，用户初始程序即在该核心级线程上运行，我们称之为主核心线程，当需要创建新的线程去运行并行任务时，主核心线程运行的程序安排一个线程创建系统调用产生一个新的核心级线程，并说明要执行的过程段或函数及初始数据，同时也要为该线程提供一个用户栈空间，线程的用户栈空间都在进程用户虚空间区，除第一个线程栈空间在线程初始化时按约定预留外，其他线程栈空间由用户态运行程序自行分配。内核建好核心级线程的线程控制块 TCB 等管理数据结构，为线程分配一个核心栈空间。线程在运行用户程序时在用户栈上工作，当中断或自陷进入操作系统核心运行时，则转到核心栈上工作。

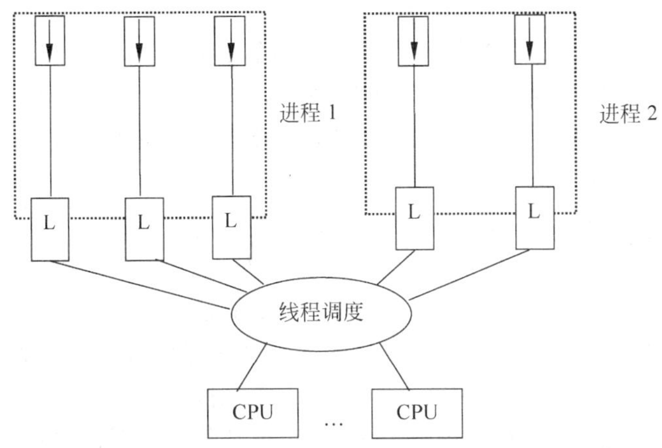

图 4-7　由操作系统内核支持实现进程内多线程示意图

4.4.2　用户级线程实现

用户级多线程结构通过线程库来实现。线程库利用轻进程（LWP）实现对用户多线程的管理。用户线程在线程库只是一个简单的数据结构和一个堆栈，核心不知道它。线程库维护的主要的数据结构包括用户线程结构、用户态堆栈、各种队列和 LWP 池。

1．用户线程调度

用户线程调度在 LWP 池中实现。每个线程都有一个从 0 到无穷大的一个调度优先级，LWP 总是从最高优先级的就绪队列中调度一个用户线程运行。当某一个线程就绪时，它插入到对应的就绪队列中，并唤醒 LWP 池中的一个空闲 LWP，去选择一个最高优先级的线程运行。如果 LWP 池空，则在就绪队列中等待。

线程调度允许抢占，如果就绪线程的优先级比活动队列中的线程的任一个的优先级高， 则要实施抢占。从活动队列中找到一个优先级最低的线程，向所对应的 LWP 发信号，LWP 接收到信号后，重新调度优先级最高的线程，则原来的线程被抢占。

当正运行的线程被某一个同步变量阻塞，或退出，或被暂停时，要释放对应的 LWP，则该 LWP 从就绪队列中调度新的线程去运行，如果没有可运行的线程，则等待在 idle 变量上，放入 LWP 池中睡眠。

2．用户线程同步

用户线程同步机制有以下几种。

互斥锁：一次只有一个线程能获得锁，常用于临界区的互斥执行。

条件变量：用于使一个线程等待某一个条件为真，必须与互斥锁一起使用。

信号灯：信号灯是一个非负的整数计数器，当一个线程进入时其值减少，当一个线程退出时其值增加。

多读者，单写者锁：允许多个线程同时读某个共享对象，但当一个线程要写该对象时，要阻塞读线程。

同步机制可用于同一进程的各个线程之间，此时同步变量在进程的常规内存空间中分配，对核心是透明的，每个同步类型都支持几种实现方法，程序员在变量初始化时选择实现方法。

同步机制也可用于不同进程的各个线程之间，在变量初始化时标志为共享，在共享的映射文件中分配。当线程被阻塞时，核心对其处理，挂在核心的睡眠队列中，此时线程同执行系统调用一样，与 LWP 捆绑在一起，LWP 不能去执行其他的线程。

3．多线库支持线程的特点

使用多线库实现用户级多线程，有如下几方面优势。

（1）为用户提供一个简明的同步并行编程环境。如多线库可为用户提供同步 I/O 请求，而多线库利用操作系统的异步 I/O 支持，当请求 I/O 的线程在同步等待 I/O 完成时多线库向操作系统发送异步 I/O，在操作系统收到异步 I/O 请求后控制返回多线库程序时，多线库把请求同步 I/O 的线程阻塞，调用其他线程运行，等到操作系统 I/O 完成后，多线库才将阻塞的线程变到就绪状态，多线库把用户的同步请求转化成了对操作系统的异步请求。

（2）开销小。线程的建立、结束、切换，管理都由多线库完成，多线库是在用户态执行的程序，所有管理数据结构也都在用户空间中，无须操作系统内核作任何支持，因此没有运行模式切换的开销。

（3）无须改变操作系统核心，用户级线程完全由用户级库程序实现，实现非常方便。

（4）用户可以根据应用的并行度，申请足够多的线程而无须考虑系统资源限制。当然，使用用户态多线库来实现多线程也有以下几点问题。

① 不能做到进程内线程在多处理机上真正并行运行，如果系统拥有多个处理机，操作系统调度进程占用处理机运行，它无法感知进程内多线程存在，因而无法让用户进程在多个处理机上同时运行。

② 如果用户级线程发系统调用或中断进入操作系统，因某种原因其所在进程被阻塞，多线库无法知道刚运行的线程被阻塞，因此也不能够转到其他线程运行。这种情形下可以对多线库加以改进，让可能引起阻塞的系统调用统统经过多线库转发，使多线库能够根据情况调度其他线程运行。

4.4.3 核心级线程实现

1．内核线程的调度

内核的调度单位是内核线程。内核线程分成三种类型， 一种是在进程中执行的 LWP；第二种是执行内核功能的特殊线程，完成调页、换进换出或为流服务的后台服务等功能；最后一种是空闲线程，当 CPU 没有就绪线程可运行时，执行空闲线程，每个 CPU 有一个对应的空闲线程，由 CPU 结构的指针指向。

系统一般支持分时类、系统类和实时类三种调度类型，每一调度类型有各自的调度策略。系统采用全局优先级模型，优先级从 0 到 159。

调度时，内核线程处于三种状态：阻塞态、就绪态和运行态。被阻塞的线程都挂在某一个同步对象的睡眠队列上，当该对象被释放时，睡眠队列中最高优先级的线程将被唤醒，进入就绪态。如果本线程的优先级高于相关联的 CPU 上正在运行的线程的优先级，则要进行“抢占”。

抢占分成用户级抢占和核心级抢占两种。用户级抢占采用“lazy”方式，直到线程从核心态返回到用户态时才实施。当有用户级抢占时，在 CPU 结构中设置一个标志位，当线程从陷入或系统调用返回用户态时，检查此标志，调用调度函数进行切换。核心级抢占立即进行，当有核心级抢占时，在 CPU 结构中设置一个标志，同一个 CPU 的抢占可立即调用调度函数来进行，不同 CPU 的抢占要通过 CPU 之间发送中断来实现。

2．同步机制

核心提供的核心内部使用的同步对象与在用户级库中为多线程结构的应用程序提供的同步对象是很类似的。提供两套类似的同步机制，可以使得用户线程在用户级库中进行同步互斥控制，而不用占用核心的资源，由核心来管理，以提高系统效率。

3．核心支持线程的特点

利用操作系统核心支持的核心级线程实现有如下几个特点。

（1）可以支持进程内多线程在多处理机上真正并行执行，这是因为核心处理机调度程序是以线程为单位，调度程序可以选定同一进程的线程同时占用处理机。具体实现时，调度程序选取同一进程中的就绪线程，把它们放在各处理机等待运行的位置，通过机间中断通知各处理机运行线程切换程序，各处理机即可切换到同一进程的线程上运行。

（2）不会出现用户级线程实现方式下，线程在内核被阻塞但多线库调度器一无所知。由于线程调度在核内进行，如果线程因等事件阻塞，核心即可将处理机切换到其他线程上运行。

（3）核心级线程比用户级线程开销要大一些。这是因为核心级线程的管理都由内核程序进行，需要用户态到核心态的切换。

（4）核心级线程占用系统空间及资源，并不适合用户根据其任务的并行度来创建相应多的核心级线程。如果用户任务并行度很高，为每个并行任务申请一个核心级线程，若系统中存在许多这种用户，系统资源会马上消耗光。

Chapter 4 Interprocess Communication and Multi-Threading

Processes executing concurrently in the operating system may be either independent processes or cooperating processes. A process is independent if it cannot affect or be affected by the other processes executing in the system. Any process that does not share data with any other process is independent. A process is cooperating if it can affect or be affected by the other processes executing in the system. Clearly, any process that shares data with other processes is a cooperating process.

4.1 Interprocess Communication

4.1.1 Shared-Memory Systems

Interprocess communication using shared memory requires communicating processes to establish a region of shared memory. Typically, a shared-memory region resides in the address space of the process creating the shared-memory segment. Other processes that wish to communicate using this shared-memory segment must attach it to their address space. Recall that, normally, the operating system tries to prevent one process from accessing another process's memory. Shared memory requires that two or more processes agree to remove this restriction. They can then exchange information by reading and writing data in the shared areas. The form of the data and the location are determined by these processes and are not under the operating system's control. The processes are also responsible for ensuring that they are not writing to the same location simultaneously.

4.1.2 Message-Passing Systems

Message passing provides a mechanism to allow processes to communicate and to synchronize their actions without sharing the same address space and is particularly useful in a distributed environment, where the communicating processes may reside on different computers connected by a network. For example, a chat program used on the World Wide Web could be designed so that chat participants communicate with one another by exchanging messages.

A message-passing facility provides at least two operations: send(message) and receive (message). Messages sent by a process can be of either fixed or variable size. If only fixed-sized messages can be sent, the system-level implementation is straightforward. This restriction, however, makes the task of programming more difficult. Conversely, variable-sized messages require a more complex system-level implementation, but the programming task becomes simpler. This is a common kind of tradeoff seen throughout operating system design.

If processes P and Q want to communicate, they must send messages to and receive messages from each other; a communication link must exist between them. This link can be implemented in a variety of ways.

1. Direct Communication

Under direct communication, each process that wants to communicate must explicitly name the recipient or sender of the communication. In this scheme, the send() and receive() primitives are defined as below:

(1) send (P , message) —Send a message to process P;

(2) receive (q, message) —Receive a message from process Q.

A communication link in this scheme has the following properties:

(1) A link is established automatically between every pair of processes that want to communicate. The processes need to know only each other's identity to communicate;

(2) A link is associated with exactly two processes;

(3) Between each pair of processes, there exists exactly one link.

This scheme exhibits symmetry in addressing, that is, both the sender process and the receiver process must name the other to communicate. A variant of this scheme employs asymmetry in addressing. Here, only the sender names the recipient; the recipient is not required to name the sender. In this scheme, the send() and receive() primitives are defined as follows:

(1) send (P , message) —Send a message to process P.

(2) receive (id, message) —Receive a message from any process; the variable id is set to the name of the process with which communication has taken place.

The disadvantage in both of these schemes (symmetric and asymmetric) is the limited modularity of the resulting process definitions. Changing the identifier of a process may necessitate examining all other process definitions. All references to the old identifier must be found, so that they can be modified to the new identifier. In general, any such hard-coding techniques, where identifiers must be explicitly stated, are less desirable than techniques involving indirection, as described next.

2. Indirect Communication

With indirect communication, the messages are sent to and received from mailboxes, or ports. A mailbox can be viewed abstractly as an object into which messages can be placed by processes and from which messages can be removed. Each mailbox has a unique identification. For example, POSIX message queues use an integer value to identify a mailbox. In this scheme, a process can communicate with some other process via a number of different mailboxes. Two processes can communicate only if the processes have a shared mailbox, however. The send() and receive() primitives are defined as follows:

(1) send (A , message) —Send a message to mailbox A;

(2) receive (A, message) —Receive a message from mailbox A.

In this scheme, a communication link has the following properties:

(1) A link is established between a pair of processes only if both members of the pair have a shared mailbox.

(2) A link may be associated with more than two processes.

(3) Between each pair of communicating processes, there may be a number of different links, with each link corresponding to one mailbox.

Now suppose that processes P_1, P_2, and P_3 all share mailbox A. Process P_1 sends a message to A, while both P_2 and P_3 execute a receive() from A. Which process will receive the message sent by P_1? The answer depends on which of the following methods we choose.

(1) Allow a link to be associated with two processes at most.

(2) Allow at most one process at a time to execute a receive() operation.

(3) Allow the system to select arbitrarily which process will receive the message (that is, either P_2 or P_3, but not both, will receive the message). The system also may define an algorithm for selecting which process will receive the message (that is, round robin where processes take turns receiving messages). The system may identify the receiver to the sender.

A mailbox may be owned either by a process or by the operating system. If the mailbox is owned by a process (that is, the mailbox is part of the address space of the process), then we distinguish between the owner (who can only receive messages through this mailbox) and the user (who can only send messages to the mailbox). Since each mailbox has a unique owner, there can be no confusion about who should receive a message sent to this mailbox. When a process that owns a mailbox terminates, the mailbox disappears. Any process that subsequently sends a message to this mailbox must be notified that the mailbox no longer exists.

In contrast, a mailbox that is owned by the operating system has an existence of its own. It is independent and is not attached to any particular process. The operating system then must provide a mechanism that allows a process to do the following:

(1) Create a new mailbox;

(2) Send and receive messages through the mailbox;

(3) Delete a mailbox.

The process that creates a new mailbox is that mailbox's owner by default. Initially, the owner is the only process that can receive messages through this mailbox. However, the ownership and receiving privilege may be passed to other processes through appropriate system calls. Of course, this provision could result in multiple receivers for each mailbox.

4.1.3 pipe

pipes provide a mechanism suitable for this kind of communication. A pipe is made of two file descriptors. The first one represents the pipe’s output. The second one represents the pipe’s input. Pipes are created by the system call pipe:

```
val pipe : unit -> file_descr * file_descr
```

The call returns a pair (fd_in, fd_out) where fd_in is a file descriptor open in read mode on the pipe's output and fd_out is file descriptor open in write mode on the pipe's input. The pipe itself is an internal object of the kernel that can only be accessed via these two descriptors. In particular, it has no name in the file system.

A pipe behaves like a queue (first-in, first-out). The first thing written to the pipe is the first thing read from the pipe. Writes (calls to write on the pipe's input descriptor) fill the pipe and block when the pipe is full. They block until another process reads enough data at the other end of the pipe and return when all the data given to write have been transmitted. Reads (calls to read on the pipe's output descriptor) drain the pipe. If the pipe is empty, a call to read blocks until at least a byte is written at the other end. It then returns immediately without waiting for the number of bytes requested by read to be available.

4.1.4 Socket Communication

A socket is defined as an endpoint for communication. A pair of processes communicating over a network employ a pair of sockets—one for each process. A socket is identified by an IP address concatenated with a port number. In general, sockets use a client–server architecture. The server waits for incoming client requests by listening to a specified port. Once a request is received, the server accepts a connection from the client socket to complete the connection. Servers implementing specific services (such as telnet, ftp, and http) listen to well-known ports (a telnet server listens to port 23, an ftp server listens to port 21, and a web, or http, server listens to port 80). All ports below 1024 are considered well known; we can use them to implement standard services.

When a client process initiates a request for a connection, it is assigned a port by the host computer. This port is some arbitrary number greater than 1024. For example, if a client on host X with IP address 146.86.5.20 wishes to establish a connection with a web server (which is listening on port 80) at address 161.25.19.8, host X may be assigned port 1625. The connection will consist of a pair of sockets: (146.86.5.20:1625) on host X and (161.25.19.8:80) on the web server. The packets traveling between the hosts are delivered to the appropriate process based on the destination port number.

4.2 Multiprocessor Systems

Although single-processor systems are most common, multiprocessor systems (also known as parallel systems or tightly coupled systems) are growing in importance. Such systems have two or more processors in close communication, sharing the computer bus and sometimes the clock, memory, and peripheral devices.

The multiple-processor systems in use today are of two types. Some systems use asymmetric multiprocessing, in which each processor is assigned a specific task. A master processor controls the system; the other processors either look to the master for instruction or have predefined tasks. This scheme defines a master-slave relationship. The master processor schedules and allocates work to the slave processors.

The most common systems use symmetric multiprocessing (SMP), in which each processor performs all tasks within the operating system. SMP means that all processors are peers; no master-slave relationship exists between processors. The benefit of this model is that many processes can run simultaneously —N processes can run if there are N CPUs—without causing a significant deterioration of performance. However, we must carefully control I/O to ensure that the data reach the appropriate processor. Also, since the CPUs are separate, one may be sitting idle while another is overloaded, resulting in inefficiencies. These inefficiencies can be avoided if the processors share certain data structures. A multiprocessor system of this form will allow processes and resources—such as memory—to be shared dynamically among the various processors and can lower the variance among the processors.

The difference between symmetric and asymmetric multiprocessing may result from either hardware or software. Special hardware can differentiate the multiple processors, or the software can be written to allow only one master and multiple slaves. For instance, Sun's operating system SunOS Version 4 provided asymmetric multiprocessing, whereas Version 5 (Solaris) is symmetric on the same hardware.

A recent trend in CPU design is to include multiple compute cores on a single chip. In essence, these are multiprocessor chips. Two-way chips are becoming mainstream, while N-way chips are going to be common in high-end systems. Aside from architectural considerations such as cache, memory, and bus contention, these multi-core CPUs look to the operating system just as N standard processors.

4.3 Threads

A thread is a basic unit of CPU utilization; it comprises a thread ID, a program counter, a register set, and a stack. It shares with other threads belonging to the same process its code section, data section, and other operating-system resources, such as open files and signals. A traditional (or heavyweight) process has a single thread of control. If a process has multiple threads of control, it can perform more than one task at a time.

4.3.1 Motivation

Many software packages that run on modern desktop PCs are multithreaded. An application typically is implemented as a separate process with several threads of control. A web browser might have one thread display images or text while another thread retrieves data from the network, for example. A word processor may have a thread for displaying graphics, another thread for responding to keystrokes from the user, and a third thread for performing spelling and grammar checking in the background.

In certain situations, a single application may be required to perform several similar tasks. For example, a web server accepts client requests for web pages, images, sound, and so forth. A busy web server may have several (perhaps thousands) of clients concurrently accessing it. If the web server ran as a traditional single-threaded process, it would be able to service only one client at a time. The amount of time that a client might have to wait for its request to be serviced could be enormous.

One solution is to have the server run as a single process that accepts requests. When the server receives a request, it creates a separate process to service that request. In fact, this process-creation method was in common use before threads became popular. Process creation is time consuming and resource intensive, as was shown in the previous chapter. If the new process will perform the same tasks as the existing process, why incur all that overhead? It is generally more efficient to use one process that contains multiple threads. This approach would multithread the web-server process. The server would create a separate thread that would listen for client requests; when a request was made, rather than creating another process, the server would create another thread to service the request.

4.3.2 Multithreading Models

Our discussion so far has treated threads in a generic sense. However, support for threads may be provided either at the user level, for user threads, or by the kernel, for kernel threads. User threads are supported above the kernel and are managed without kernel support, whereas kernel threads are supported and managed directly by the operating system. Virtually all contemporary operating systems—including Windows XP, Linux, Mac OS X, Solaris, and Tru64 UNIX (formerly Digital UNIX)—support kernel threads.

Ultimately, there must exist a relationship between user threads and kernel threads. In this section, we look at three common ways of establishing this relationship.

1. many-to-one model

The many-to-one model maps many user-level threads to one kernel thread. Thread management is done by the thread library in user space, so it is efficient; but the entire process will block if a thread makes a blocking system call. Also, because only one thread can access the kernel at a time, multiple threads are unable to run in parallel on multiprocessors. Green threads—a thread library available for Solaris—uses this model.

2. one-to-one model

The one-to-one model maps each user thread to a kernel thread. It provides more concurrency than the many-to-one model by allowing another thread to run when a thread makes a blocking system call; it also allows multiple threads to run in parallel on multiprocessors. The only drawback to this model is that creating a user thread requires creating the corresponding kernel thread. Because the overhead of creating kernel threads can burden the performance of an application, most implementations of this model restrict the number of threads supported by the system. Linux, along with the family of Windows operating systems—including Windows 95, Windows 98, Windows NT, Windows 2000, and Windows XP—implement the one-to-one model.

3. many-to-many model

The many-to-many model multiplexes many user-level threads to a smaller or equal number of kernel threads. The number of kernel threads may be specific to either a particular application or a particular machine (an application may be allocated more kernel threads on a multiprocessor than on a uniprocessor). Whereas the many-to-one model allows the developer to create as many user threads as she wishes, true concurrency is not gained because the kernel can schedule only one thread at a time. The one-to-one model allows for greater concurrency, but the developer has to be careful not to create too many threads within an application (and in some instances may be limited in the number of threads she can create). The many-to-many model suffers from neither of these shortcomings: Developers can create as many user threads as necessary, and the corresponding kernel threads can run in parallel on a multiprocessor. Also, when a thread performs a blocking system call, the kernel can schedule another thread for execution.

习　　题

1．信箱通信是一种________通信方式。

A．直接通信　　B．间接通信　　C．低级通信　　D．信号量

2．用信箱实现通信时，应有________和________两条基本原语

3．简述进程通信的三种基本方式。

第5章 内存管理

内存是计算机系统的重要组成部分。内存管理，是指软件运行时对计算机内存资源的分配和使用的技术。其最主要的目的是如何高效、快速地分配，并且在适当的时候释放和回收内存资源。内存管理的基本方法包括分区式管理、分页式管理、分段式管理和段页式管理等多种分配管理方案。

5.1 重 定 位

前面介绍过计算机程序运行的基本原理。计算机程序运行时，组成程序的指令一定要放在计算机的内存中，计算机的控制器只能从内存中读取指令，分析并执行指令，如图 5-1 所示。那么多道程序运行时，被系统接纳，并创建了进程的程序（在宏观上，处于运行状态的程序），都需要放入计算机的内存。如何把内存分配给它们？当程序运行结束后，怎样把内存收回？这是内存管理的主要内容。

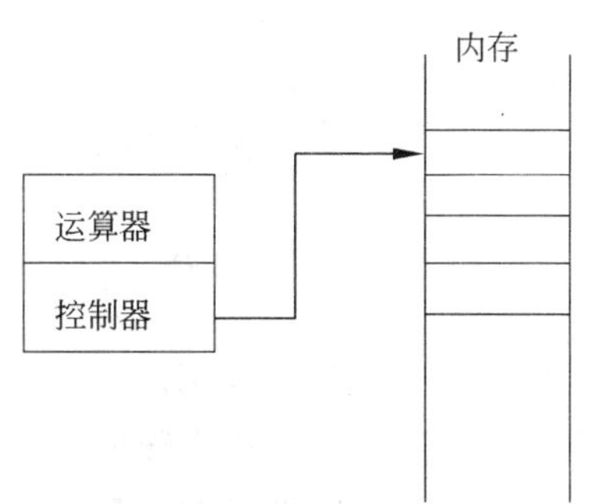

图 5-1 计算机程序的执行

在了解内存管理的相关知识之前，先学习几个相关的概念。

（1）物理地址：就是内存的地址，是以字节为单位，对内存单元的编址。

（2）逻辑地址：用户源程序经过编译或汇编后形成的目标指令代码的编址。

（3）地址空间：地址的编址范围。

（4）物理地址空间：内存地址的编址范围。也就是计算内存的编址范围，它是由实际的物理内存的大小决定的。

（5）逻辑的地址空间：用户程序指令的编址范围，是由程序的大小决定的。

（6）重定位：是把程序的逻辑地址空间变换成内存中的实际物理地址空间的过程，也就是说在装入时对目标程序中指令和数据的修改过程。

重定位这个概念是内存管理中非常重要的概念。重定位是实现多道程序在内存中同时运行的基础。

程序是静态的，它是为解决某种问题而编写的指令代码的集合。程序被写出来后，经过编译链接或者汇编后，形成目标指令代码顺序，按照顺序的逻辑地址编址。程序要想运行，必须被装入内存，装入内存的地址是由操作系统决定的。程序的逻辑地址与内存的物理地址一般是不一样的。程序中的指令可能会根据自己的逻辑地址存取数据，程序被装入内存后，地址发生了变化，所以需要重定位，调整程序中的地址才能使程序正常运行。

如图 5-2 所示，程序中有一条指令，LOAD1,2500，其含义是将 2500 号单元的数据送到 1 号寄存器中。这里的 2500 号单元，是逻辑地址空间中的地址单元。当程序装入内存后，被装在了从 10000 号地址单元开始的物理地址空间中。那么当程序执行到 LOAD1,2500 指令时就不能直接使用 2500 这个地址，而是应该在这个地址基础上加上重定位的起始地址。即 2500+10000，这个将逻辑地址转换成真实物理地址的过程就是重定位。

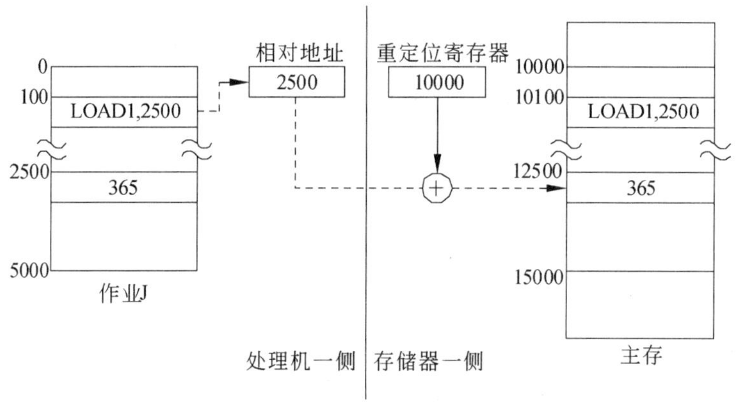

图 5-2　重定位

重定位有两种，分别是动态重定位与静态重定位。

（1）静态重定位：即在程序装入内存的过程中完成，是指在程序开始运行前，程序中的各个地址有关的项均已完成重定位，地址变换通常是在装入时一次完成的，以后不再改变，故成为静态重定位。

（2）动态重定位：它不是在程序装入内存时完成的，而是 CPU 每次访问内存时 由动态地址变换机构（硬件）自动进行把相对地址转换为绝对地址。动态重定位需要软件和硬件相互配合完成。

5.2　分区式管理

运行一个程序，要通过操作系统的命令接口或程序调用接口，向操作系统提交一个作业，操作系统接纳这个作业，要为它创建进程，为它分配内存，程序装入内存，才能运行。怎样分配内存，程序怎样装入内存，是本节讨论的主要话题。

5.2.1 单一连续区分配

早期的单用户、单任务系统中，一般采用单一连续区的分配方式。这是一种简单的、朴素的存储器分配方案。它的基本思想就是把内存分成两个区，系统程序区和用户程序区，如图 5-3 所示。系统程序区存放操作系统程序，大部分操作系统程序常驻内存；用户程序区分配给用户程序，程序运行结束，释放用户程序区，另外有作业提交之后，在把用户程序区分配给新的作业。这种分配方案，简单，容易实现，早期的单用户的操作系统，CPM、MS-DOS 都是采用这样的内存分配方案。

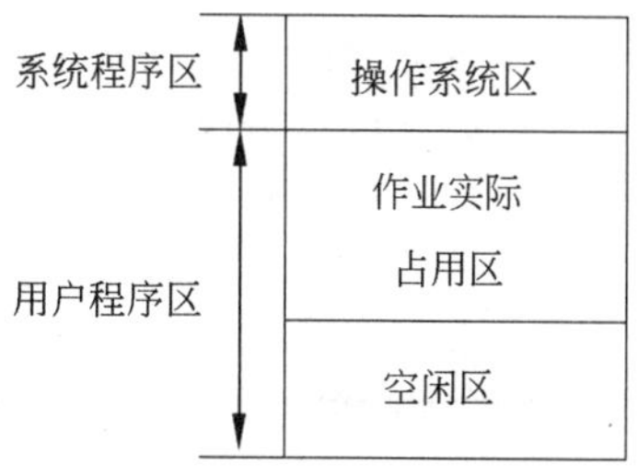

图 5-3 连续区分配

单一连续区分配方案，不能运行大于用户程序区的作业，小的作业不能占满用户程序区，但剩余的部分也不能被利用，只能暂时空闲。所以这种分配方案内存的利用率不是很高。

在单一连续区的分配方案中要设置一些存储器保护机构，防止用户程序访问系统程序区，防止破坏、修改操作系统程序，造成不良后果。

5.2.2 固定分区分配

在多道程序的系统中，单一连续区的分配方式显然是不能满足要求的。一种简单的改进的分配方案就是采用分区的方法，把内存化成若干个连续的区域，每个连续区叫做一个分区，这种分配方案叫做分区式分配方案。分区式分配是能满足多道程序设计的最简单的存储管理技术，它允许几个作业共享主存空间，这几个作业被装入不同的分区中，每个分区可装入一个作业，这样可以在内存中同时放几道作业，实现多道程序运行。

在分区式分配方案中，分区个数是固定的，叫做固定式分区分配；分区的个数不固定的可变式分区方案。这一小节只要讨论固定式分区分配方案，在下一小节中再讨论可变式分区分配方案。

1. 固定式分区

操作系统
200MB
200MB
200MB
200MB

操作系统
100MB
100MB
200MB
400MB

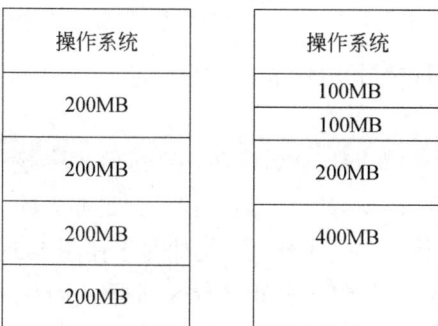

图 5-4 固定式分区

固定式分区是在处理作业之前存储器就已经被划分成固定个数的分区，每个分区的大小可以相同，也可以不同，如图 5-4 所示。但是，一旦划分好分区后，主存储器中的分区的个数就固定了，且每个分区的大小不再不变。

分区大小相同，看起来内存分配均衡，好像比较公平。例如内存中的用户程序区有 800MB，平均分成四个分区，每个分区 200MB。但在这种情况下，大于 200MB 的大作业不能够运行；而小作业也要占用一个分区，如果作业很小，假设只有 10MB，分区剩余 190MB 空闲，浪费比较严重。

分区大小不同，虽然分区个数固定，但是分区的大小可以不均分，设置一些小的分区，一些中等的分区，再设置一些大的分区。如果新到的作业较大就分配到较大的分区，作业较小就分配到较小的分区，这样既可以运行相对较大一些的作业，又能使较小的作业不会占用较大的分区，造成剩余"零头"较大，浪费较大的现象。

2．分区表

固定式分区方案的实现通过分区分配表或者分区分配链表来实现。分区分配表是一个二维表，每行描述一个分区，分区的个数固定，分区表的行数就是固定的。分区表中包括每个分区的分区号、起始地址、分区大小、占用大小和备注等信息。

例如，内存分区如图 5-5（a）所示，则分区表如图 5-5（b）。

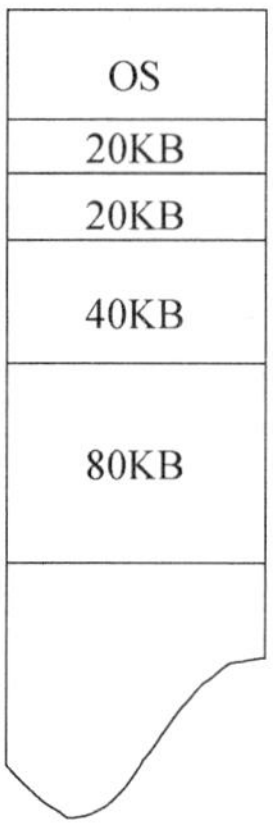

OS
20KB
20KB
40KB
80KB

(a) 内存分区

分区号	起始地址	分区大小	占用大小	备注
1	40KB	20KB	0KB	
2	60KB	20KB	5KB	
3	80KB	40KB	30KB	
4	120KB	80KB	0KB	
5	200KB	…	…	
6	…	…	…	

(b) 分区表

图 5-5　内存分区与分区表

上例中是以分区大小不等为例的。分区大小相等的分配方式也可以通过这样的分区表来管理，只是分区大小那一列的数值都是相等的。

在分区表中可以通过"占用大小"这一列来判断分区是否已分配，当这一列的值为 0 时，说明这个分区是空闲的。这一列如果有大于 0 的值，这个值一定小于等于分区大小那一列的值，它们的差值是这个分区的空闲区域的大小，也叫内存的"零头"，上例中 2 号分区的"零头"是 15KB，3 号分区的"零头"是 10KB。零头越大说明内存浪费越多，内存的利用率越低。

固定分区分配方案中，分配内存和回收内存可以通过维护分区表来实现。

3．分区分配

采用固定分区方案是比较容易分配内存的。当有一个新的作业进来时，在分区表中找一个大小比较合适的空闲分区，分配给它，使内存"零头"尽量小。如果找不到足够大的空闲分区，则这个作业暂时无法分配内存空间，系统将调度另一个作业。例如对于在图 5-5（b）所示的分区表，如果新来一个作业 J_1 大小为 15KB，1 号分区和 4 号分区都是空闲区，把 J_1 分配到 1 号分区，"零头"是 5KB，比较小，所以 J_1 分配到 1 号分区，分配后的分区表如表 5-1 所示。

表 5-1　分配后的分区表

分区号	起始地址	分区大小	占用大小	备注
1	40KB	20KB	15KB	
2	60KB	20KB	5KB	
3	80KB	40KB	30KB	
4	120KB	80KB	0KB	
5	200KB	…	…	
6	…	…	…	

4．分区回收

当一个作业运行结束后，需要把它使用的内存收回来。采用固定分区方案，内存的回收还是比较容易的。当有一个作业运行结束后，在分区表中找到对应的分区所在行，把占用大小清零即可。例如对于在图 5-5（b）所示的分区表，如果 3 号分区的作业运行结束了，只需把 3 号分区那一行的占用大小为改成 0KB 即可。回收分区后的分区分配表如表 5-2 所示。

表 5-2　回收分区后的分区分配表

分区号	起始地址	分区大小	占用大小	备注
1	40KB	20KB	15KB	
2	60KB	20KB	5KB	
3	80KB	40KB	0KB	
4	120KB	80KB	0KB	
5	200KB	…	…	
6	…	…	…	

5.2.3　可变式分区

固定式分区分配方案，不利于大作业的运行，有时也分配的分区还会产生很大“零头”。与固定式分区分配方案对应的是可变式分区分配方案，它的基本思想是不预先划分内存空间，而是在作业装入时根据作业的实际需要动态地划分内存空间。这样分区的个数是不固定的，当没有作业提交时整个用户程序区只有一个大的空闲分区，可以运行一个比较大的作业；当有一个作业 A 被接纳了，根据 A 作业的实际大小，划出一个分区，大小同 A 作业一样大，把它分配给作业 A，另一个分区是剩下的部分，还是空闲分区，如图 5-6 所示。

1．分区表

管理动态的分区分配，系统应设置两张表，一个记录已分配分区的表，一个记录空闲分区的表。由于分区个数不固定，所以两张表的行数都是不固定的。初始的时候，已分配分区表是空的，空闲分区表只有一行。

假设系统中连续提交了 A、B、C、D 四个作业，系统动态的生成四个已分配分区和一个空闲分区。已分配分区表和空闲分区表如图 5-7（b）、（c）所示。

图 5-6　可变式分区

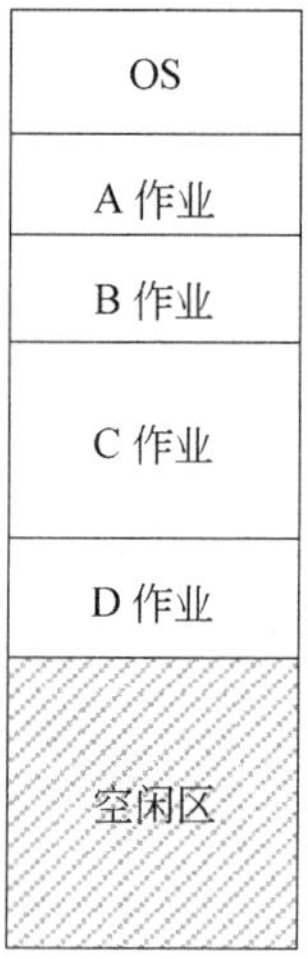

OS
A 作业
B 作业
C 作业
D 作业
空闲区

(a) 内存分区

分区号	起始地址	分区大小	占用大小
1	40KB	10KB	10KB
2	50KB	20KB	20KB
3	70KB	40KB	40KB
4	110KB	20KB	20KB
5	…	…	空表目

(b) 分配分区表

分区号	起始地址	分区大小	状态
1	130KB	1000KB	空闲
2	…	…	空表目

(c) 空闲分区表

图 5-7　可变式分区管理

假设过了一段时间，A 作业和 C 作业运行结束，系统收回它们所占内存，这时的系统状态变为如图 5-8 所示。

OS
空闲区
B 作业
空闲区
D 作业
空闲区

(a) 内存分区

分区号	起始地址	分区大小	占用大小
1	…	…	空表目
2	50KB	20KB	20KB
3	…	…	空表目
4	110KB	20KB	20KB
5	…	…	空表目

(b) 分配分区表

分区号	起始地址	分区大小	状态
1	40KB	10KB	空闲
2	70KB	40KB	空闲
3	130KB	1000KB	空闲
4	…	…	空表目

(c) 空闲分区表

图 5-8　A、C 作业运行结束的状态

看得出来，当系统运行一段时间后，用户程序区会产生许多的空闲区和已分配分区，它们会分散分布在用户程序区。分区的分配和回收要比固定式分区分配方法复杂一些。

空闲分区也可以组织成链表的形式，叫做空闲分区链。为了实现对空闲分区的分配和链接，在每个分区的起始部分，设置一些用于控制分区分配的信息，以及用于链接各分区的前向指针，在分区尾部则设置一后向指针，通过前、后向指针将所有空闲分区链成一个双向链表，如图 5-9 所示。为了检索空闲分区方便，在分区尾部重复设置状态位和分区大小表目，当分区分配出去以后，把状态位由“0”改为“1”。

2．分区分配算法

为把一个新作业装入内存，要按照一定的算法找到一个合适的空闲分区，分配给这个新作业。常用的分配算法有四种。

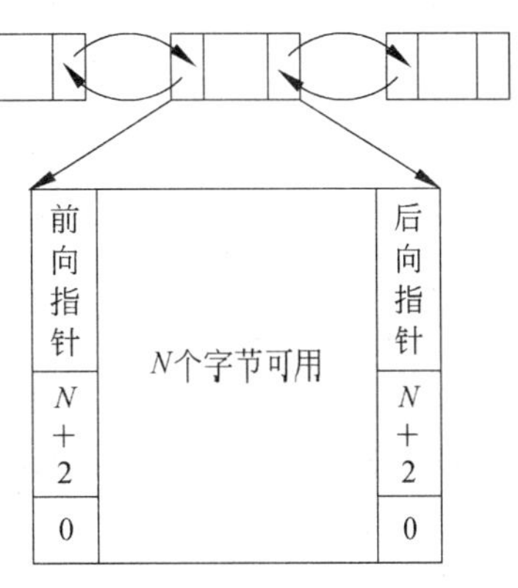

图 5-9　空闲分区链表

1）首次适应算法（First Fit）

首次适应算法要求空闲分区按首址递增的次序排序（队列），当进程申请大小为 u.size 的内存时，系统从空闲区表的第一个表目开始查询，直到首次找到等于或大于 u.size 的空闲区 m。如果 m.size-u.size，大于等于系统规定的最小分区的大小 size，则从 m 中划出大小为 u.size 的分区分配给进程，余下的部分 m.size-u.size，仍作为一个空闲区留在空闲区表中，但要修改其首址和大小。

如果多出来的部分 m.size-u.size，小于系统规定的最小分区的大小 size，则把多出来的部分 m.size-u.size，看成是分区的“零头”，把整个分区 m 分配给进程，分区存在 m.size-u.size 的“零头”，在空闲分区表中把 m 分区置成空表目。

首次适应分配算法的流程图如图 5-10 所示。

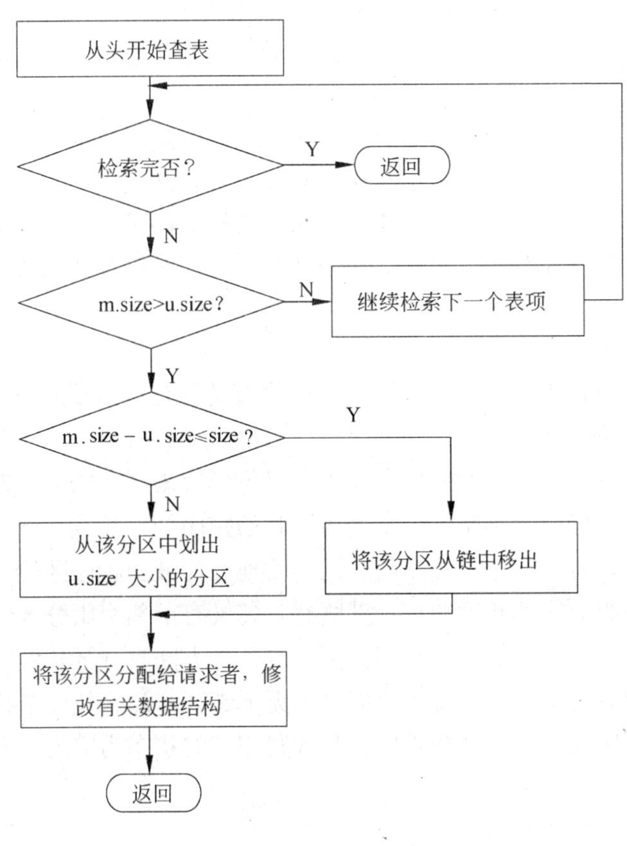

图 5-10　首次适应分配算法流程图

2）循环首次适应算法（Next Fit）

首次适应算法每次都是从内存的低地址部分开始查找的，所以内存的低地址部分使用比较频繁，出现小的、比较零碎的空闲分区的机会比较多，内存的使用不够均衡。在首次适应算法基础上稍加改动就是循环首次适应算法。循环首次适应算法的做法是不在每次从队首开始查找，而是从上次找到的空闲区的下一个空闲区开始查找，直到找到第一个能满足要求的空闲区，如果找到最后一个空闲区（队尾），还没找到要求的空闲区，则返回队首，从头再来查找。

为实现该算法，应设置一个起始查寻指针，以指示下一次起始查寻的空闲分区，采用循环查找的方式。该算法能使内存中的空闲分区分布得更均匀，减少查找空闲分区的开销，但这会导致缺乏大的空闲分区。

3）最佳适应算法（Best Fit）

最佳适应算法是从全部空闲区中找出能满足作业要求的且大小最小的空闲分区，这种方法能使碎片尽量小。为实现此算法，空闲分区表（空闲区链）中的空闲分区要按大小从小到大进行排序，自表头开始查找到第一个满足要求的自由分区分配。每次分配时，找到的空闲区表中能够满足作业需求的最小的空闲区，尽量不分割在空闲区，保留大的空闲区，便于大作业的装入。但会造成许多小的空闲区。

4）最坏适应算法（Worst Fit）

最坏适应算法其实“不坏”，它只是每次都找最大的空闲分区分配给作业，是这个空闲分区分配一部出去后，剩余的空闲区最大。这样做主要是考虑使剩余部分的空间不至于太小，仍可用来装入其他作业，被其他作业利用。

最坏适应分配算法要扫描整个空闲分区或链表，总是挑选一个最大的空闲分区分割给作业使用。该算法的实现方式是将所有的空闲分区按其容量从大到小的顺序形成一空闲分区链，查找时只要看第一个分区能否满足作业要求。这样做的优点是可使剩下的空闲分区不至于太小，产生碎片的几率最小，对中、小作业有利，同时该算法查找效率很高。缺点是存储器中缺乏大的空闲分区。

3. 分区回收算法

可变分区分配方式下，当回收内存分区时，应检查是否有与回收区相邻的空闲分区，若有，则应合并成一个空闲区。

相邻可能有上邻空闲区、下邻空闲区、既上邻又下邻空闲区、既无上邻又无下邻空闲区。若有上邻空闲区，只修改上邻空闲区长度为回收的空闲分区长度与原上邻区长度之和即可；若有下邻空闲区，修改这个下邻空闲分区的首地址为被回收分区的地址，长度为下邻空闲分区的长度和回收分区的长度即可；若既有上邻又有下邻空闲分区，修改上邻空闲分区的长度为上邻空闲分区长度加下邻空闲分区长度再加上回收分区长度之和，再把下邻空闲分区的状态改为空表目即可；若既无上邻区又无下邻区，那么找一个状态为空表目的一行，记下该回收分区区的起始地址和长度，且改写相应的状态为空闲，表明该行指示了一个空闲分区。

回收时的各种情况如图 5-11 所示。

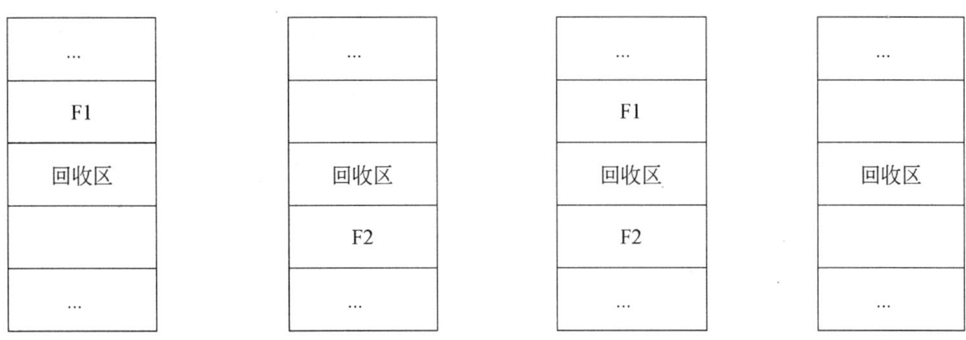

图 5-11　内存回收时的各种情况

4．紧缩

采用可变式分区的管理方式，有时会出现这样状况：系统中有三个空闲分区，大小分别为 36KB、24KB、74KB，现在内存的总的空闲区域是 134KB，但是系统接到进程 5，大小为 80KB 的作业，这个作业的大小比每个空闲区都大，无法为它分配内存。这时可以通过移动内存中的进程，把空闲分区连在一起，再为 80KB 的作业分配内存，就能够是作业运行了。这种移动内存中的进程，把空闲分区连在一起的做法叫做“紧凑”，在有些教科书中也叫“拼接”或者叫“程序搬家”，如图 5-12 所示。

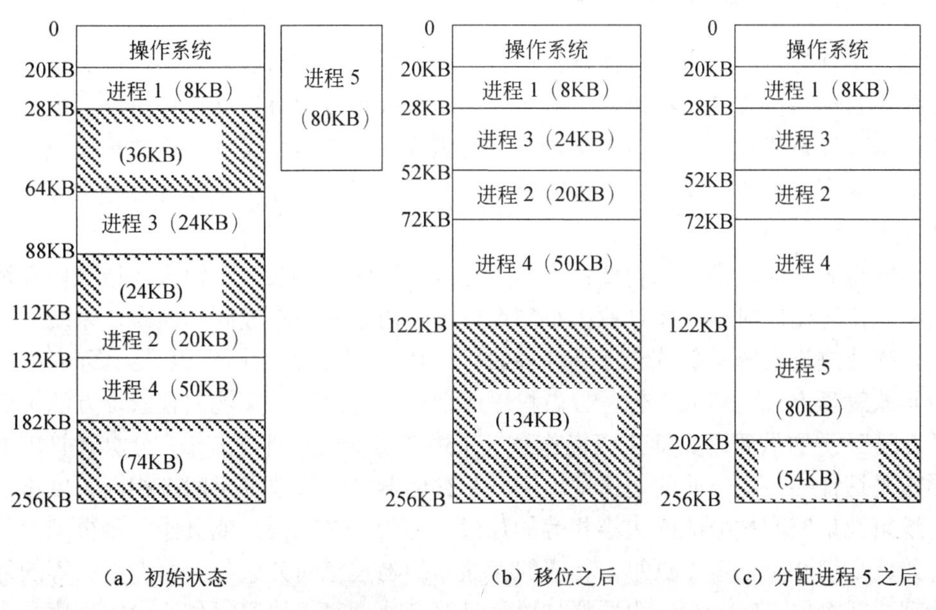

图 5-12　可重定位分区的紧缩

紧凑的开销很大，它不仅要修改被移动进程的地址信息，还要复制进程空间，所以如果不是很必要，尽量不做紧凑，仅在接纳作业，且每个空闲分区单独均不能满足要求，但所有空闲分区之和能够满足要求时，才做一次紧凑。

紧凑功能的分配算法如图 5-13 所示。

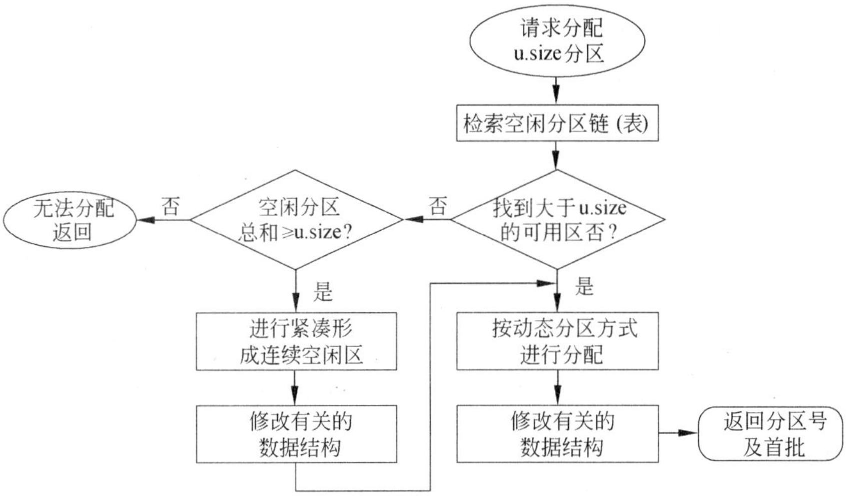

图 5-13　具有紧凑功能的分区分配算法

5.3　分页式管理

5.3.1　分页的基本工作原理

回顾一下计算机的基本工作原理，计算机程序运行时，组成程序的指令一定是要放在计算机的内存中；计算机的控制器只能从内存中读取指令，分析并执行指令；计算机取出指令后，指令计数器自动指向下一条指令。根据程序执行的顺序性，系统将作业的全部信息一次装入内存，并使程序连续存放。所以分区式管理，作业一定要占有整个分区，程序要连续存放。

采用固定式分区，小作业也要占用一个分区，有时“零头”会很大，造成内存浪费；采用可变式，开始的时候，按作业大小划定分区，内存利用率较高，但是，系统运行一段时间后，作业进进出出，系统会出现许多的“碎片”，内存的利用率也受到影响。

分页式管理方式，解放思想，打破程序连续存放的限制，实现内存管理方法上的一次重要“突破”。分页式管理目前主流操作系统采用较多的内存管理方法。分页式管理的基本工作原理是这样的：将作业的逻辑地址空间和存储器的物理地址空间按相同长度进行等量划分，逻辑地址空间被分成的大小相等的片段，称为页（Page）或页面，各页编上号码，从 0 开始，如第 0 页、第 1 页等。相应的存储器的物理空间分成与页面大小相等的片段，称为物理块或页框（Frame），也同样加以编号第 0 块、第 1 块等。作业中的程序装入内存时，按照作业的页数分配物理块，分配的物理块可以连续也可以不连续，如图 5-14 所示。

1. 页表

在分页式管理系统中，进程的每个页面离散地分配到了内存中的任意物理块中。进程的正确执行必须保证能够找到每个页面对应的物理块，为此，系统为每个进程建立一张页面与物理块的映射表，简称为页表，如图 5-15 所示。

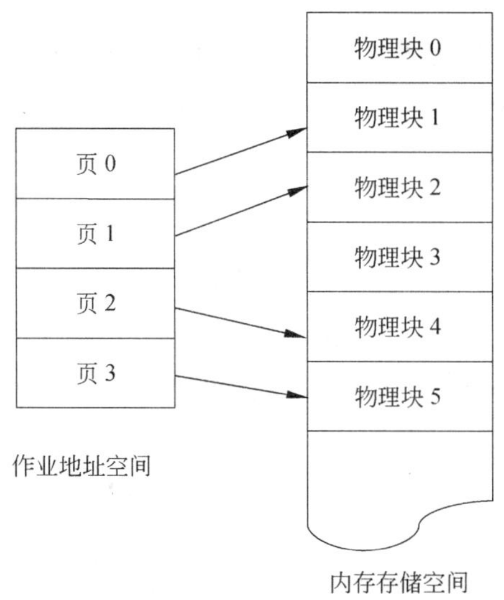

图 5-14　分页管理示意图

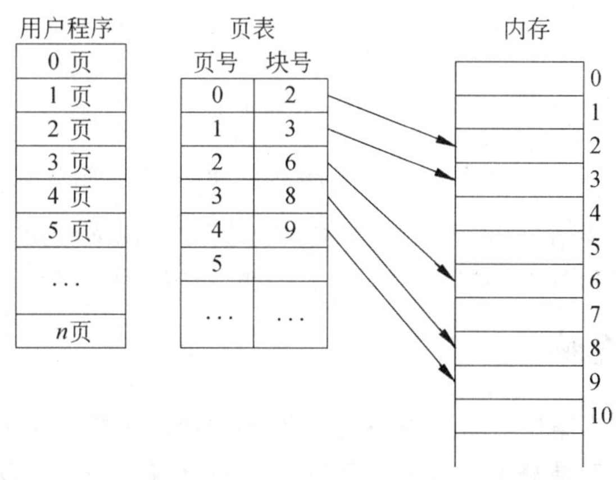

图 5-15　页表示意图

进程的逻辑地址空间内的所有的页，依次在页表中占据一个表项，其中记录了相应页在内存中对应的物理块号。进程执行时，通过查找页表，找到每页在内存中的物理块号。页表通常放在内存中设置的页表区，系统为每一个进程提供一个页表，页表的始地址存放在进程的 PCB 中。

掌握分页式管理的基本工作原理，还应注意以下几个问题。

（1）页内“零头”：每个作业的大小很难正好是页面长度的整数倍，所以最后一页经常装不满一块而形成了不可利用的“零头”，称之为“页内零头”。

（2）页面大小：设计页面的大小很重要，页面的大小偏小，产生页内的“零头”就小，有利于提高内存的利用率，但是会使作业的页面过多，导致页表过长，占用大量内存。如果页面偏大，可能导致页内“零头”较大。因此页面的大小要选择适中，根据内存容量和地址访问机构决定。通常页面大小定义成 2 的整数次幂，例如 2^{10} 或 2^{12} 等。

2．地址结构

分页系统的逻辑地址和物理地址可以分解成两部分，逻辑地址分成页号、页内偏移量（页内地址），分别记为 P，d；物理地址可以分成物理块号、物理块内位移（物理块内地址），记为 b，d。假设页面长度为 L，逻辑地址为 A，那么可通过下列公式计算页号 P 和页内偏移量 d：

$$P=\text{int}\left[\frac{A}{L}\right]$$

$$d=[A]\ \text{mod}\ L$$

其中，int 是整除函数，mod 是取余函数。例如页面大小为 1024B，逻辑地址 2153，用上述公式求得 P=2，d=105，即逻辑地址为 2153 的那一行指令，在第 2 页，页内的偏移量为 105。

从上述公式可以看出，如果页面的大小 L，是 2 的整数次幂（例如是 n 次幂）。对于逻辑地址 A 的二进制值的后 n 位为页内偏移量（页内地址）；前面的位数组成的二进制数就是这个逻辑地址的页号。例如页面大小为 1024B，即 2^{10}，则 0～9 位为页内偏移量，10～31 位为页号 P，如图 5-17 所示。页号占 22 位，地址空间最多允许有 4MB 个页面。

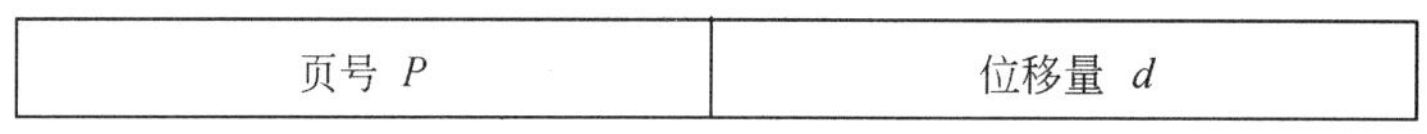

图 5-16　逻辑地址

在执行计算机的指令时，移位操作占的指令周期数要比乘除法指令的周期数少得多，移位操作指令比计算操作指令执行快得多，所以，如果页面大小是 2 的整数次幂，可以通过移位的方式获得页号和页内偏移量，可以提高地址转换的速度。

5.3.2　动态地址变换

在分区式内存管理方案中，程序在内存中连续存放，逻辑地址到物理地址的转换靠重定位寄存器就可以了，但是分页式管理，程序可以不连续存放了，这个地址变换变得复杂了。为实现进程的逻辑地址到物理地址的变换，系统中必须设置地址变换机构，来完成逻辑地址到物理地址的转换过程。

系统设置一个页表寄存器 PTR（Page-Table Register），其中存放页表在内存的始地址和页表长度。当进程要访问某个逻辑地址中的数据时，分页地址变换机构会自动将逻辑地址分为页号和页内偏移量两部分，再以页号为索引去检索页表，查找操作有硬件执行。在执行检索之前，先将页号 P 与页表寄存器中的页表长度相比，如果页号大于或等于页表长度，则本次访问的逻辑地址已超越了进程的地址空间，系统会产生一个越界中断；如果页号大小于页表长度，将页表始地址与页 P 号和表项长度的乘积相加，得到该表项在页表中的位置，从页表中可以得到该页的物理块号 f，装入物理地址寄存器，同时将逻辑地址的页内偏移量 d 也送入物理地址寄存器，得到物理地址，实现逻辑地址到物理地址的转换。地址变换机构如图 5-17 所示。

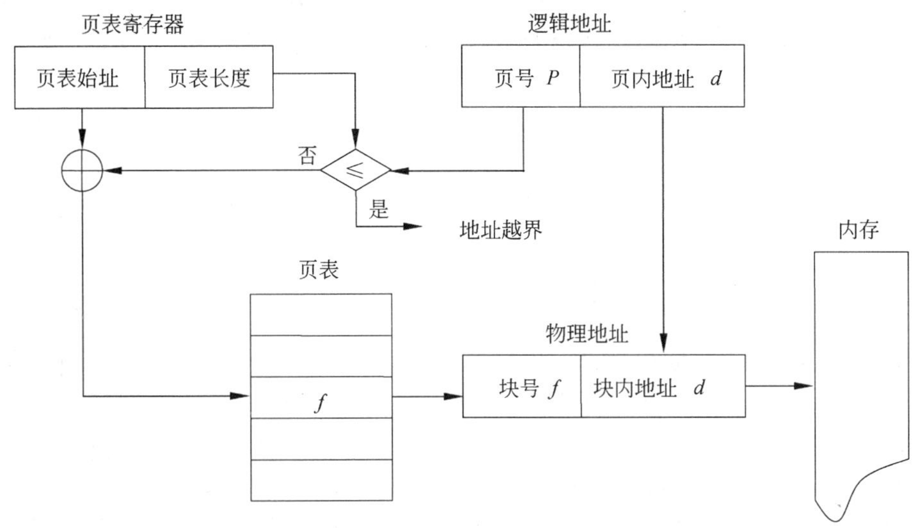

图 5-17　地址变换机构

例如，设页长为 1KB，程序的逻辑地址字长为 16 位，页表如图 5-18 所示。在执行指令 MOV r1，[2500]指令，即把 2500 号地址的数值送到 r1 寄存器。

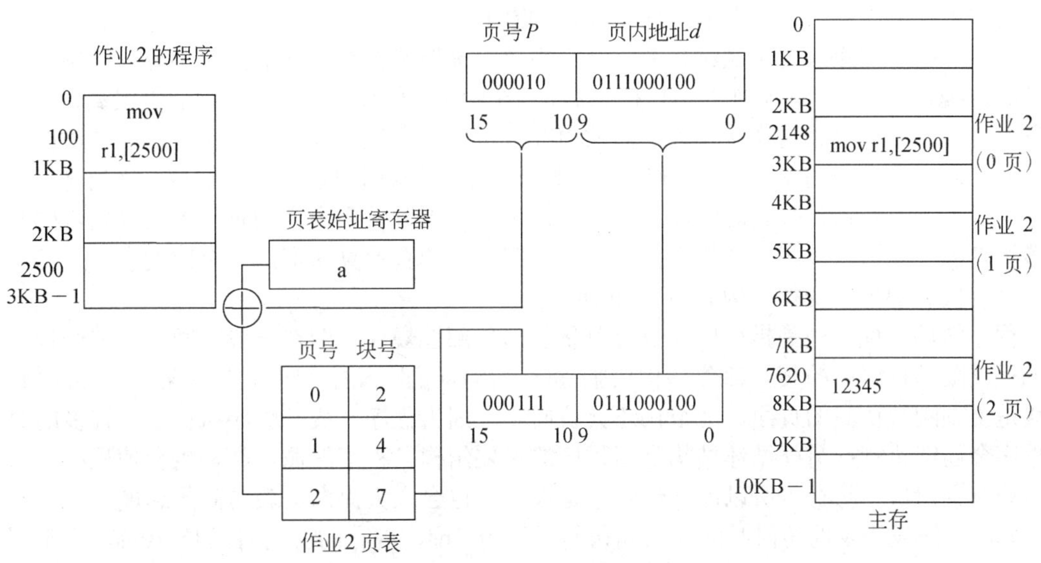

图 5-18　地址变换过程示例

首先，取出程序地址字 2500 送逻辑地址寄存器，由硬件分离出页号 P 和页内地址 W，因为页长为 1KB，所以页内地址占 10 位（0～9 位），页号占 6 位（10～15 位），取出逻辑地址寄存器中的高 6 位即为页号，低 10 位即为页内地址。

当然我们通过计算可以得到 P=2，W=452。根据页号 P=2，硬件自动查该进程的页表，找到第 2 页对应的块号为 7，将块号送到物理地址寄存器的高 10 位中。将逻辑地址寄存器

中的 W 值 452 复制到物理地址寄存器的低 10 位中，从而形成内存地址 7×1024+452=7620。

把上述地址变换过程写成公式转换的形式可以设页面长度为 L，程序的逻辑地址为 A，则，把 A 写成如下形式

$$A=P\times L+d$$

其中 P 是页号，d 是页内偏移量。

在页表中查询 p 页对应块号是 f，那么，物理地址 M 可以写成

$$M=f\times L+d$$

例 1 页面大小为 1024B，已知页表见表 5-3，求逻辑地址 2153 的物理地址。

表 5-3 页表

页号	块号
0	4
1	6
2	3
3	8

解：先求出页号和页内偏移量 P、d

P=int（2153/1024）=2，d=2153 mod 1024 =105

可以写成：

$$A=2153=2\times 1024+105$$

查询页表可知，对应的块号 f=3，物理地址为

$$M=3\times 1024+105=3177$$

5.3.3 快表

采用分页式内存管理方式，CPU 取指令时要先查询页表，才能知道指令所在的内存的物理位置，页表也是放在内存的，所以取指令的操作要访问两次内存。可见，这种方式增加了访问内存的次数，而页表的访问是非常频繁的。

在计算机存储系统的层次结构中，介于中央处理器和内存之间的高速小容量存储器。叫做高速缓存（Cache），又称“联想存储器”。高速缓存存取速度快，但是造价很高，一般容量不会很大。像页表这样被频繁访问的内存，可以在高速缓存中存有副本，经常访问的页表放在高速缓存中的副本称为“快表”。做逻辑地址转换时可以先查询快表，找到物理块号，可以大大节约访问页表的时间。如果在快表中找不到再到内存的页表中去查找，同时将该页表的表项写入快表，若快表已满，则按照一定的置换算法淘汰一个旧的表项。

具有快表的地址变换机构如图 5-19 所示。

程序的局部性原理是指程序在执行时呈现出局部性规律，即在一段时间内，整个程序的执行仅限于程序中的某一部分。相应地，执行所访问的存储空间也局限于某个内存区域。也就是说刚刚被访问的页面可能很快还会访问到。因为程序一般是顺序执行的，许多的时候程序会连续执行；另外循环结构是写程序的重要结构，程序的循环结构也会使程序会在一定的范围内反复执行。所以即使快表不是很大，也会大大提高页表访问的速度。

例 2 假设快表的访问时间检索和访问时间为 2ns，内存的访问时间是 10ns，由于程序的局部性，一般快表的访问命中率达到 90%。计算一下，有无快表的访问速度。

无快表的情况下，100 万条指令的访问时间是

$$2\times 10\times 10^{6}(\text{ns})=20\text{s}$$

有快表的情况下，如果在快表中找到页表表项，再访问内存的时间开销是 12ns；如果再快表中没找到相应表项，再访问页表、访问内存，需要时间开销为 22ns。

100 万条指令的访问时间是

$$12\times 90\%\times 10^{6}(\text{ns})+22\times 10\%\times 10^{6}(\text{ns})=13\text{s}$$

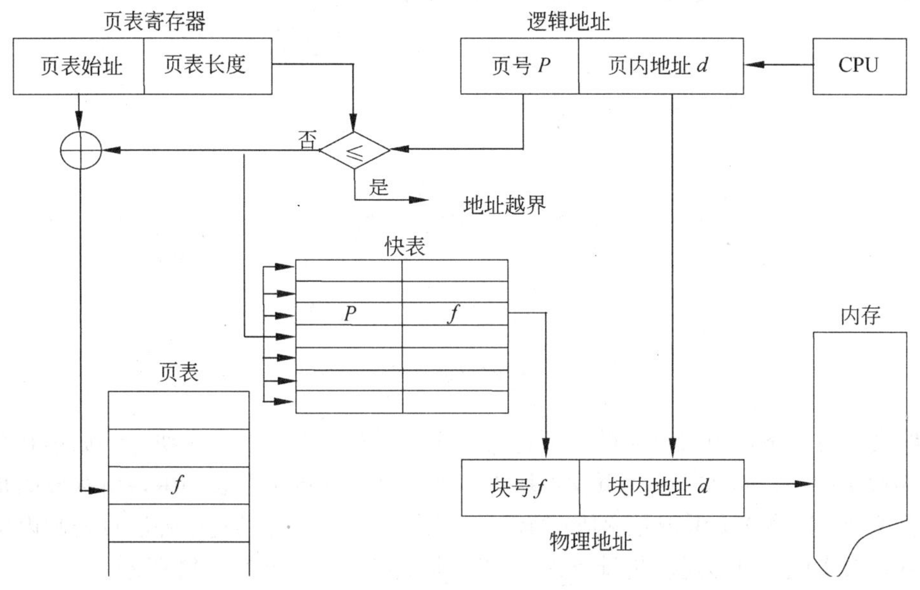

图 5-19　利用快表的地址变换机构

5.3.4　两级和多级页表

现代的大多数计算机系统，都支持非常大的逻辑地址空间（2^{32}~2^{64}）。在这样的环境下，页表就变得非常大，要占用相当大的内存空间。例如，对于一个具有 32 位逻辑地址空间的分页系统，规定页面大小为 4KB 即 2^{12} B，则在每个进程页表中的页表项可达 1 兆个之多。又因为每个页表项占用一个字节，故每个进程仅仅其页表就要占用 4KB 的内存空间，而且还要求是连续的。可以采用这样两个方法来解决这一问题：

（1）采用离散分配方式来解决难以找到一块连续的大内存空间的问题；

（2）只将当前需要的部分页表项调入内存，其余的页表项仍驻留在磁盘上，需要时再调入。

采用多级页表管理方式，需要将页表进行分页，并将各页表页分别放到不同的内存块中。为这些页表再建立一张页表，称为外层页表，从而形成两级页表，此时的逻辑地址结构如图 5-20 所示。

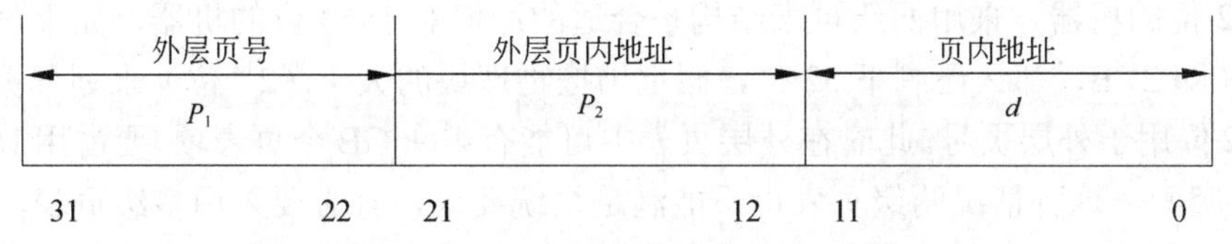

图 5-20　两级页表的逻辑地址结构

为了地址变换实现上的方便，在地址变换机构中增设一个外层页表寄存器，用于存放外层页表始地址，利用访问外层页号 P_1，经过地址变换得到的不是物理块号，而是内层页表的首地址，这个首地址与外侧页内地址 P_2 相加，得到内层页表的表项，在这个表项中得

到物理块号 b，b 与页内地址 d 组成物理地址，如图 5-21 所示。

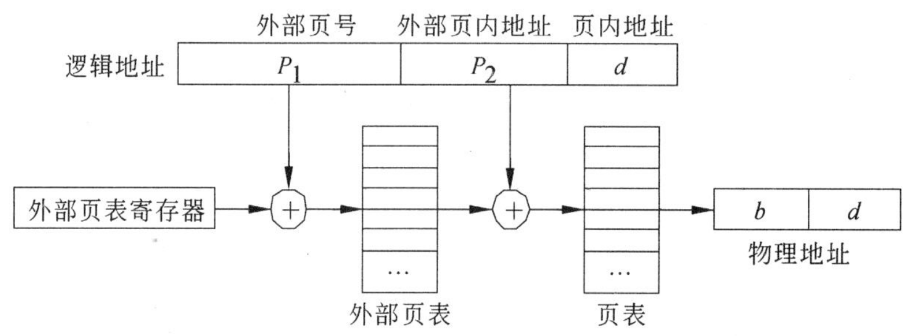

图 5-21　两级页表的地址变换

两级页表方式下，内层页表的表项记录对应页码在内存中的物理块号，如第 0 页存放在 1 页物理块中；1 页存放在 4 页物理块中。外层页表的每个页表项中，所存放的是内层页表的首地址，如第 0 页页表是存放在第 1011 页物理块中。可以利用外层页表和内层页表这两级页表来实现从进程逻辑地址到内存物理地址的变换，如图 5-22 所示。

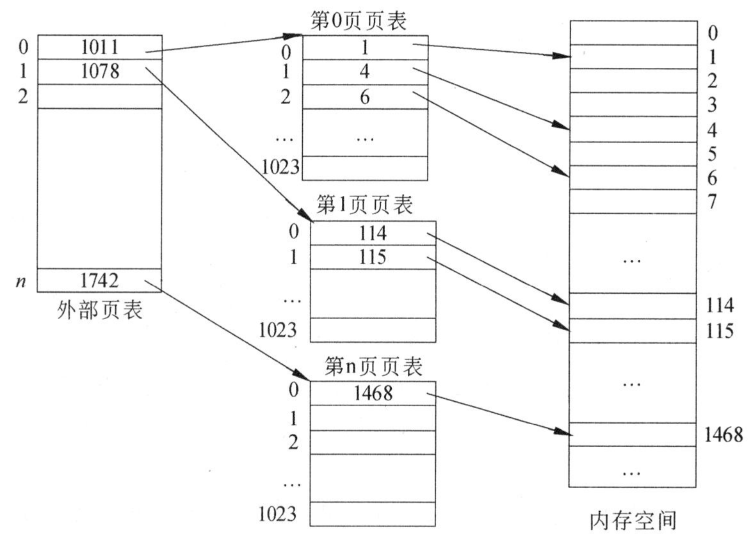

图 5-22　两级页表的地址变换示例

对于 32 位的机器，采用两级页表结构是合适的；但对于 64 位的机器，如果页面大小仍采用 4KB 即 2^{12}B，那么还剩下 52 位，假定仍按物理块的大小（2^{12} 位）来划分页表，则将余下的 42 位用于外层页号。此时在外层页表中可能有 4096GB 个页表项，要占用 16384GB 的连续内存空间。这种情况两级页表也不能满足系统要求，有必要采用多级页表，将外层页表再进行分页，也是将各分页离散地装入到不相邻接的物理块中，再利用第 2 级的外层页表来映射它们之间的关系。

对于 64 位的计算机，如果要求它能支持 2^{64}（=1844744TB）规模的物理存储空间，则即使是采用三级页表结构也是难以办到的；所以在有些支持 64 位机的操作系统中，直接寻址的内存空间减少到 45 位长度，寻址到 2^{45}，利用三级页表结构实现分页存储管理。

5.4 分段式管理

5.4.1 分段的基本工作原理

分页管理方式是许多操作系统采取的内存管理方式，管理手段比较成熟，内存的利用率也得到了提高。但是分页管理方式中，页的划分是按照约定的页面的大小来划分的，是一种等分的方式，与程序逻辑没有关系，那么对于作业的公共程序段的共享使用，不够方便。分段管理的基本思想是每个作业的地址空间按照自身的逻辑关系划分若干个段，例如主程序段、子程序段、数据段等，每个段都有自己的名字。每个段有一个段号，用段号代替段名，每个段都从 0 还是独立编址，段内地址连续。段的长度由相应的逻辑信息组的长度决定，段的长度一般互不相等，不像分页管理那样，每个页面的大小是一样的。

由于每个段独立编址，所以不能单纯给出一个逻辑地址数值，来指定一条指令。描述一条指令的地址，要有段号和一个逻辑地址共同指定，即（n，A），n 是段号，A 是段内的逻辑地址。因此说分段管理的逻辑地址空间是二维的；与分页管理不一样，分页管理的逻辑地址空间是一维的，页号是一维的逻辑地址空间的一部分。

分段管理中的段的大小不一致，不能像分页管理那样用页和块来对应。分配内存时，为每个段分配一个连续的存储空间，一般采用可变式分区的管理方式，每个段可以占用一个分区。段间可以不连续。通过段表来标识各段在内存中的位置。

1．段表

在分段式管理系统中，要为每一个段分配一个连续的内存分区，段与段之间可以不连续，段在内存中的位置通过段表来描述。进程的每个段在段表中占有一个表项，其中记录该段在内存中的起始地址（基址），和段的长度，如图 5-23 所示，段表一般也存储在内存中，执行中的进程通过查找段表，找到每个段在内存中的位置，实现逻辑段到物理内存的映射。

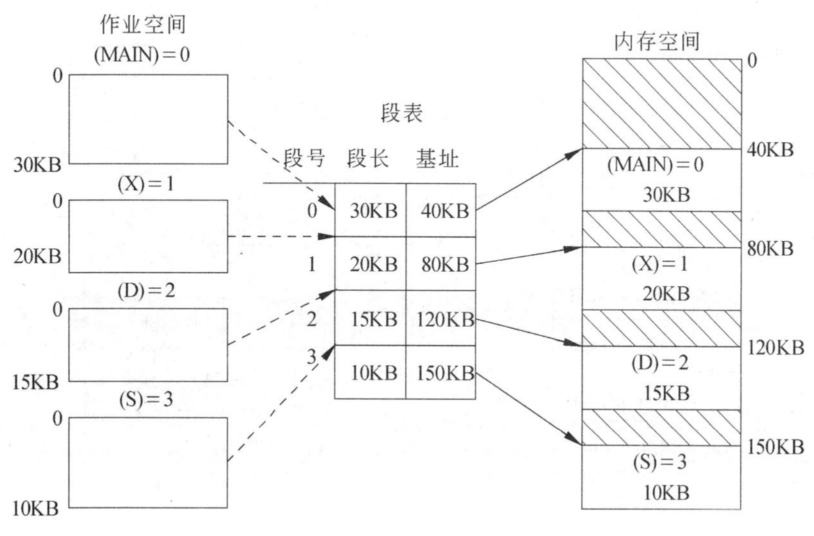

图 5-23　段表

2．地址结构

分段存储器管理系统中，逻辑地址分为段号和段内偏移量（段内地址），如图 5-24 所示。以 32 位地址空间为例，来说明在分段存储管理系统中的地址结构。

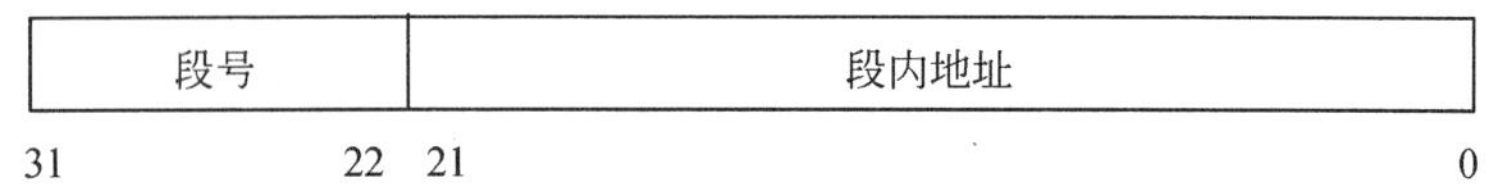

图 5-24　分段管理的逻辑地址结构

图 5-24 中，0～21 位共 22 位用来表示段内地址，22～31 位共 10 位来表示段号。在这样的地址结构中，允许一个作业最多可分为 2^{10} 个段，每段的最大长度为 2^{22} 即 4MB。

5.4.2　地址变换

为了实现分段存储管理系统中的逻辑地址到物理地址的变换，系统设置了段表寄存器用来存储当前运行进程的段表的始址和段表长度。在进行地址变换时，首先比较逻辑地址中的段号和段表寄存器中的段表长度，如果段号大于或等于段表长度，则访问越界，产生一个越界中断，有系统处理。如果没有越界，则用段号和段表寄存器中的段表始地址检索段表，按照始地址和段号找到该逻辑地址所在的段在段表中的位置，从中读出该段的始址，然后比较逻辑地址中的段内地址和段长，如果段内地址大于或等于段长，则越界。如果没有越界，则用该段的始址加上段内地址得到要访问的物理地址，地址变换过程如图 5-25 所示。

像分页系统一样，段表也是放在内存中，取得一条指令或数据，需要访问两次内存，为了提高地址变换的速度，也可以采用高速缓存作为快表，把最近一段时间常用的段表项放在快表中，每次先查快表，如果快表中没用要找的表项，则在段表中查找，找到后，进行地址变换，同时把表项存入快表。

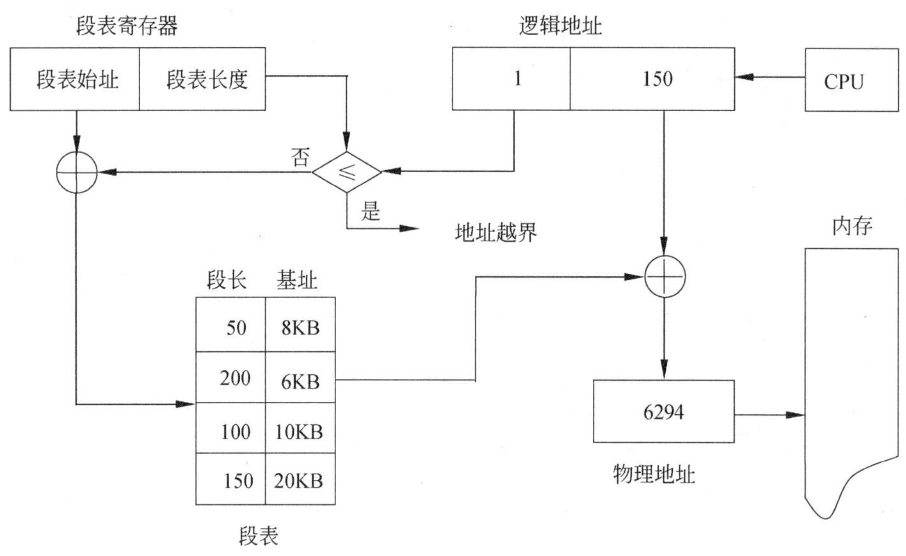

图 5-25　分段管理的地址变换过程

5.4.3 分段管理的信息共享

1．信息共享

在分段式管理系统，一个突出的优点是易于实现段的共享，即允许若干个进程共享一个或多个分段，其次对段的保护也十分简单易行。在分页系统中，虽然也能实现程序和数据的共享，但由于分页方法与程序逻辑无关，所以实现共享远不如分段系统来得方便。

通过一个例子来讨论分段系统共享问题。例如，有一个多用户系统，可同时接纳 40 个用户，它们都执行一个文本编辑文件（Text Editor）。如果文本编辑程序有 160KB 的代码段和另外 40KB 的数据区。则总共需要 8MB 的内存空间。

40 个用户使用编辑功能，它们可以共享 160KB 的代码段。为了实现程序的代码段共享，只需在每个进程的段表中为文本编辑程序设置的一个表项，它们在段表表项中指向 editor 编辑软件。分段系统共享 editor 程序，的示意图，如图 5-26 所示。

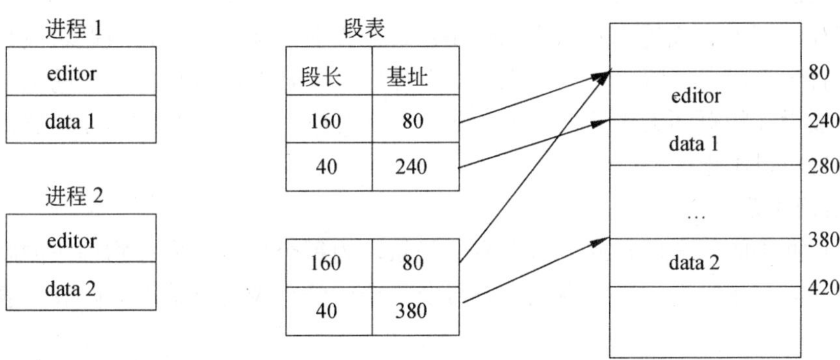

图 5-26　分段管理的段共享

2．存储保护

在分段式管理系统中，用户个分段是信息的逻辑单位，因此容易对各段实现保护。保护分为越界保护和越权保护。

1）越界保护

在地址变换过程中，需要进行段号和段表长度的比较，以及段内地址和段长的比较，如果段号小于段表长度并且段内地址小于段长，才能进行地址变换。否则产生越界中断，终止程序运行。

2）越权保护

分段式的内存管理方式和分页的存储器一样，通过在段表中设置存取控制字段，来对各段进行保护。

3．分页分段管理的主要区别

（1）页是信息的物理单位，分页是为实现离散分配方式，以消减内存的外零头，提高内存的利用率。或者说，分页仅仅是由于系统管理的需要而不是用户的需要。段则是信息的逻辑单位，它含有一组意义相对完整的信息。分段的目的是为了能更好地满足用户的需要。

（2）页的大小固定且由系统决定，由系统把逻辑地址划分为页号和页内地址两部分，是由机器硬件实现的，因而在系统中只能有一种大小的页面；而段的长度却不固定，决定于用户所编写的程序，通常由编译程序在对源程序进行编译时，根据信息的性质来划分。

（3）分页的作业地址空间是一维的，即单一的线性地址空间，程序员只需利用一个记忆符，即可表示一个地址；而分段的作业地址空间则是二维的，程序员在标识一个地址时，既需给出段名，又需给出段内地址。

5.5 段页式管理

5.5.1 段页式的基本工作原理

分页存储管理方式提高了内存的利用率，分段存储管理方式以用户程序的实际段落分段，方便用户使用，结合两者优点，将分页存储管理方式和分段存储管理方式组合在一起，形成了段页式的存储管理方式。

段页式存储管理方式的基本工作原理是：每个作业的地址空间按照逻辑关系分成若干段，每个段有一个段名，每段可以独立从 0 开始编址，每个段内再按设计的页面大小分成页；内存也按设计的页面大小分成块。为每个作业分配内存，分配给作业足够的块，内存块可以连续也可以不连续。

1. 段表及页表

系统为作业里的每个段建立一张页表，记录段内每个页面与内存物理块的映射。系统为整个作业建立一张段表，段表内记录段号及每个段的页表的首地址。段表页表的示意图见图 5-27 所示。

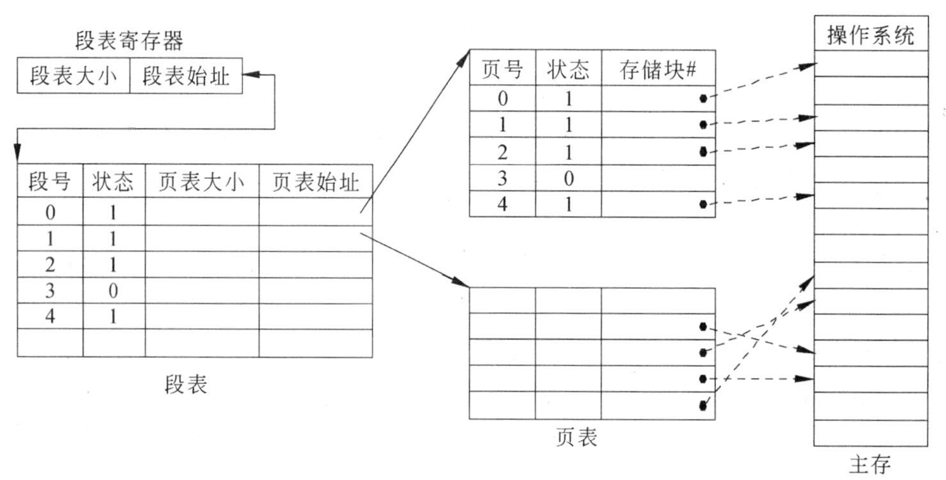

图 5-27　段页式管理示意图

2. 地址结构

根据段页式管理方式的基本工作原理，为了访问段页式的地址空间，逻辑地址分成三部分，段号 s、段内页号 P、和页内地址 d。例如一个作业分成三个段，主程序段、子程序段和数据段，分别为 15KB、8KB 和 10KB，设计页面大小为 4KB，各段分页及地址结构如图 5-28 所示。

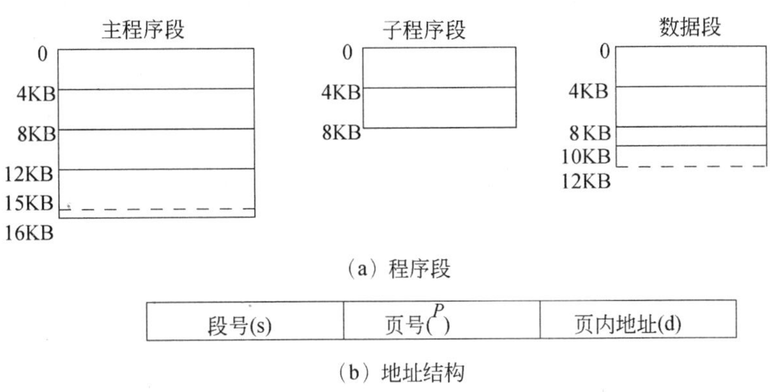

图 5-28　段页式管理示例

页面的大小为 4KB，第 1 个段，即 0 号段，有 1KB 的页内零头；第 3 段即 2 号段有 2KB 的页内零头。在页式管理方法中，每个作业最多只有一个页内零头，大小不会超过一页。在段页式管理方法中，可能有多个页内零头，页内零头的个数不超过分段的个数。

5.5.2　地址变换

段页式存储管理系统中，为了实现地址变换，配置一段表寄存器来存放段表的始址和段长。地址映射时，首先利用段号和段长进行比较，如果段号小于段长，利用段表寄存器中的段表的始址和段号求出该段的段表项在段表中的位置，从中得到该段的页表始址，并利用逻辑地址中的段内页号得到该页对应的页表项的位置，从中读出该页所对应的物理块号，把物理块号和页内地址组成物理地址。地址变换如图 5-29 所示。

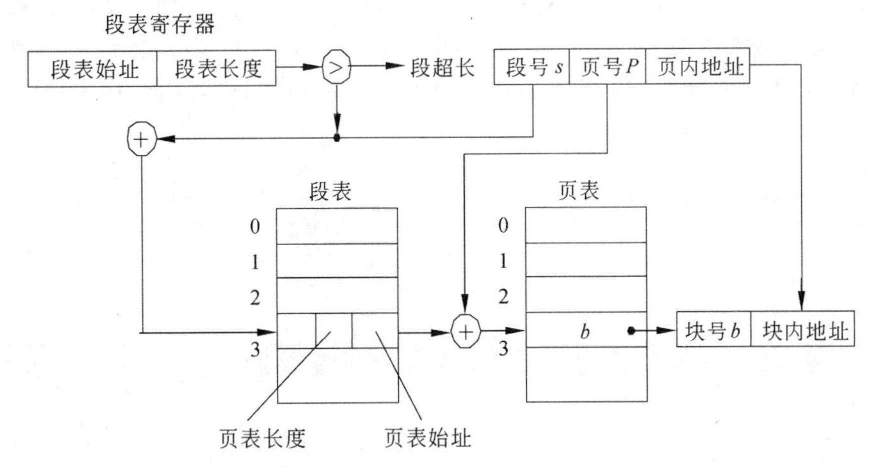

图 5-29　段页式管理的地址变换

在段页式存储管理方式中，执行一条指令需要访问三次内存。第一次访问段表，从中得到页表的位置；第二次访问页表，得出该页所对应的物理块号；第三次按照得到的物理地址访问内存。为了提高地址变换速度，可以和分页式存储管理方式及分段式存储管理方式一样，设置一高速缓存的快表，利用段号和页号去检索快表，可以快速得到物理块号。

段页式存储管理方式，地址也是二维的，因为每个段都是从 0 编址，都有自己的逻辑地址空间。描述一条指令的地址时要给出段号，还要给出段内偏移量。即（*s*，*A*）。

例 3 段页式管理系统中，已知页面大小为 1KB，段表页表如图 5-30 所示，计算下列逻辑地址（0，90），（0，2058），（1，28），（1，1038）的物理地址。

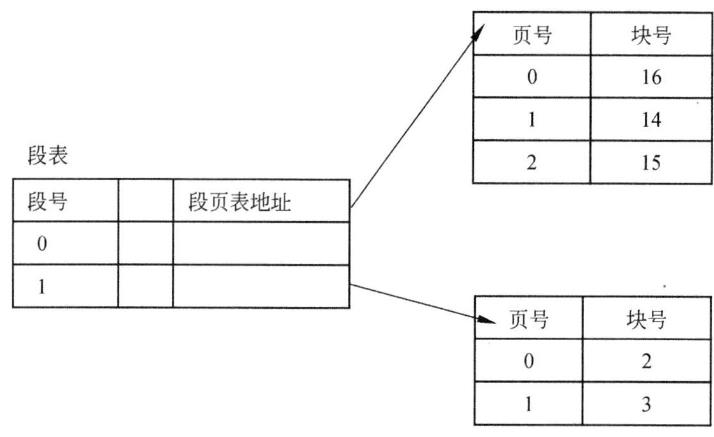

图 5-30 段页表

解：逻辑地址（0，90）处于第 0 段，在段表中 0 段指示的页表，段内偏移量 90 可以写成

$A=P\times L+d$ 的形式，即：90=0×1024+90，页号为 0，在页表中找到 0 页对应的块号为 2，物理地址 P_1 应为 2×1024+90=2138，即（0，90）对应的物理地址 P_1=2138。

同理：（0，2058），2058=2×1024+10，P_2=8×1024+10=8202（1，28），处于第 1 段，28=0×1024+28，P_3=16×1024+28=16412（1，1038），1038=1×1024+14，P_4=14×1024+14=14350。

Chapter 5 Memory Management

The main purpose of a computer system is to execute programs. These programs, together with the data they access, must be in main memory (at least partially) during execution.To improve both the utilization of the CPU and the speed of its response to users, the computer must keep several processes in memory. Many memory-management schemes exist, reflecting various approaches, and the effectiveness of each algorithm depends on the situation. Selection of a memory-management scheme for a system depends on many factors, especially on the hardware design of the system. Each algorithm requires its own hardware support.

5.1 Address Binding

Usually, a program resides on a disk as a binary executable file. To be executed, the program must be brought into memory and placed within a process. Depending on the memory management in use, the process may be moved between disk and memory during its execution. The processes on the disk are waiting to be brought into memory for execution form the input queue.

The normal procedure is to select one of the processes in the input queue and to load that process into memory. As the process is executed, it accesses instructions and datum from memory. Eventually, the process terminates, and its memory space is declared available.

Most systems allow a user process to reside in any part of the physical memory. Thus, although the address space of the computer starts at 00000, the first address of the user process need not be 00000. This approach affects the addresses that the user program can use. In most cases, a user program will go through several steps—some of which may be optional—before being executed. Addresses may be represented in different ways during these steps. Addresses in the source program are generally symbolic (such as count). A compiler will typically bind these symbolic addresses to relocatable addresses (such as "14 bytes from the beginning of this module"). The linkage editor or loader will in turn bind the relocatable addresses to absolute addresses (such as 74014). Each binding is a mapping from one address space to another.

An address generated by the CPU is commonly referred to as a **logical address**, whereas an address seen by the memory unit—that is, the one loaded into the memory-address register of the memory—is commonly referred to as a **physical address**.

The compile-time and load-time address-binding methods generate identical logical and physical addresses. However, the execution-time address-binding scheme results in differing logical and physical addresses. In this case, we usually refer to the logical address as a virtual address. We use logical address and virtual address interchangeably in this text. The set of all logical addresses generated by a program is a **logical address space**; the set of all physical addresses corresponding to these logical addresses is a **physical address space**. Thus, in the execution-time address-binding scheme, the logical and physical address spaces differ.

The run-time mapping from virtual to physical addresses is done by a hardware device called the memory-management unit (MMU), shown in Figure 5.1. We can choose from many different methods to accomplish such mapping. For the time being, we illustrate this mapping with a simple MMU scheme, which is a generalization of the base-register scheme. The base register is now called a relocation register. The value in the relocation register is added to every address generated by a user process at the time it is sent to memory. For example, if the base is at 14000, then an attempt by the user to address location 0 is dynamically relocated to location 14000; an access to location 346 is mapped to location 14346. The MS-DOS operating system running on the Intel 80x86 family of processors uses four relocation registers when loading and running processes.

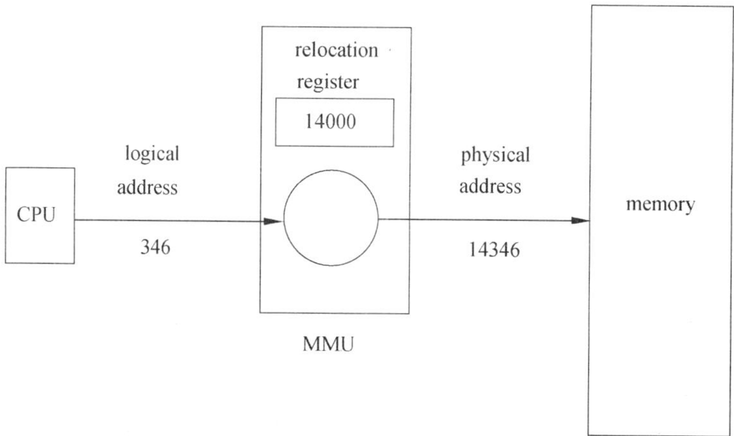

Figure 5.1 The run-time mapping

The user program never sees the real physical addresses. The program can create a pointer to location 346, store it in memory, manipulate it, and compare it with other addresses—all as the number 346. Only when it is used as a memory address (in an indirect load or store, perhaps) is it relocated relative to the base register. The user program deals with logical addresses. The memory-mapping hardware converts logical addresses into physical addresses. The final location of a referenced memory address is not determined until the reference is made.

We now have two different types of addresses: logical addresses (in the range 0 to max) and physical addresses (in the range R + 0 to R + max for a base value R). The user generates only logical addresses and thinks that the process runs in locations 0 to max. The user program supplies logical addresses; these logical addresses must be mapped to physical addresses before they are used.

The concept of a logical address space that is bound to a separate physical address space is central to proper memory management.

5.2 Continuous Memory Allocation

The main memory must accommodate both the operating system and the various user processes. We therefore need to allocate the parts of the main memory in the most efficient way possible. This section explains one common method, contiguous memory allocation.

The memory is usually divided into two partitions: one for the resident operating system and one for the user processes. We can place the operating system in either low memory or high memory. The major factor affecting this decision is the location of the interrupt vector. Since the interrupt vector is often in low memory, programmers usually place the operating system in low memory as well.

We usually want several user processes to reside in memory at the same time. We therefore need to consider how to allocate available memory to the processes that are in the input queue waiting to be brought into memory. In this contiguous memory allocation, each process is contained in a single contiguous section of memory.

One of the simplest methods for allocating memory is to divide memory into several fixed-sized partitions. Each partition may contain exactly one process. Thus, the degree of multiprogramming is bound by the number of partitions. In this multiple-partition method, when a partition is free, a process is selected from the input queue and is loaded into the free partition. When the process terminates, the partition becomes available for another process.

In the fixed-partition scheme, the operating system keeps a table indicating which parts of memory are available and which are occupied. Initially, all memory is available for user processes and is considered one large block of available memory; a hole. When a process arrives and needs memory, we search for a hole large enough for this process. If we find one, we allocate only as much memory as is needed, keeping the rest available to satisfy future requests.

As processes enter the system, they are put into an input queue. The operating system takes into account the memory requirements of each process and the amount of available memory space in determining which processes are allocated memory. When a process is allocated space, it is loaded into memory; and it can then compete for the CPU. When a process terminates, it releases its memory, which the operating system may then fill with another process from the input queue.

At any given time, we have a list of available block sizes and the input queue. The operating system can order the input queue according to a scheduling algorithm. Memory is allocated to processes until, finally, the memory requirements of the next process cannot be satisfied—that is, no available block of memory (or hole) is large enough to hold that process. The operating system can then wait until a large enough block is available, or it can skip down the input queue to see whether the smaller memory requirements of some other process can be met.

In general, at any given time we have a set of holes of various sizes scattered throughout memory. When a process arrives and needs memory, the system searches the set for a hole that is large enough for this process. If the hole is too large, it is split into two parts. One part is allocated to the arriving process; the other is returned to the set of holes. When. a process terminates, it releases its block of memory, which is then placed back in the set of holes. If the new hole is adjacent to other holes, these adjacent holes are merged to form one larger hole. At this point, the system may need to check whether there are processes waiting for memory and whether this newly freed and recombined memory could satisfy the demands of any of these waiting processes.

This procedure is a particular instance of the general dynamic storage-allocation problem, which concerns how to satisfy a request of size Pi from a list of free holes. There are many solutions to this problem. The first-fit, best-fit, and worst-fit strategies are the ones most commonly used to select a free hole from the set of available holes.

1. Allocation Algorithm

(1) First fit. Allocate the first hole that is big enough. Searching can start either at the beginning of the set of holes or where the previous first-fit search ended. We can stop searching as soon as we find a free hole that is large enough.

(2) Best fit. Allocate the smallest hole that is big enough. We must search the entire list, unless the list is ordered by size. This strategy produces the smallest leftover hole.

(3) Worst fit. Allocate the largest hole. Again, we must search the entire list, unless it is sorted by size. This strategy produces the largest leftover hole, which may be more useful than the smaller leftover hole from a best-fit approach.

Simulations have shown that both first fit and best fit are better than worst fit in terms of decreasing time and storage utilization. Neither first fit nor best fit is clearly better than the other in terms of storage utilization, but first fit is generally faster.

2. Recycling and Fragmentation

Both the first-fit and best-fit strategies for memory allocation suffer from external fragmentation. As processes are loaded and removed from memory, the free memory space is broken into little pieces. External fragmentation exists when there is enough total memory space to satisfy a request, but the available spaces are not contiguous; storage is fragmented into a large number of small holes. This fragmentation problem can be severe. In the worst case, we could have a block of free (or wasted) memory between every two processes. If all these small pieces of memory were in one big free block instead, we might be able to run several more processes.

Whether we are using the first-fit or best-fit strategy can affect the amount of fragmentation. (First fit is better for some systems, whereas best fit is better for others.) Another factor is which end of a free block is allocated. (Which is the leftover piece the one on the top or the one on the bottom?) No matter which algorithm is used, external fragmentation will be a problem.

Depending on the total amount of memory storage and the average process size, external fragmentation may be a minor or a major problem. Statistical analysis of first fit, for instance, reveals that, even with some optimization, given N allocated blocks, another 0.5 N blocks will be lost to fragmentation. That is, one-third of memory may be unusable! This property is known as the 50-percent rule.

Memory fragmentation can be internal as well as external. Consider a multiple-partition allocation scheme with a hole of 18,464 bytes. Suppose that the next process requests 18,462 bytes. If we allocate exactly the requested block, we are left with a hole of 2 bytes. The overhead to keep track of this hole will be substantially larger than the hole itself. The general approach to avoiding this problem is to break the physical memory into fixed-sized blocks and allocate memory in units based on block size. With this approach, the memory allocated to a process may be slightly larger than the requested memory. The difference between these two numbers is internal fragmentation —memory that is internal to a partition but is not being used.

3. Compaction

One solution to the problem of external fragmentation is compaction. The goal is to shuffle the memory contents so as to place all free memory together in one large block. Compaction is not always possible, however. If relocation is static and is done at assembly or load time, compaction cannot be done; compaction is possible only if relocation is dynamic and is done at execution time. If addresses are relocated dynamically, relocation requires only moving the program and data and then changing the base register to reflect the new base address. When compaction is possible, we must determine its cost. The simplest compaction algorithm is to move all processes toward one end of memory; all holes move in the other direction, producing one large hole of available memory. This scheme can he expensive.

Another possible solution to the external-fragmentation problem is to permit the logical address space of the processes to be noncontiguous, thus allowing a process to be allocated physical memory wherever the latter is available. Two complementary techniques achieve this solution: paging and segmentation

5.3 Paging

5.3.1 Basic Method

The basic method for implementing paging involves breaking physical memory into fixed-sized blocks called frames and breaking logical memory into blocks of the same size called pages.

When a process is to be executed, its pages are loaded into any available memory frames from the backing store. The backing store is divided into fixed-sized blocks that are of the same size as the memory frames.

The hardware support for paging is illustrated in Figure 5.2. Every address generated by the

CPU is divided into two parts: a page number (p) and a page offset (d). The page number is used as an index into a page table. The page table contains the base address of each page in physical memory. This base address is combined with the page offset to define the physical memory address that is sent to the memory unit. The paging model of memory is shown in Figure 5.3.

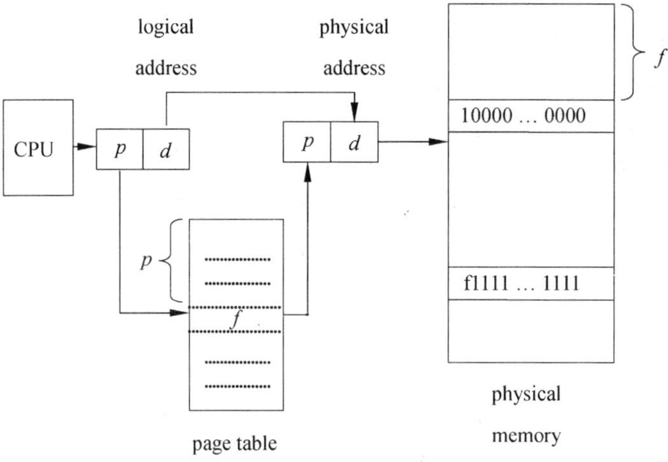

Figure 5.2 The hardware support for paging

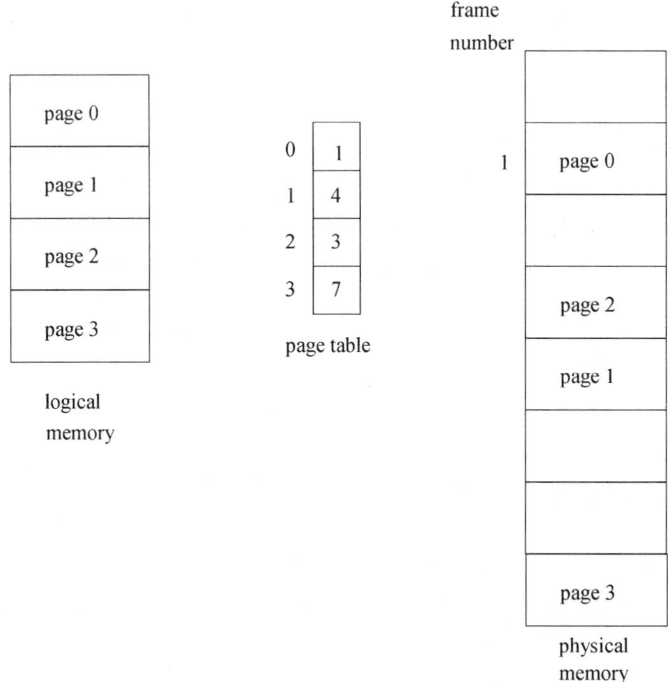

Figure 5.3 The paging model of memory

The page size (like the frame size) is defined by the hardware. The size of a page is typically a power of 2, varying between 512 bytes and 16 MB per page, depending on the computer architecture. The selection of a power of 2 as a page size makes the translation of a logical address into a page number and page offset particularly easy. If the size of logical address space is 2^m and a page size is 2^n addressing units (bytes or words), then the high-order *m–n* bits of a logical address designate the page number, and the a low-order bits designate the page offset. Thus, the logical address is as follows(Figure 5.4), where p is an index into the page table and *d* is the displacement within the page.

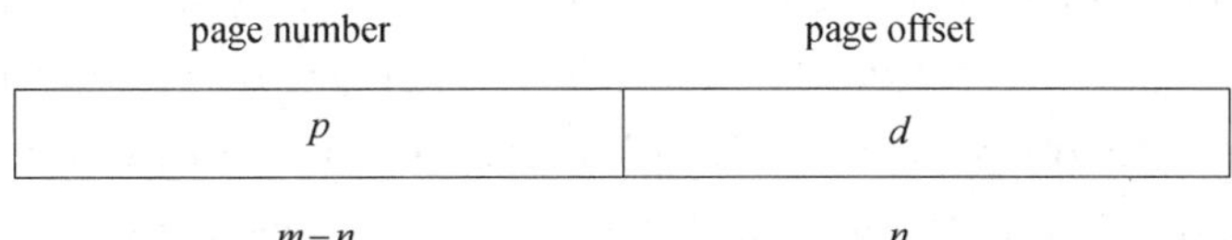

Figure 5.4 Logical address

5.3.2 Caching

The standard solution to this problem is to use a special, small, fast-lookup hardware cache, called a translation look-aside buffer (TLB). The TLB is associative, high-speed memory. Each entry in the TLB consists of two parts: a key (or tag) and a value. When the associative memory is presented with an item, the item is compared with all keys simultaneously If the item is found, the corresponding value field is returned. The search is fast; the hardware, however, is expensive. Typically, the number of entries in a TLB is small, often numbering between 64 and 1,024.

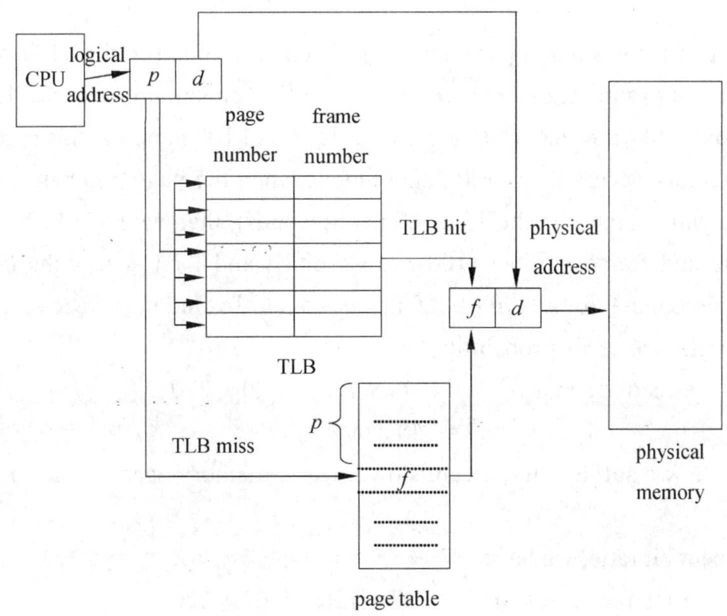

Figure 5.5 Paging hardware with TLB

The TLB is used with page tables in the following way. The TLB contains only a few of the page-table entries. When a logical address is generated by the CPU, its page number is presented to the TLB. If the page number is found, its frame number is immediately available and is used to access memory. The whole task may take less than 10 percent longer than it would if an unmapped memory reference were used.

If the page number is not in the TLB (known as a TLB miss), a memory reference to the page table must be made. When the frame number is obtained, we can use it to access memory (Figure 5.5). In addition, we add the page number and frame number to the TLB, so that they will be found quickly on the next reference. If the TLB is already full of entries, the operating system must select one for replacement. Replacement policies range from least recently used (LRU) to random. Furthermore, some TLBs allow entries to be wired down, meaning that they cannot be removed from the TLB. Typically, TLB entries for kernel code are wired down.

Some TLBs store address-space identifiers (ASIDs) in each TLB entry. An ASID uniquely identifies each process and is used to provide address-space protection for that process. When the TLB attempts to resolve virtual page numbers, it ensures that the ASID for the currently running process matches the ASID associated with the virtual page. If the ASIDs do not match, the attempt is treated as a TLB miss. In addition to providing address-space protection, an ASID allows the TLB to contain entries for several different processes simultaneously. If the TLB does not support separate ASIDs, then every time a new page table is selected (for instance, with each context switch), the TLB must be flushed (or erased) to ensure that the next executing process does not use the wrong translation information. Otherwise, the TLB could include old entries that contain valid virtual addresses but have incorrect or invalid physical addresses left over from the previous process.

The percentage of times that a particular page number is found in the TLB is called the hit ratio. An 80-percent hit ratio means that we find the desired page number in the TLB 80 percent of the time. If it takes 20 nanoseconds to search the TLB and 100 nanoseconds to access memory, then a mapped-memory access takes 120 nanoseconds when the page number is in the TLB. If we fail to find the page number in the TLB (20 nanoseconds), then we must first access memory for the page table and frame number (100 nanoseconds) and then access the desired byte in memory (100 nanoseconds), for a total of 220 nanoseconds. To find the effective memory-access time, we weight each case by its probability:

$$
\begin{aligned}
\text{effective access time} &= 0.08 \times 120 + 0.20 \times 220 \\
&= 140 \text{ nanoseconds}
\end{aligned}
$$

In this example, we suffer a 40-percent slowdown in memory-access time (from 100 to 140 nanoseconds).

For a 98-percent hit ratio, we have

$$
\begin{aligned}
\text{effective access time} &= 0.98 \times 120 + 0.02 \times 220 \\
&= 122 \text{ nanoseconds}
\end{aligned}
$$

This increased hit rate produces only a 22 percent slowdown in access time.

5.3.3 Hierarchical Paging

Most modern computer systems support a large logical address space (2^{32} to 2^{64}). In such an environment, the page table itself becomes excessively large. For example, consider a system with a 32-bit logical address space. If the page size in such a system is 4 KB (2^{12}), then a page table may consist of up to 1 million entries ($2^{32}/2^{12}$). Assuming that each entry consists of 4 bytes, each process may need up to 4 MB of physical address space for the page table alone. Clearly, we would not want to allocate the page table contiguously in main memory. One simple solution to this problem is to divide the page table into smaller pieces. We can accomplish this division in several ways.

One way is to use a two-level paging algorithm, in which the page table itself is also paged (Figure 5.6). Remember our example of a 32-bit machine with a page size of 4 KB. A logical address is divided into a page number consisting of 20 bits and a page offset consisting of 12 bits. Because we page the page table, the page number is further divided into a 10-bit page number and a 10-bit page offset. Thus, a logical address is as follows.

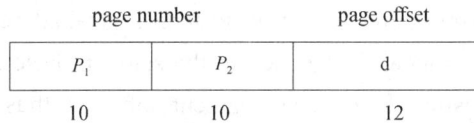

page number		page offset
P_1	P_2	d
10	10	12

Figure 5.6 Logical address

where P_1 is an index into the outer page table and P_2 is the displacement within the page of the outer page table. Because address translation works from the outer page table inward, this scheme is also known as a forward-mapped page table.

5.4 Segmentation

Segmentation is a memory-management scheme that supports this user view of memory. A logical address space is a collection of segments. Each segment has a name and a length. The addresses specify both the segment name and the offset within the segment. The user therefore specifies each address by two quantities: a segment name and an offset. (Contrast this scheme with the paging scheme, in which the user specifies only a single address, which is partitioned by the hardware into a page number and an offset, all invisible to the programmer.)

For simplicity of implementation, segments are numbered and are referred to by a segment number, rather than by a segment name. Thus, a logical address consists of a two tuple:

$\langle$segment-number, offset$\rangle$

Normally, the user program is compiled, and the compiler automatically constructs segments reflecting the input program.

A Compiler might create separate segments for the following:

(1) The code;

(2) Global variables;

(3) The heap, from which memory is allocated;

(4) The stacks used by each thread;

(5) The standard C library;

Although the user can now refer to objects in the program by a two-dimensional address, the actual physical memory is still, of course, a one-dimensional sequence of bytes. Thus, we must define an implementation to map two-dimensional user-defined addresses into one-dimensional physical addresses. This mapping is effected by a segment table. Each entry in the segment table has a segment base and a segment limit. The segment base contains the starting physical address where the segment resides in memory, whereas the segment limit specifies the length of the segment.

The use of a segment table is illustrated in Figure 5.7. A logical address consists of two parts: a segment number, s, and an offset into that segment, The segment number is used as an index to the segment table. The offset d of the logical address must be between 0 and the segment limit. If it is not, we trap to the operating system (logical addressing attempt beyond end of segment). When an offset is legal, it is added to the segment base to produce the address in physical memory of the desired byte. The segment table is thus essentially an array of base—limit register pairs.

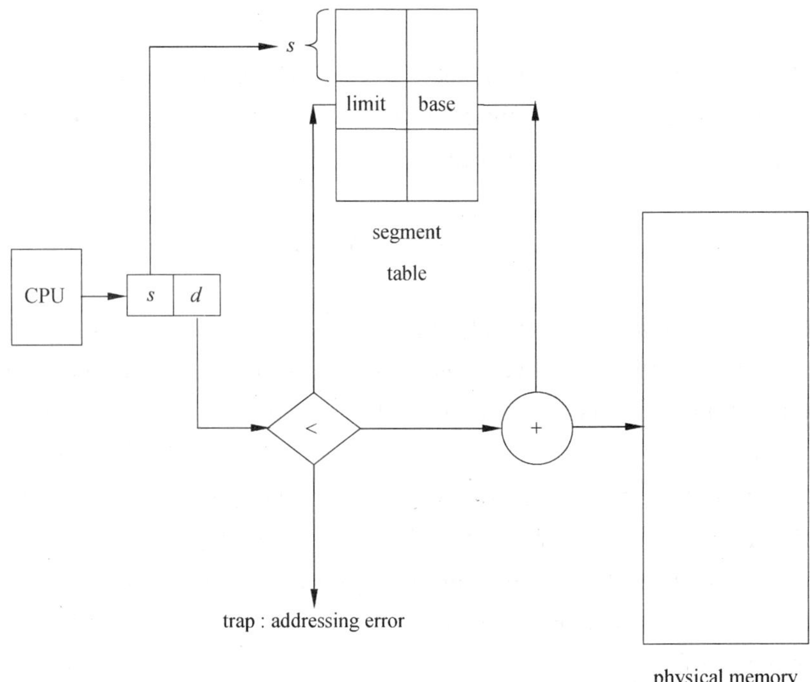

Figure 5.7 Segmentation hardware

As an example, consider the situation shown in Figure 5.8. We have five segments numbered from 0 through 4. The segments are stored in physical memory as shown. The segment table has a separate entry for each segment, giving the beginning address of the segment in physical memory (or base) and the length of that segment (or limit). For example, segment 2 is 400 bytes long and begins at location 4300. Thus, a reference to byte 53 of segment 2 is mapped onto location 4301) + 53 = 4353. A reference to segment 3, byte 852, is mapped to 3200 (the base of segment 3) + 852 = 4052. A reference to byte 1222 of segment 0 would result in a trap to the operating system, as this segment is only 1,000 bytes long.

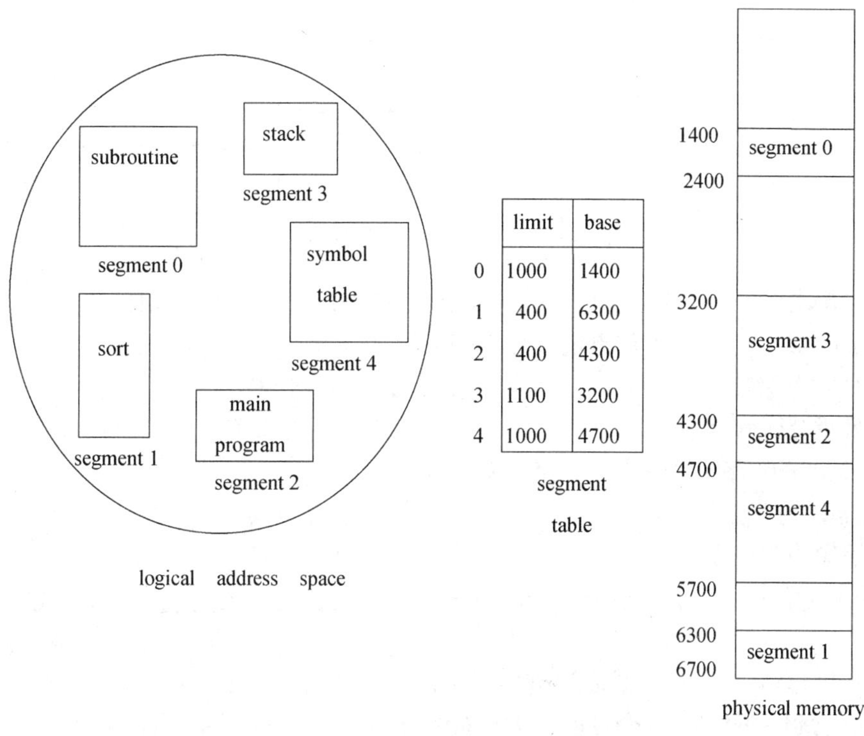

Figure 5.8 Example of segmentation

习 题

一、选择题

1、在以下存储管理方案中，不适用于多道程序设计系统的是_________。

A．单用户连续分配　　B．固定式分区分配

C．可变式分区分配　　D．页式存储管理

2．动态重定位技术依赖于____________。

A．重定位装入程序　B．重定位寄存器　C．地址机构　D．目标程序

3．设内存的分配情况如图所示。若要申请一块 40KB 的内存空间，若采用最佳适应算法，则所得到的分区首址为___________。

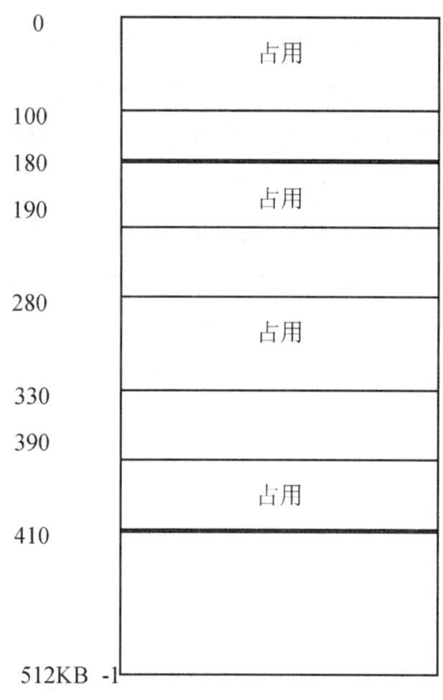

A. 100KB　　B. 190KB　　C. 330KB　　D. 410KB

4. 很好地解决了“零头”问题的存储管理方法是____________。

A. 页式存储管理　　B. 段式存储管理

C. 多重分区管理　　D. 可变式分区管理

5. 在可变式分区存储管理中的紧凑技术可以____________。

A. 集中空闲区　　B. 增加主存容量

C. 缩短访问周期　　D. 加速地址转换

6. 分区管理中采用“最佳适应”分配算法时，宜把空闲区按________次序登记在空闲区表中。

A. 长度递增　　B. 长度递减　　C. 地址递增　　D. 地址递减

7. 在固定分区分配中，每个分区的大小是__________。

A. 相同　　B. 随作业长度变化

C. 可以不同但预先固定　　D. 可以不同但根据作业长度固定

8. 采用段式存储管理的系统中，若地址用 24 位表示，其中 8 位表示段号，则允许每段的最大长度是___________。

A. 2^{24}　　B. 2^{16}　　C. 2^{8}　　D. 2^{32}

9. 首次适应算法（FF）是一种_________算法。

A. 页面淘汰　　B. 内存分配　　C. 进程调度　　D. 作业调度

10. 把作业地址空间中使用的逻辑地址变成内存中物理地址的过程称为___________。

A. 重定位　　B. 物理化　　C. 逻辑化　　D. 加载

11. 首次适应算法的空闲区是_________。

A. 按地址递增顺序连在一起　　B. 始端指针表指向最大空闲区

C. 按大小递增顺序连在一起　　D. 寻找从最大空闲区开始

12．在分页系统环境下，程序员编制的程序，其地址空间是连续的，分页是由________完成的。

A．程序员　　B．编译地址　　C．用户　　D．系统

13．在段页式存储管理系统中，内存等分成___①___，程序按逻辑模块划分成若干___②___。

A．块　B．基址　C．分区　D．段　E．页号　F．段长

14．在可变式分区分配方案中，某一作业完成后，系统收回其主存空间，并与相邻空闲区合并，为此需修改空闲区表，造成空闲区数减 1 的情况是________。

A．无上邻空闲区，也无下邻空闲区　　B．有上邻空闲区，但无下邻空闲区

C．有下邻空闲区，但无上邻空闲区　　D．有上邻空闲区，也有下邻空闲区

15．在可变式分区分配方案中，某一作业完成后，系统收回其主存空间，并与相邻空闲区合并，为此需修改空闲区表，造成空闲区数加 1 的情况是________。

A．无上邻空闲区，也无下邻空闲区　　B．有上邻空闲区，但无下邻空闲区

C．有下邻空闲区，但无上邻空闲区　　D．有上邻空闲区，也有下邻空闲区

16．采用可变式分区的内存分配方法时，怎样才能运行一个比每个空闲区都大的作业？例如系统中只有三个空闲区，分别为 10KB、20KB、15KB，这时有一个大小为 30KB 的作业被提交，采用________使其运行呢？

A．生产者-消费者算法　　B．“紧凑”的方法

C．银行家算法　　D．小作业优先算法

17．在没有快表的情况下，段页式系统每访问一次数据，需要访问内存的次数为________。

A．一次　　B．两次　　C．三次　　D．四次

18．某页式存储管理系统中的地址结构如图所示，则________。

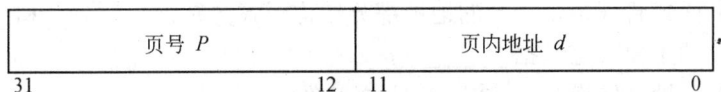

页号 P	页内地址 d
31　　　　12	11　　　　0

图 5-31　逻辑地址

A．页的大小为 1KB，最多有 8MB 页

B．页的大小为 2KB，最多有 4MB 页

C．页的大小为 4KB，最多有 1MB 页

D．页的大小为 8KB，最多有 2MB 页

19．在段页式管理系统中，作业的地址空间________.

A．由页号．页内地址组成，是一维的

B．由页号．页内地址组成，是二维的

C．由段号．段内地址组成，段内地址再分页，它是二维的

D．由段号．段内地址组成，段内地址再分页，它是三维的

20．某段表的内容如下。

段号	段首址	段长度
0	120KB	40KB
1	760KB	30KB
2	480KB	30KB
3	370KB	20KB

一逻辑地址为（2,154），它对应的物理地址为________。

A．120KB+2　　　　B．480KB+154

C．30KB+154　　　　D．2+480KB

21．在一个页式存储管理系统中，页表内容如下所示。

页号	块号
0	2
1	1
2	6
3	3
4	7

若页的大小为4KB，则地址转换机构将逻辑地址0转换成的物理地址为________。

A．8192　　B．4096　　C．2048　　D．1024

22. 如果一个程序为多个进程所共享，那么该程序的代码在执行的过程中不能被修改，即程序应该是________。

A．可执行码　　B．可重入码　　C．可改变码　　D．可再现码

二、填空题

1．将作业地址空间中的逻辑地址转换为主存中的物理地址的过程称为________。

2．分区分配中的存储保护通常采用________方法。

3．在页式和段式管理中，指令的地址部分结构形式分别为___①___和___②___。

4．段表表目的主要内容包括________。

5．把___①___地址转换为___②___地址的工作称为地址映射。

6．重定位的方式有___①___和___②___两种

7．分区管理中采用“首次适应”分配算法时，应将空闲区按________次序登记在空闲区表中。

8．页表表目的主要内容包括________。

9．主存中一系列物理存储单元的集合称为________。

10．静态重定位在___①___时进行；而动态重定位在中___②___时进行。

三、综合题

1．画图并解释重定位的地址变换过程。

2．写出分区式内存管理中内存分配算法的算法描述，以及内存回收算法的描述。

3．页式管理中程序的页，必须在内存中的连续块中存放，对吗？请简单解释。

4．分页管理彻底解决了内存的“零头”问题，对吗？请简单解释。

5．下表给出了某系统中的空闲分区表，系统采用可变式分区存储管理策略。现有以下作业序列：96KB、20KB、200KB。若用首次适应算法和最佳适应算法来处理这些作业序列，试问哪一种算法可以满足该作业序列的请求，为什么？

分区号	大小	起始地址
1	32KB	100KB
2	10KB	150KB
3	5KB	200KB
4	218KB	220KB
5	96KB	530KB

6．在某系统中，采用固定分区分配管理方式，内存分区（单元字节）情况如表所示。现有大小为 1KB、9KB、33KB、121KB 的多个作业要求进入内存，试画出它们进入内存后的空间分配情况，并说明主存浪费有多大？

分区号	大小	起始地址
1	8KB	20KB
2	32KB	28KB
3	120KB	60KB
4	332KB	180KB

7．有一页式系统，其页表存放在主存中。

（1）如果对主存的一次存取需要 1.5μs，试问实现一次页面访问的存取时间的多少？

（2）如果系统加有快表，平均命中率为 85%，当页表在快表中时，其查找时间忽略为 0，试问此时的存取时间为多少？

8．若在一分页存储管理系统中，某作业的页表如下所示。已知页面大小为 1024B，试将逻辑地址 1011、2148、3000、4000、5012 转化为相应的物理地址。

页号	块号
0	2
1	3
2	1
3	6

9．在一分页存储管理系统中，逻辑地址长度为 16 位，页面大小为 4096B，现有一逻辑地址为 2F6AH，且第 0、1、2 页依次存入在物理块 5、10、11 中，问相应的物理地址为多少？

10．在一分页存储管理系统中，逻辑地址长度为 16 位，页面大小为 1KB，某作业共有 5 页，第 0，1，2，3，4 页依次存放在物理块 1，4，6，7，9 中。

（1）将逻辑地址 1086 转换为相应的物理地址。

（2）将逻辑地址 6020 转换为相应的物理地址。

第6章 虚拟存储器

虚拟存储器，是指具有请求调入功能和置换功能，能从逻辑上对内存容量加以扩充的一种存储器系统。就是利用请求调入和置换的方法将一定的外存容量模拟成内存，同时对程序的调入调出内存的方式进行管理，从而得到一个比实际内存容量大得多的内存空间，使得程序的运行不受内存大小的限制。

6.1 虚拟存储器概述

6.1.1 虚拟存储器的引入

计算机程序运行时，组成程序的指令一定是要放在计算机的内存中才能执行；计算机的控制器从内存中读取指令，分析并执行指令；计算机取出指令后，指令计数器自动指向下一条指令。程序执行具有顺序性。所以一般情况下，系统将程序一次装入内存，并使程序连续存放。所以一个作业中程序的大小受到实际的计算机内存的限制。

假设一台处理器有 32 位地址线，那么其最大寻址空间就是 2^{32}，即为 4GB。但是，如果这台计算机只配备了 1GB 的内存，有 3GB 的一大段的寻址空间就没有实际存储器与其对应了，或者说，就浪费了。那么能否既不需要扩展实际存储器，又要充分利用处理器的寻址空间呢？可以，就是采用虚拟存储技术来实现虚拟存储器。在虚拟存储器的概念之下，程序员在设计程序时，完全可以不顾及实际内存有多少，只要不超过计算机处理器寻址空间即可。

分页式存储管理方式，打破程序连续存放的限制，实现内存管理方法上的一次重要“突破”。虚拟存储器的技术，打破了程序一次性全部放入内存的限制，实现了内存管理方法上的第二次重要“突破”。

所谓虚拟存储器，是指具有请求调入功能和置换功能，能从逻辑上对内存容量加以扩充的一种存储器系统。就是利用请求调入和置换的方法将一定的外存容量模拟成内存，同时对程序的调入调出内存的方式进行管理，从而得到一个比实际内存容量大得多的内存空间，使得程序的运行不受内存大小的限制。

虚拟存储器具有以下特征。

（1）多次性。多次性是指一个作业被分成多次调入内存运行，亦即在作业运行时没有必要将其全部装入，只需将当前要运行的那部分程序和数据装入内存即可。以后每当要运行到尚未调入的那部分程序时，再将它调入。多次性是虚拟存储器最重要的特征，任何其他的存储管理方式都不具有这一特征。因此，也可以认为虚拟存储器是具有多次性特征的存储器系统。

（2）对换性。对换性是指允许在作业的运行过程中进行换进、换出，亦即，在进程运行期间，允许将那些暂不使用的程序和数据，从内存调至外存的对换区（换出），待以后需要时再将它们从外存调至内存（换进）；甚至还允许将暂时不运行的进程调至外存，待它们重新具备运行条件时再调入内存。换进和换出能有效地提高内存利用率。可见，虚拟存储器具有对换性特征。

（3）虚拟性。虚拟性是指能够从逻辑上扩充内存容量，使用户所看到的内存容量远大于实际内存容量。这是虚拟存储器所表现出来的最重要的特征，也是实现虚拟存储器的最重要的目标。

值得说明的是，虚拟性是以多次性和对换性为基础的，或者说，仅当系统允许将作业分多次调入内存，并能将内存中暂时不运行的程序和数据换至盘上时，才有可能实现虚拟存储器；而多次性和对换性又必须建立在离散分配的基础上。

6.1.2 交换技术

分区式内存管理方式，对作业的大小有严格的限制。作业运行时，系统将作业的全部信息一次装入内存，并一直驻留内存，直至运行结束。当作业大于内存的空闲分区时，作业无法被接收、运行。为充分利用计算机的内存资源，可以采用覆盖和交换技术，使较大的作业也能够在系统中运行。

覆盖和交换技术是实现虚存管理的最基本的方法，让几个程序段共享一段内存空间，通过覆盖和交换，把暂时不需要运行的程序段调出内存，腾出内存空间，把将要运行的程序段调入内存执行。

1．覆盖技术

所谓覆盖技术，是指同一内存分区可以被不同的程序段重复使用的技术。通常一个作业由若干个功能上相互独立的程序段组成，作业在一次运行时，也只是用到其中的几段，有些程序段是不会同时使用的。让那些不会同时运行的程序段交替使用同一个内存分区，实现部分内存的共享，这就是覆盖技术。被多个程序段共享使用的内存段称为覆盖区；共享使用覆盖区的程序段称为覆盖段。

假设系统中有一个作业 J 由 6 个程序段组成，如图 6-1（a）所示，可以看出，主程序是一个独立段，它调用子程序 1 和子程序 2，且子程序 1 与子程序 2 是互斥被调用的两个段。在子程序 1 执行过程中，它调用子程序 11，而子程序 2 执行过程中调用子程序 21 和子程序 22，显然子程序 21 和子程序 22 也是互斥调用的。因此可以为作业 J 建立如图 6-1（b）所示的覆盖结构：主程序段是作业 J 的常驻内存段，而其余部分组成覆盖段。根据上述分析，子程序 1 和子程序 2 组成覆盖段 0，子程序 11、子程序 21 和子程序 22 组成覆盖段 1。相应覆盖区的大小应为每个覆盖段中最大的覆盖段的大小。

覆盖技术的主要特点是打破了必须将一个作业的全部信息装入内存后才能运行的限制，在一定程度上解决了小内存运行大作业的矛盾，为后续虚拟存储器概念的建立打下基础。

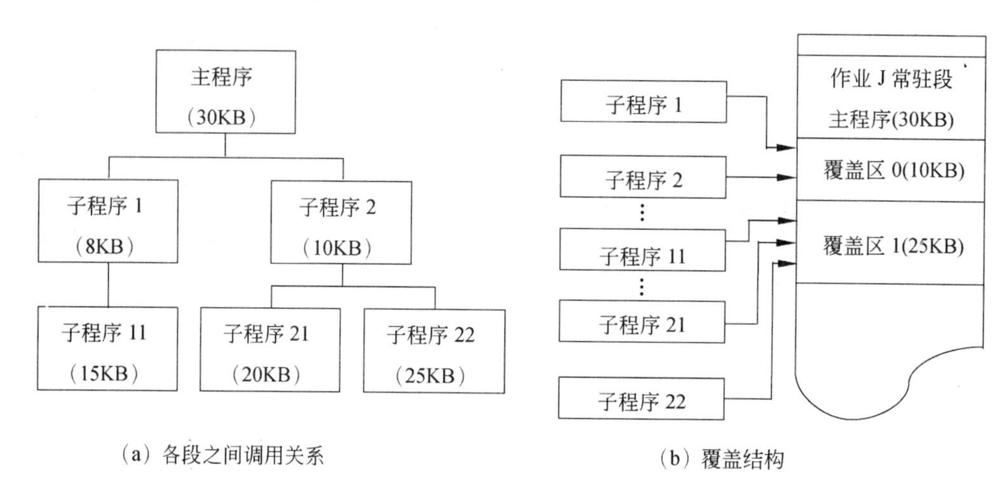

图 6-1 覆盖示例

2. 交换技术

所谓交换（又叫做对换）就是指系统根据需要把内存中暂时不能运行的进程或者暂时不用的程序和数据，部分或者全部移到外存，以便腾出足够的内存空间，再把已具备运行条件的进程或进程所需要的程序和数据，移到相应的内存区，并使其投入运行，如图 6-2 所示。

具有交换功能的操作系统，通常把外存分为文件区和交换区，文件区用于存放文件，交换区用于存放从内存中换出的作业。为了能对交换区中的空闲盘块进行管理，在系统中应配置相应的数据结构，以记录外存的使用情况。其形式与内存在动态分区分配方式中所用数据结构相似，同样可以用空闲分区表或空闲分区链来管理交换区。

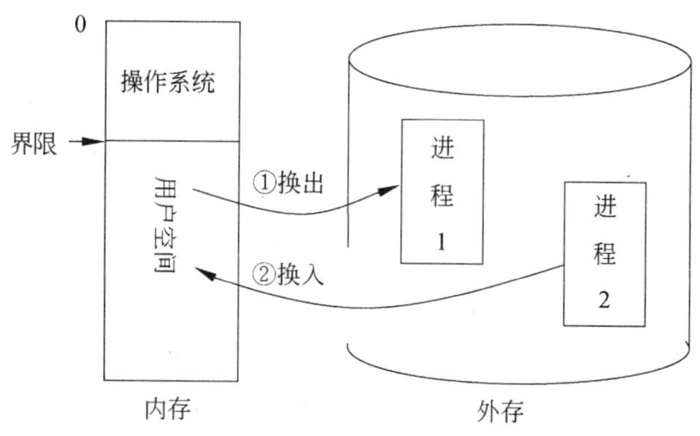

图 6-2 对换两个进程

每当一进程由于创建子进程而需要更多的内存空间，但又无足够的内存空间等情况发生时，系统应将某进程换出。其过程是：系统首先选择处于阻塞状态且优先级最低的进程作为换出进程，将该进程的程序和数据传送到磁盘的交换区上。若传送过程未出现错误，便可回收该进程所占用的内存空间，并对该进程的进程控制块做相应的修改。

系统定时地查看所有进程的状态，从中找出“就绪”状态但已换出的进程，将其中换出时间（换出到磁盘上）最久的进程作为换入进程，将之换入。直至已无可换入的进程或无可换出的进程为止。

6.2 请求页式管理

实现虚拟存储器的最重要的方法是请求页式管理和请求段式管理。本节从程序的局部性原理出发，讨论请求页式管理的实现原理及方法。

6.2.1 程序的局部性原理

早在 1968 年，Denning.P 就曾提出局部性原理：程序在执行时将呈现出局部性规律，即在一较短的时间内，程序的执行仅局限于某个部分；相应地，它所访问的存储空间也局限于某个区域。他提出了下述几个论点。

（1）程序执行时，除了少部分的转移和过程调用指令外，在大多数情况下仍是顺序执行的。该论点也在后来的许多学者对高级程序设计语言(如 FORTRAN 语言、Pascal 语言)及 C 语言规律的研究中被证实。

（2）过程调用将会使程序的执行轨迹由一部分区域转至另一部分区域，但经研究看出，过程调用的深度在大多数情况下都不超过 5。这就是说，程序将会在一段时间内都局限在这些过程的范围内运行。

（3）程序中存在许多循环结构，这些虽然只由少数指令构成，但是它们将多次执行。

（4）程序中还包括许多对数据结构的处理，如对数组进行操作，它们往往都局限于很小的范围内。

局限性还表现在下述两个方面。

（1）时间局限性。如果程序中的某条指令一旦执行，则不久以后该指令可能再次执行；如果某数据被访问过，则不久以后该数据可能再次被访问。产生时间局限性的典型原因是在程序中存在着大量的循环操作。

（2）空间局限性。一旦程序访问了某个存储单元，在不久之后，其附近的存储单元也将被访问，即程序在一段时间内所访问的地址，可能集中在一定的范围之内，其典型情况便是程序的顺序执行。

基于局部性原理，应用程序在运行之前，没有必要全部装入内存，仅须将那些当前要运行的少数页面或段先装入内存便可运行，其余部分暂留在盘上。程序在运行时，如果它所要访问的页已调入内存，便可继续执行下去；但如果程序所要访问的页尚未调入内存（称为缺页），此时程序应利用操作系统所提供的请求调页功能，将它们调入内存，以使进程能继续执行下去。如果此时内存已满，无法再装入新的页，则还须再利用页的置换功能，将内存中暂时不用的页调至盘上，腾出足够的内存空间后，再将要访问的页调入内存，使程序继续执行下去。

6.2.2 工作集

根据程序的局部性原理，一般情况下，进程在一段时间内总是集中访问一些页面，这些页面称为活跃页面，如果分配给一个进程的页帧数太少了，使该进程所需的活跃页面不能全部装入内存，则进程在运行过程中将频繁发生中断，这种现象叫做“抖动”。系统出现“抖动”的现象会耗费大量的时间用来将页面调入调出，就会大大降低系统的工作效率。如果能为进程提供与活跃页面数相等的页帧数，则可以减少缺页中断次数。

一个页面置换算法的好坏与进程运行的页面走向有很大的关系。虚拟存储系统的有效操作依赖于程序中访问的局部化程度。局部化程度愈突出，缺页的情况愈少，进程运行效率愈高。

工作集就是一个进程在某一小段时间内访问页面的集合。它是程序局部性的近似表示。

操作系统监督每个进程的工作集并给它分配足够工作集所需的内存块。若有足够多的额外块，就可装入并启动另外的进程。如果工作集的大小增加了，超出可用块的总数，操作系统要选择一个进程让它挂起，把它原来占的块分给别的进程。

这种工作集策略可防止抖动，同时保持尽可能高的多道程序度，从而使 CPU 的利用最优。实现工作集模型的困难是怎样保持工作集的轨迹。

工作集：一个运行进程在 $t-w$ 到 t 这个时间间隔内所访问的页的集合称为该进程在时间 t 的工作集，记为：

$$W(t,w)$$

W（t,w）为工作集尺寸：工作集中包含的页面数。w 为对于给定的访问序列选取定长的区间，称为工作集窗口。

6.2.3 缺页中断

中断是指计算机在执行程序的过程中，当出现异常情况或特殊请求时，计算机停止现行程序的运行，转向对这些异常情况或特殊请求的处理，处理结束后再返回现行程序的间断处，继续执行原程序。缺页中断就是要访问的页不在主存，需要操作系统将其调入主存后再进行访问。

请求页式管理在作业或进程开始执行之前，都不把作业或进程的程序段和数据段一次性地全部装入内存，而只装入被认为是经常反复执行和调用的工作集部分。其他部分则在执行过程中动态装入。请求页式管理的调入方式是，当需要执行某条指令而又发现它不在内存时或当执行某条指令需要访问其他的数据或指令时．这些指令和数据不在内存中，从而发生缺页中断，系统将外存中相应的页面调入内存。

请求页式管理的地址变换过程与静态页式管理时的相同，也是通过页表查出相应的页面号之后，由页面号与页内相对地址相加而得到实际物理地址，但是，由于请求页式管理只让进程或作业的部分程序和数据驻留在内存中，因此，在执行过程中，不可避免地会出现某些虚页不在内存中的问题。

关于虚页不在内存时的处理，涉及到两个问题。第一，采用何种方式把所缺的页调入内存。第二，如果内存中没有空闲页面时，把调进来的页放在什么地方。也就是说，采用什么样的策略来淘汰已占据内存的页。还有，如果在内存中的某一页被淘汰，且该页曾因程序的执行而被修改，则显然该页是应该重新写到外存上加以保存的。而那些未被访问修改的页、因为外存已保留有相同的副本，写回外存是没有必要的。因此，在页表中还应增加一项以记录该页是否曾被改变。

这个问题可以通过用扩充页表的方法解决。即与每个虚页号相对应，除了页面号之外，再增设该页是否在内存的状态位以及该页在外存中的副本起始地址。

请求分页的页表机制，它是在纯分页的页表机制上增加若干项而形成的。是实现请求分页管理的重要的数据结构。扩充后的页表如图 6-3 所示。

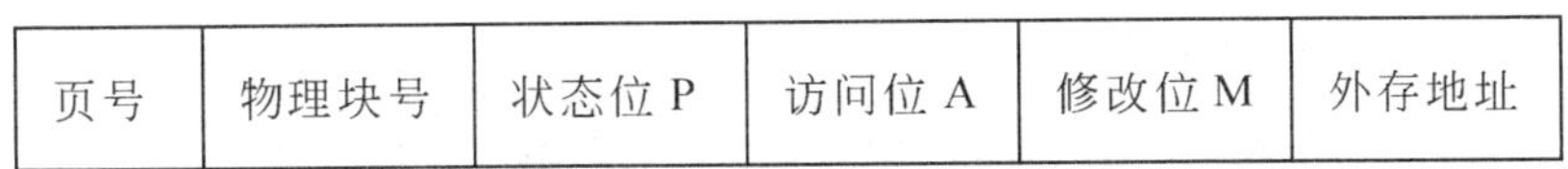

页号	物理块号	状态位 P	访问位 A	修改位 M	外存地址

图 6-3　扩充后的页表

各字段说明如下。

状态位 P：用于指示该页是否已调入内存，供程序访问时判断是否产生应该缺页中断。

访问位 A：用于记录本页在一段时间内被是否访问过，或记录本页最近已有多长时间未被访问，供选择换出页面时参考。

修改位 M：表示该页在调入内存后是否被修改过。由于内存中的每一页都在外存上保留一份副本，因此，若未被修改，在置换该页时就不需再将该页写回到外存上，以减少系统的开销和启动磁盘的次数；若已被修改，则必须将该页重写到外存上，以保证外存中所保留的始终是最新的副本。简言之，M 位供置换页面时，判断是否需要写回到磁盘。

外存地址：用于指出该页在外存上的地址，通常是磁盘的扇区号，供调入该页时寻找外村上的页面时使用。

每当用户程序要访问的页面尚未调入内存时便产生一次缺页中断，以请求操作系统将所缺的页面调入内存。缺页中断是一种特殊的中断，与一般的中断相比，有着明显的区别。

（1）指令执行期间，产生和处理中断信号。通常，CPU 是在一条指令执行完后，才检查是否有中断请求到达。若有，便去响应，否则，继续执行下一条指令。然而，缺页中断是在指令执行期间，发现所要访问的指令或数据不在内存时所产生和处理的。

（2）一条指令在执行期间，可能产生多次缺页中断。在图 6-4 中给出了一个例子。如在执行一条指令 COPY A TO B 时，可能要产生 6 次缺页中断，其中指令本身跨了两个页面，A 和 B 又分别各是一个数据块，也都跨了两个页面。基于这种特征，系统中的硬件机构能保存多次中断时的状态，并保证最后能返回到中断前产生缺页中断的指令处继续执行。

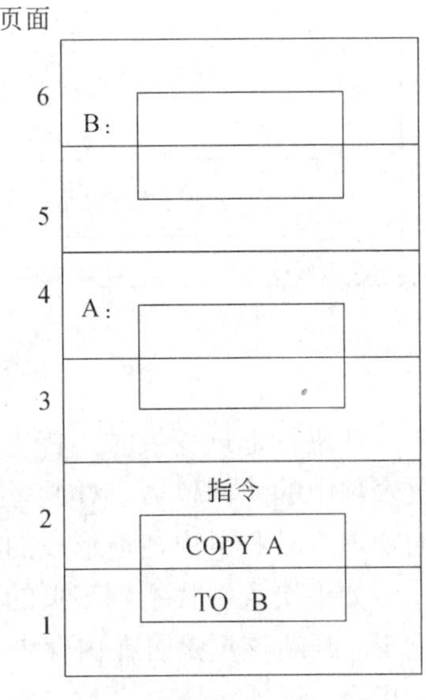

图 6-4　产生 6 次缺页的指令

6.2.4　地址变换

请求分页系统中的地址变换机构，同样是在纯分页地址变换机构的基础上发展形成的。在分页系统地址变换机构的基础上，为实现虚拟存储器增加了某些功能，如产生和处理缺页中断、页面的调入调出以及页面置换等。图 6-5 给出了请求分页系统中的地址变换过程。

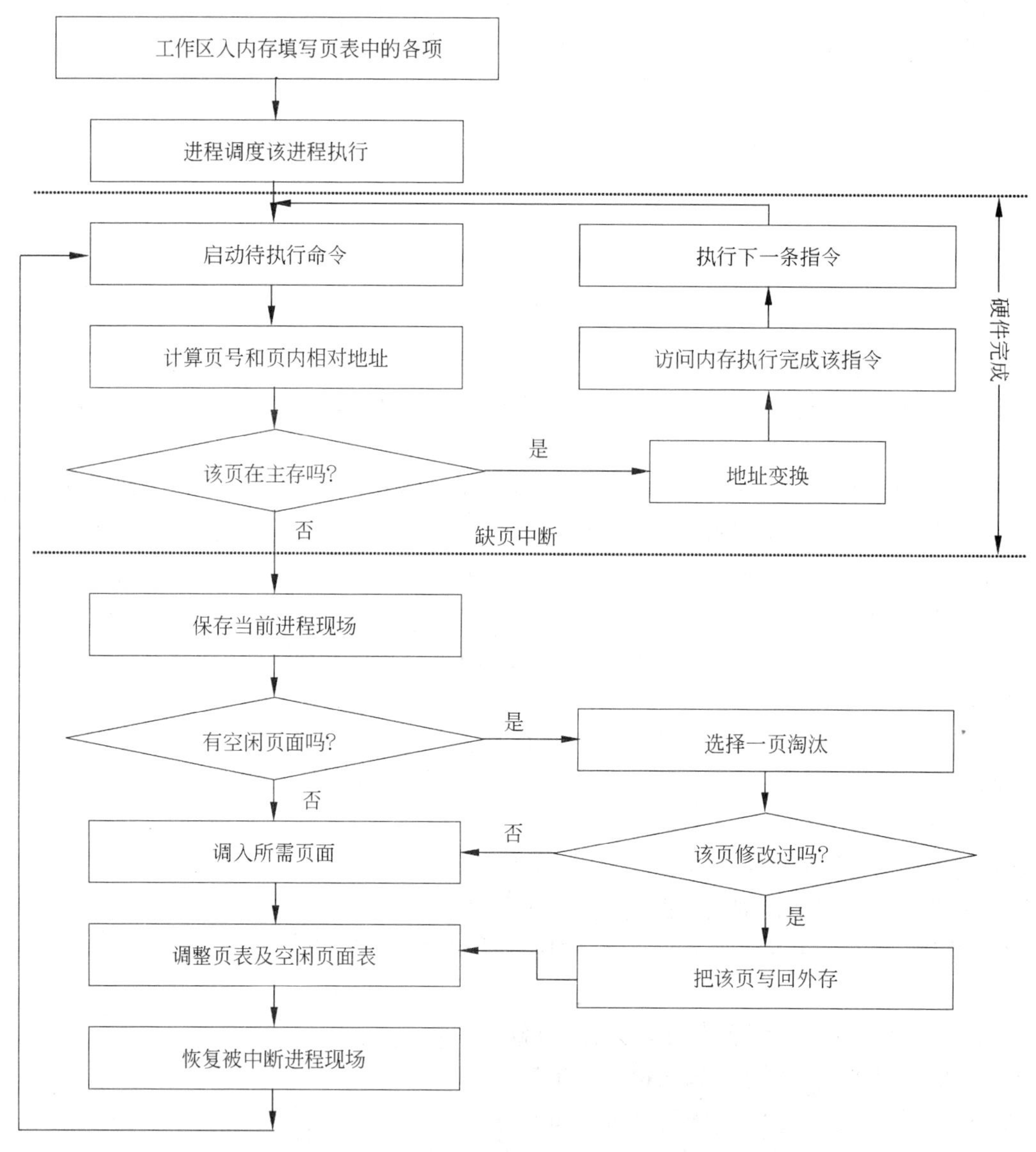

图 6-5　请求页式系统的地址变换过程

在进行地址变换时，首先去检索快表，试图从中找出所要访问的页。若找到，便修改页表项中的访问位 A。对于写指令，还须将修改位 M 置换成“1”，然后利用页表项中给出的物理块号和页内地址形成的物理地址，地址变换过程到此结束。

如果快表中找不到该页的页表项，则到页表中查找对应页表项，判断页表项中的状态位 P，判断该页是否在内存中，若该页面已在内存中，便修改页表项中的访问位 A。对于写指令，还须将修改位 M 置换成“1”，并将此表项写入快表，然后利用页表项中给出的物理块号和页内地址形成的物理地址，地址变换过程到此结束；如果判断状态位 P，发现该页面不在内存中，则产生缺页中断，在外存中找到对应页面，再看该进程是否有内存的空闲内存块，如果有则调入页面，如果没有选择一页淘汰。再把对应页面调入内存。

选择页面淘汰出内存时要遵循一定的置换算法，选择淘汰页。选出淘汰页后判断该页面的修改位 M，判断该页面是否被修改过，如果被修改过，将这一页写回外存，如果没有被修改过，不用写回外存，可用调入页面直接覆盖。

在动态页管理的流程中，有关地址变换部分是由硬件自动完成的。当硬件变换机构发现所要求的页不在内存时，产生缺页中断信号，由中断处理程序做出相应的处理。中断处理程序是由软件实现的。除了在没有空闲页面时要按照置换算法选择出被淘汰页面之外，还要从外存读入所需要的虚页。这个过程要启动相应的外存和涉及到文件系统。因此，请求页式管理是一个十分复杂的处理过程，内存的利用率的提高是以牺牲系统开销的代价换来的。

6.2.5 物理块的分配

根据程序的局部性原理和工作集理论，进程在一段时间内总是集中访问一些页面。所以应用程序在运行之前，没有必要全部装入内存，仅须将那些当前要运行的少数页面装入内存便可运行，其余部分暂留在盘上。程序在运行时，如果它所要访问的页已调入内存，便可继续执行下去；如果程序所要访问的页尚未调入内存，程序会利用操作系统所提供的请求调页功能，将它们调入内存，以使进程能继续执行下去。

把当前要运行的少数页面装入内存，需要为进程分配一定量的物理块数。这个块数的多少很重要，分配太少了，很容易出现系统“抖动”的现象。

1．最小物理块数

最小物理块数是指能保证进程正常运行所需的最小物理块数。当系统为进程分配的物理块数少于此值时，进程将无法运行。进程应获得的最少物理块数与计算机的硬件结构有关，取决于指令的格式、功能和寻址方式。对于某些简单的机器，若是单地址指令且采用直接寻址方式，则所需的最少物理块数为两块，其中，一块是用于存放指令的页面，另一块则是用于存放数据的页面。如果该机器允许间接寻址时，则至少要求有三个物理块。对于某些功能较强的机器，至少要求有 6 个物理块，因为其指令长度可能是两个或多于两个字节，因而其指令本身有可能跨两个页面，且源地址和目标地址所涉及的区域也都可能跨两个页面，如同前面讨论的中断机构中可能发生 6 次中断一样，涉及 6 个页面，至少分配 6 个物理块。

2．物理块的分配策略

在请求分页系统中，可采取两种内存分配策略，即固定和可变分配策略。在进行置换时，也可采取两种策略，即全局置换和局部置换。于是可组合出以下三种适用的策略。

请求分页系统支持虚拟存储器，可以采用两种内存块分配策略，固定分配和可变分配。

（1）固定分配（Fixed Allocation）策略是分配给进程的内存块数是固定的，并在最初装入时（即进程创建时）确定块数。分给每个进程的内存块数基于进程类型（交互式、批处理型、应用程序型等），或者由程序员或系统管理员提出的建议。当进程执行过程中出现缺页时，只能从分给该进程的内存块中进行页面置换。

（2）可变分配（Variable Allocation）策略允许分给进程的内存块数随进程的活动而改

变。如果一个进程在运行过程中持续缺页率太高，这就表明该进程的局部化行为不好，需要给它分配另外的内存块，以减少它的缺页率。如果一个进程的缺页率特别低，就可以减少分配的内存块，但不要显著增加缺页率。

可变分配策略的功能更强，但需要操作系统估价出各活动进程的行为，这就增加了操作系统的软件开销，并且依赖于处理器平台所提供的硬件机制。

内存块分配的另一个重要问题是页面置换范围。多个进程竞争内存块时，可以把页面置换分为两种主要类型，全局置换和局部置换。

（1）全局置换（Global Replacement）允许一个进程从全体存储块的集合中选取置换块，尽管该块当前已分给其他进程，但还是能强行剥夺。

（2）局部置换（Local Replacement）是每个进程只能从分给它的一组块中选择置换块。

采用局部置换策略，分给进程的块数是不能变更的。采用全局置换策略，一个进程可以只从分给其他进程的块里挑选。这样，如果没有别的进程挑选它的块，那么分给该进程的块数就增加了。可以由一个核心进程专门负责页面置换工作。

全局置换算法存在的一个问题是，程序无法控制自己的缺页率。一个进程在内存中的一组页面不仅取决于该进程的页面走向，而且也取决于其他进程的页面走向。因此，相同程序由于外界环境不同会造成执行上的很大差别。使用局部置换算法就不会出现这种情况，一个进程在内存中的页面仅受本进程页面走向的影响。

3．物理块分配算法

1）平均分配算法

这是将系统中所有可供分配的物理块，平均分配给各个进程。 例如，当系统中有100个物理块，有5个进程在运行时，每个进程可分得20个物理块。这种方式貌似公平，但实际上是不公平的，因为它未考虑到各进程本身的大小。如有一个进程其大小为200页，只分配给它 20 个块，这样，它必然会有很高的缺页率；而另一个进程只有 10 页，却有 10个物理块闲置未用。

2）按比例分配算法

这是根据进程的大小按比例分配物理块的算法。如果系统中共有 n 个进程，每个进程的页面数为 S_i，则系统中各进程页面数的总和为：

$$S = \sum_{i=1}^{n} S_i$$

又假定系统中可用的物理块总数为 m，则每个进程所能分到的物理块数为 b_i，将有：

$$b_i = \frac{S_i}{S} \times m$$

b_i 应该取整，它必须大于最小物理块数。

3）考虑优先权的分配算法

在实际应用中，为了照顾到重要的、紧迫的作业能尽快地完成，应为它分配较多的内存空间。通常采取的方法是把内存中可供分配的所有物理块分成两部分：一部分按比例地分配给各进程；另一部分则根据各进程的优先权，适当地增加其相应份额后，分配给各进程。在有的系统中，如重要的实时控制系统，则可能是完全按优先权的大小来为各进程分配其物理块的。

6.2.6 调页策略

1．何时调入页面

1）预调页策略

如果进程的许多页是存放在外存的一个连续区域中，则一次调入若干个相邻的页，会比一次调入一页更高效些。但如果调入的一批页面中的大多数都未被访问，则又是低效的。可采用一种以预测为基础的预调页策略，将那些预计在不久之后便会被访问到的页面预先调入内存。如果预测较准确，那么，这种策略显然是很有吸引力的。但遗憾的是，目前预调页的成功率仅约 50%。故这种策略主要用于进程的首次调入时，由程序员指出应该先调入哪些页。

2）请求调页策略

当进程在运行中需要访问某些部分程序和数据时，若发现其所在的页面不在内存，便立即提出请求，由 OS 将其所需的页面调入内存。由请求调页策略所确定调入的页，是一定会被访问的，再加上请求调页策略比较易于实现，故在目前的虚拟存储器中大多采用此策略。但这种策略每次仅调入一页，故须花费较大的系统开销，增加了磁盘 I/O 的启动频率。

2．从何处调入页面

在请求分页系统中的外存分为两部分：用于存放文件的文件区和用于存放对换页面的对换区。通常，由于对换区是采用连续分配方式，而事件是采用离散分配方式，故对换区的磁盘 I/O 速度比文件区的高。这样，每当发生缺页请求时，系统应从何处将缺页调入内存，可分成如下三种情况。

（1）系统拥有足够的对换区空间，这时可以全部从对换区调入所需页面，以提高调页速度。为此，在进程运行前，便须将与该进程有关的文件，从文件区拷贝到对换区。

（2）系统缺少足够的对换区空间，这时凡是不会被修改的文件，都直接从文件区调入；而当换出这些页面时，由于它们未被修改而不必再将它们换出，以后再调入时，仍从文件区直接调入。但对于那些可能被修改的部分，在将它们换出时，便须调到对换区，以后需要时，再从对换区调入。

（3）Linux 方式。Linux 使用请求调页把可执行映像装入进程虚拟内存中。每当一个命令被执行时，包含该命令的文件被打开，它的内容被映射到进程虚拟空间。这是通过修改描述进程内存映射的数据结构来完成的，被称作“内存映射”。然后，只有映像的开始部分被实际装入物理内存，映像其余部分留在磁盘上。随着进程的执行，它会产生页故障，Linux 使用进程内存映射以决定映像的哪一部分被装入内存去执行。

3．页面调入过程

每当程序所要访问的页面未在内存时，便向 CPU 发出一缺页中断，中断处理程序首先保留 CPU 环境，分析中断原因后，转入缺页中断处理程序。该程序通过查找页表，得到该页在外存的物理块后，如果此时内存能容纳新页，则启动磁盘 I/O 将所缺之页调入内存，然后修改页表。如果内存已满，则须先按照某种置换算法从内存中选出一页准备换出；如果该页未被修改过，可不必将该页写回磁盘；但如果此页已被修改，则必须将它写回磁盘，然后再把所缺的页调入内存，并修改页表中的相应表项，置其存在位为“1”，并将此页表项写入快表中。在缺页调入内存后，利用修改后的页表，去形成所要访问数据的物理地址，再去访问内存数据。

6.3　页面置换算法

在进程运行过程中，若其所要访问的页面不在内存而需把它们调入内存，但内存已无空闲空间时，为了保证该进程能正常运行，系统必须从内存中调出一页程序或数据送磁盘的对换区中。但应将哪个页面调出，须根据一定的算法来确定。通常，把选择换出页面的算法称为页面置换算法（Page-Replacement Algorithms）。置换算法的好坏，直接影响的系统性能。

6.3.1　最佳置换算法

最佳（Optimal）置换算法是由 Belady 于 1966 年提出的一种理论上的算法。其所选择的被淘汰页面，将是以后永不使用的，或许是在最长（未来）时间内不再被访问的页面。采用最佳置换算法，通常可保证获得最低的缺页率。

假定系统为某进程分配了三个物理块，并考虑有以下的页面号引用串：

7，0，1，2，0，3，0，4，2，3，0，3，2，1，2，0，1，7，0，1

进程运行时，先将 7，0，1 三个页面装入内存。以后，当进程要访问页面 2 时，将会产生缺页中断。此时 OS 根据最佳置换算法，看 7，0，1 这三个页面哪一个是最晚使用到的，选择页面 7 予以淘汰。这是因为页面 0 将作为第五个被访问的页面，页面 1 是第 14 个被访问的页面，而页面 7 则要在第 18 次页面访问时才需要调入。

然后，访问页面 0 时，因它已在内存而不必产生缺页中断。接着，当进程访问页面 3 时，又将引起页面 1 被淘汰；因为，它在现有的 1，2，0 三个页面中，页面 1 将是以后最晚才被访问的。图 6-6 示出了采用最佳置换算法时的置换图。可以看出，采用最佳置换算法发生了 9 次缺页中断，发生了 6 次页面置换。

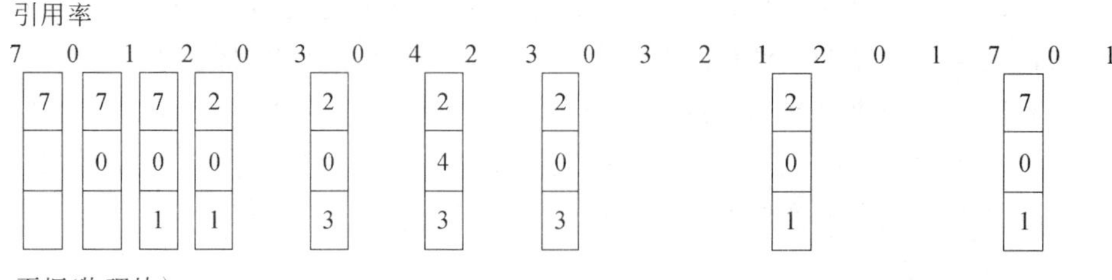

图 6-6　利用最佳页面置换算法时的置换图

最佳置换算法是一种理想化得的算法，它具有较好的性能，但是实际上却是不可实现的。因为这种算法需要预先知道页面的走向次序，可实际上，程序中有分支结构，页面的实际走向是不能事先确定的。所以这种算法在实际的应用中是不可行的。

6.3.2 先进先出页面置换算法

先进先出（FIFO）页面置换算法是最早出现的置换算法。该算法总是淘汰最先进入内存的页面，即选择在内存中驻留时间最久的页面予以淘汰。该算法实现简单，只需把一个已调入内存的页面，按先后次序链接成一个队列，并设置一个指针，称为替换指针，使它总是指向最老的页面。但该算法与进程实际运行的规律不相适应，因为在进程中，有些页面经常被访问，例如，含有全局变量、常用函数、例程等的页面，FIFO 算法并不能保证这些页面不被淘汰。

图 6-7 中所示的这组页面走向，采用 FIFO 置换算法发生了 15 次缺页中断，发生了 12 次页面置换。

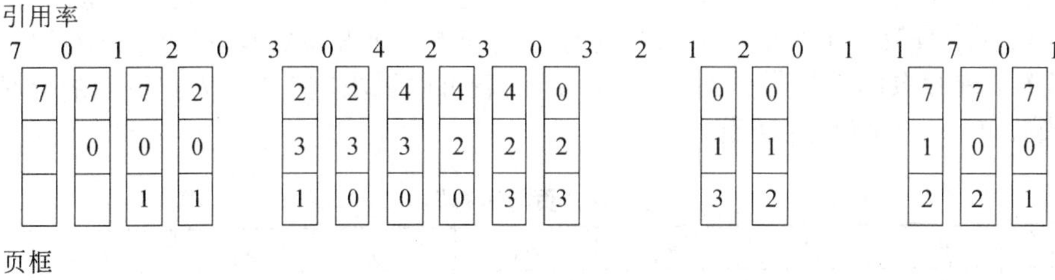

图 6-7 利用 FIFO 置换算法时的页面置换图

6.3.3 最近最久未使用置换算法

FIFO 置换算法性能之所以较差，是因为它所依据的条件是各个页面调入内存的时间，而页面调入先后并不能反映页面的使用情况。最近最久未使用（LRU）的页面置换算法，是根据页面调入内存后的使用情况进行决策的。因为根据程序的局部性原理，刚刚被访问过页面，可能很快还被访问到。由于无法预测各个页面将来的使用情况，只能利用“最近的过去”作为“最近的将来”的近似，因此，LRU 置换算法是选择最近最久未使用的页面予以淘汰。该算法赋予每个页面一个访问字段，用来记录一个页面自上次被访问以来所经历的时间 t，当须淘汰一个页面时，选择现有页面中其 t 值最大的，即最近最久未使用的页面予以淘汰。

利用 LRU 算法对上例进行页面置换的结果如图 6-8 所示。当进程第一次对页面 2 进行访问时，由于页面 7 是最近最久未被访问的，故将它置换出去。当进程第一次对页面 3 进行访问时，第 1 页成为最近最久未使用的页，将它换出。可以看出，前 5 个时间的图像与最佳置换算法时的相同，但这并非是必然的结果。因为，最佳置换算法是从“向后看”的观点出发的，即它是根据以后各页的使用情况；而 LRU 算法则是“向前看”的，即根据各页以前的使用情况来判断，而页面过去和未来的走向之间并无必然的联系。

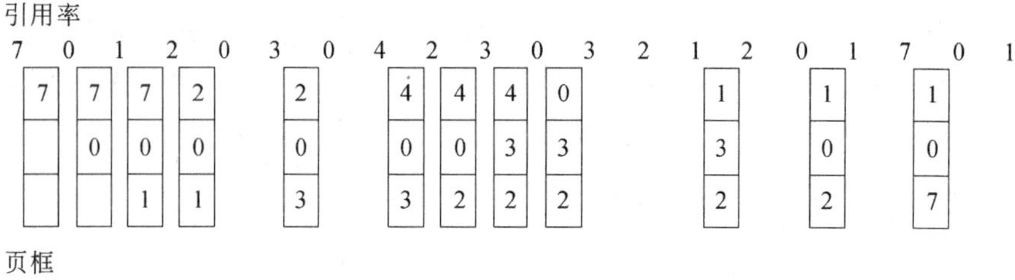

图 6-8 利用 LRU 置换算法时的页面置换图

1．寄存器

为了记录某进程在内存中各页的使用情况，须为每个在内存中的页面配置一个移位寄存器，可表示为

$$R=R_{n-1}R_{n-2}R_{n-3}\ldots R_2R_1R_0$$

当进程访问某物理块时，要将相应寄存器的 R_{n-1} 位置换成 1。此时，定时信号将每隔一定时间（例如 100ms）将寄存器右移一位。如果我们把 n 位寄存器的数看作是一个整数，那么，具有最小数值的寄存器所对应的页面，就是最近最久未使用的页面。表 6-1 示出了某进程在内存中具有 4 个页面，为每个内存页面配置一个 8 位寄存器时的 LRU 访问情况。这里，把 4 个内存页面的序号分别定为 1～4。由表可以看出，第 3 个内存页面的 R 值最小，当发生缺页时，首先将它置换出去。

表 6-1 页与寄存器

实页 R	R_7	R_6	R_5	R_4	R_3	R_2	R_1	R_0
1	0	1	0	1	0	0	1	0
2	1	0	1	0	1	1	0	0
3	0	0	0	0	0	1	0	0
4	0	1	1	0	1	0	1	1

2．栈

可利用一个特殊的栈来保存当前使用的各个页面的页面号。每当进程访问某页面时，便将该页面的页面号从栈中移出，将它压入栈顶。因此，栈顶始终是最新被访问页面的编号，而栈底则是最近最久未使用页面的页面号。假定现有一进程所访问的页面的页面号序列为：

4，7，0，7，1，0，1，2，1，2，6

随着进程的访问，栈中页面号的变化情况如图 6-9 所示。在访问页面 6 时发生了缺页，此时页面 4 是最近最久未被访问的页，应将它置换出去。

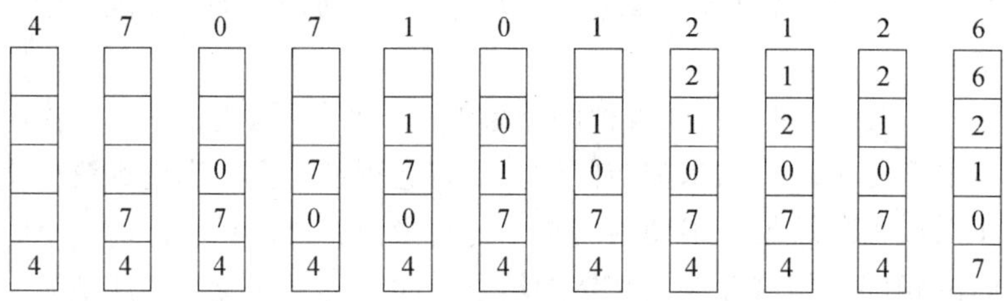

图 6-9 用栈保存当前使用页面时栈的变化情况

6.3.4 Clock 置换算法

LRU 算法是较好的一种算法，但由于它要求有较多的硬件支持，故在实际应用中，大多采用 LRU 的近似算法。Clock 算法就是用得较多的一种 LRU 近似算法。

1. 简单的 Clock 置换算法

当采用简单的 Clock 算法时，只需为每页设置一位访问位，再将内存中的所有页面都通过链接指针链接成一个循环的队列。当某页被访问时，其访问位被置 1。置换算法在选择一页淘汰时，只需检查页的访问位。如果是 0，就选择该页面换出；若为 1，则重新将它置 0，暂不换出，而给该页第二次驻留内存的机会，再按照 FIFO 算法检查下一个页面。当检查到队列中最后一个页面时，若其访问位仍为 1，则再返回到队首去检查第一个页面。图 6-10 给出了该算法的流程和示例。由于该算法是循环地检查各个页面的使用情况，故称为 Clock 算法。但因该算法只有一位访问位，只能用它表示该页是否已经使用过，而置换时是将未使用过的页面换出去，故又把该算法称为最近最久未使用算法 NRU（Not Recently Used）。

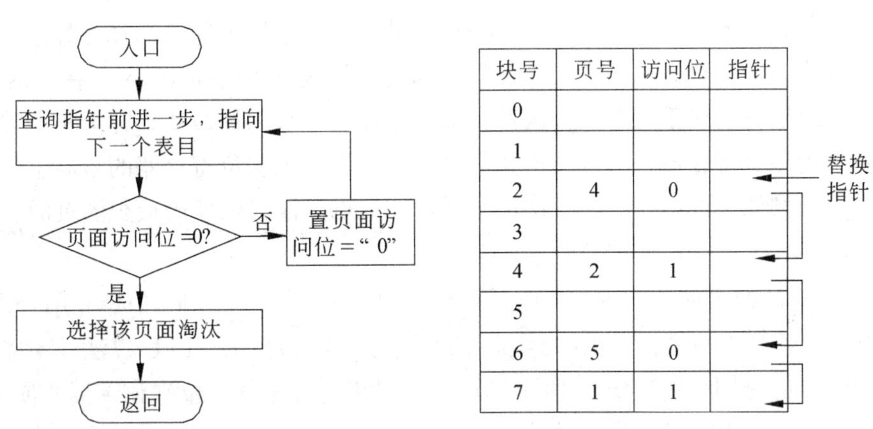

图 6-10　简单 Clock 置换算法的流程和示例

2. 改进型 Clock 置换算法

在将一个页面换出时，如果该页已被修改过，便须将该页重新写回到磁盘上；但如果该页未被修改过，则不必将它拷回磁盘。在改进型 Clock 算法中，除须考虑页面的使用情况外，还须再增加一个因素，即置换代价，这样，选择页面换出时，既要是未使用过的页面，又要是未被修改过的页面。把同时满足这两个条件的页面作为首选淘汰的页面。

由访问位 A 和修改位 M 可以组合成下面四种类型的页面。

（1）1 类（A=0, M=0）：表示该页最近既未被访问，又未被修改，是最佳淘汰页。

（2）2 类（A=0, M=1）：表示该页最近未被访问，但已被修改，并不是很好的淘汰页。

（3）3 类（A=1, M=0）：最近已被访问，但未被修改，该页有可能再被访问。

（4）4 类（A=1, M=1）：最近已被访问且被修改，该页可能再被访问。

其执行过程可分成以下三步。

（1）从指针所指示的当前位置开始，扫描循环队列，寻找 A=0 且 M=0 的 1 类页面，将所遇到的第一个页面作为所选中的淘汰页。在第一次扫描期间不改变访问位 A。

（2）如果第一步失败，即查找一周后未遇到 1 类页面，则开始第二轮扫描，寻找 A=0 且 M=1 的 2 类页面，将所遇到的第一个这类页面作为淘汰页。在第二轮扫描期间，将所有扫描过的页面的访问位都置 0。

（3）如果第二步也失败，亦即未找到 2 类页面，则将指针返回到开始的位置，并将所有的访问位复 0。然后重复第一步，如果仍失败，必要时再重复第二步，此时就一定能找到被淘汰的页。

6.3.5 其他置换算法

还有许多其他进行页面置换的算法。例如，最少使用置换算法、页面缓冲算法等。在此我们对这两种置换算法稍加介绍。

1．最少使用置换算法

在采用最少使用（Least Frequently Used，LFU）置换算法时，应为在内存中的每个页面设置一个移位寄存器，用来记录该页面被访问的频率。该置换算法选择在最近的时期使用最少的页面作为淘汰页。由于存储器具有较高的访问速度，例如 100ns，在 1ms 时间内可能对某页面连续访问成千上万次，因此，通常不能直接利用计数器来记录某页被访问的次数，而是采用移位寄存器方式。每次访问某页时，便将该移位寄存器的最高位置 1，再每隔一定时间（例如 100ms）右移一次，这样，在最近一段时间使用最少的页面$\sum R_i$最小的页。

LRU 置换算法的页面访问图与 LRU 置换算法的访问图完全相同，或者说，利用这样一套硬件既可以实现 LRU 算法，又可实现 LFU 算法。应该指出，LFU 算法并不能真正反映出页面的使用情况，因为在每一时间间隔内，只是用寄存器的一位来记录页的使用情况，因此访问一次和访问 10000 次是等效的。

2．页面缓冲算法

虽然 LRU 和 Clock 置换算法都比 FIFO 算法好，但它们都需要一定的硬件支持，并需付出较多的开销，而且，置换一个已修改的页比置换未修改页的开销要大，而页面缓冲算法（Page Buffering Algorithm，PBA）则既可改善分页系统的性能，又可采用一种较简单的置换策略。VAX/VMS 操作系统便是使用页面缓冲算法。它采用了前述的可变分配和局部置换方式，置换算法采用的是 FIFO。该算法规定将一个被淘汰的页放入两个链表中的一个，即如果页面未被修改，就将它直接放入空闲链表中；否则，便放入已修改页面的链表中。须注意的是，这时页面在内存中并不做物理上的移动，而只是将页表中的表项移到上述两个链表之一中。

空闲页面链表，实际上是一个空闲物理块链表，其中的每个物理块都是空闲的，因此，可在其中装入程序或数据。当需要读入一个页面时，便可利用空闲物理块链表中的第一个物理块来装入该页。当有一个未被修改的页要换出时，实际上并不将它换出内存，而是把该未被修改的页所在的物理块挂在自由页链表的末尾。类似地，在置换一个已修改的页面时，也将其所在的物理块挂在修改页面链表的末尾。利用这种方式可使已被修改的页面和未被修改的页面都仍然保留在内存中。当该进程以后再次访问这些页面时，只需花费较小的开销，使这些页面又返回到该进程的驻留集中。当被修改的页面数目达到一定值时，例如 64 个页面，再将它们一起写回到磁盘上，从而显著地减少了磁盘 I/O 的操作次数。一个较简单的页面缓冲算法已在 MACH 操作系统中实现了，只是它没有区分已修改页面和未修改页面。

6.4 请求段式管理

6.4.1 段表机制

在请求分段式管理中所需的主要数据结构是段表。由于在应用程序的许多段中，只有一部分段装入内存，其余的一些段仍留在外存上，故须在段表中增加若干项，以供程序在调进、调出时参考。如图 6-11 所示。

段名	段长	段的基址	存取方式	访问字段 A	修改位 M	存在位 P	增补位	外存始址

图 6-11　扩充后的段表

在段表项中，除了段名（号）、段长、段在外存中的始地址外，还增加了以下诸项。

（1）存取方式：用于标识本分段的存取属性只是执行、只读，还是允许读/写。

（2）访问字段 A：其含义与请求分页的相应字段相同，记录该段被访问的频繁程度。

（3）修改位 M：用于表示该页在进入内存后是否已被修改过，供置换页面时参考。

（4）存在位 P：指示本段是否已调入内存，供程序访问时参考。

（5）增补位：这是请求分段式管理中所特有的字段，用于表示本段在运行过程中是否做过动态增长。

（6）外存始址：指示本段在外存中的起始地址，即起始盘块号。

6.4.2 缺段中断机制

在请求分段系统中，每当发现运行进程所要访问的段尚未调入内存时，便由缺段中断机构产生一缺段中断信号，进入 OS 后由缺段中断处理程序将所需的段调入内存。缺段中断机构与缺页中断机构类似，它同样需要在一条指令的执行期间产生和处理中断，以及在一条指令执行期间，可能产生多次缺段中断。但由于分段是信息的逻辑单位，因而不可能出现一条指令被分割在两个分段中和一组信息被分割在两个分段中的情况。缺段中断的处理过程如图 6-12 所示。由于段不是定长的，这使对缺段中断的处理比对缺页中断的处理复杂。

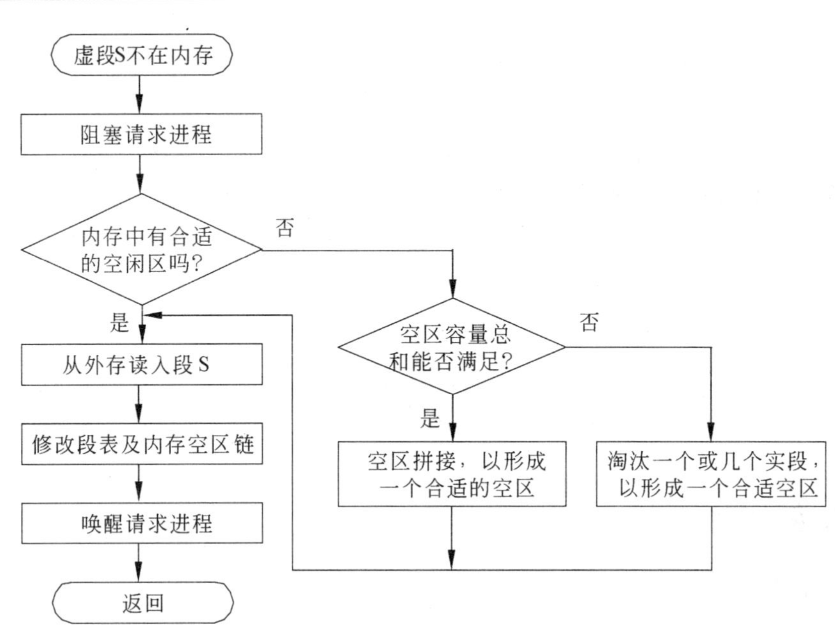

图 6-12 缺段中断的处理过程

6.4.3 地址变换机制

请求分段系统中的地址变换机构在分段系统地址变换机构的基础上形成的。因为被访问的段并非全在内存，所以在地址变换时，若发现所要访问的段不在内存，必须先将所缺的段调入内存，并修改段表，然后才能再利用段表进行地址变换。为此，在地址变换机构中又增加了某些功能，如缺段中断的请求及处理等。图 6-13 给出请求分段系统的地址变换过程。

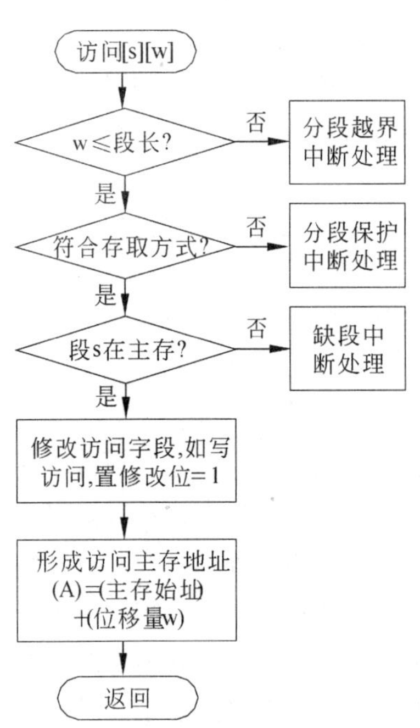

图 6-13 请求段式系统的地址变换

6.4.4 分段的共享与保护

1. 共享段表

为实现分段共享，可在系统中配置一张共享段表，所有各共享段都在共享段表中占有一表项。表项中记录了共享段的段号、段长、内存地址、存在位等信息，并记录了共享此分段的每个进程的情况。共享段表如图 6-14 所示。

共享进程计数 count。非共享段仅为一个进程所需要。当进程不再需要该段时，可立即释放该段，由系统回收该段所占用的空间。而共享段是为多个进程所需要的，当某进程不再需要时，系统并不回收该段所占内存区，仅当所有共享该段的进程全都不再需要它时，才由系统回收该段所占内存区。为了记录有多少个进程需要共享该分段，特设置了一个整型变量 count。

存取控制字段。对于一个共享段，应给不同的进程以不同的存取权限。例如，对于文件主，通常允许他读和写；而对其他进程，则可能只允许读，甚至只允许执行。

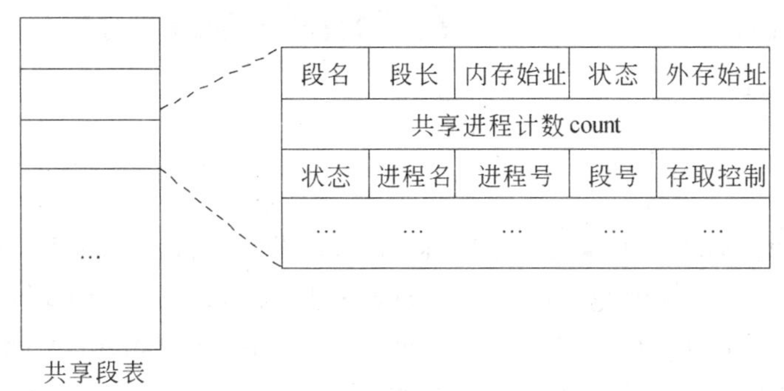

图 6-14　共享段表项

段号：对于一个共享段，不同的进程可以各用不同的段号去共享该段。

2．共享段的分配与回收

1）共享段的分配

在为共享段分配内存时，对第一个请求使用该共享段的进程，由系统为该共享段分配一物理区，再把共享段调入该区，同时将该区的始址填入请求进程的段表的相应项中，还须在共享段表中增加一表项，填写有关数据，把 count 置为 1 之后，当又有其他进程需要调用该共享段时，由于该共享段已被调入内存，故此时无须再为该段分配内存，而只需在调用进程的段表中，增加一表项，填写该共享段的物理地址；在共享段的段表中，填上调用进程的进程名、存取控制等，再执行 count:=count＋1 操作，表明有两个进程共享该段。

2）共享段的回收

当共享此段的某进程不再需要该段时，应将该段释放，包括撤销在该进程段表中共享段所对应的表项，以及执行 count:= count –1 操作。若结果为 0，则须由系统回收该共享段的物理内存，取消在共享段表中该段所对应的表项，表明此时已没有进程使用该段；否则（减 1 结果不为 0），则只是取消调用者进程在共享段表中的有关记录。

3．分段保护

1）越界检查

在段表寄存器放有段表长度信息；同样，在段表中也为每个段设置有段长字段。在进行存储访问时，首先将逻辑地址空间的段号与段表长度进行比较，如果段号等于或大于段表长度，将发出地址越界中断信号；其次，还要检查段内地是否等于或大于段表长度，将发出地址越界中断信号，从而保证了每个进程只能在自己的地址空间内运行。

2）存取控制检查

（1）只读，只允许进程对该段中的程序或数据进行读访问。

（2）只执行，只允许进程调用该段去执行，不准读该段的内容，也不允许对该段执行写操作。

（3）读/写，允许进程对该段进行读/写访问。

对于共享段而言，存取控制就显得尤为重要，因而对不同的进程，应赋予不同的读写权限。这时，既要保证信息的安全性，又要满足运行需要。

Chapter 6 Virtual Memory

Virtual memory is a technique that allows the execution of processes that are not completely in memory. One major advantage of this scheme is that programs can be larger than physical memory. Further, virtual memory abstracts main memory into an extremely large, uniform array of storage, separating logical memory as viewed by the user from physical memory. This technique frees programmers from the concerns of memory-storage limitations. Virtual memory also allows processes to share files easily and to implement shared memory. In addition, it provides an efficient mechanism for process creation. Virtual memory is not easy to implement, however, and may substantially decrease performance if it is used carelessly. In this chapter, we discuss virtual memory in the form of demand paging and examine its complexity and cost.

6.1 Virtual Memory Concept

6.1.1 Background

The requirement that instructions must be in physical memory to be executed seems both necessary and reasonable; but it is also unfortunate, since it limits the size of a program to the size of physical memory. In fact, an examination of real programs shows us that, in many cases, the entire program is not needed. For instance, consider the following:

(1) Programs often have code to handle unusual error conditions. Since these errors seldom, if ever, occur in practice, this code is almost never executed;

(2) Arrays, lists, and tables are often allocated more memory than they actually need. An array may be declared 100 by 100 elements, even though it is seldom larger than 10 by 10 elements. An assembler symbol table may have room for 3,000 symbols, although the average program has less than 200 symbols;

(3) Certain options and features of a program may be used rarely.

Even in those cases where the entire program is needed, it may not all be needed at the same time.

Virtual memory involves the separation of logical memory as perceived by users from physical memory. This separation allows an extremely large virtual memory to be provided for programmers when only a smaller physical memory is available. Virtual memory makes the task of programming much easier, because the programmer no longer needs to worry about the amount of physical memory available; she can concentrate instead on the problem to be programmed.

The virtual address space of a process refers to the logical (or virtual) view of how a process is stored in memory. Typically, this view is that a process begins at a certain logical address say, address 0—and exists in contiguous memory. In fact physical memory may be organized in page frames and that the physical page frames assigned to a process may not be contiguous. It is up to the memory-management unit (MMU) to map logical pages to physical page frames in memory.

6.1.2 Shared Pages and Swapping

1. Shared Pages

An advantage of paging is the possibility of sharing common code. This consideration is particularly important in a time-sharing environment. Consider a system that supports 40 users, each of whom executes a text editor. If the text editor consists of 150 KB of code and 50 KB of data space, we need 8,000 KB to support the 40 users. If the code is reentrant code (or pure code), however, it can be shared, as shown in Figure 6.1. Here we see a three-page editor—each page 50 KB in size (the large page size is used to simplify the figure) —being shared among three processes. Each process has its own data page.

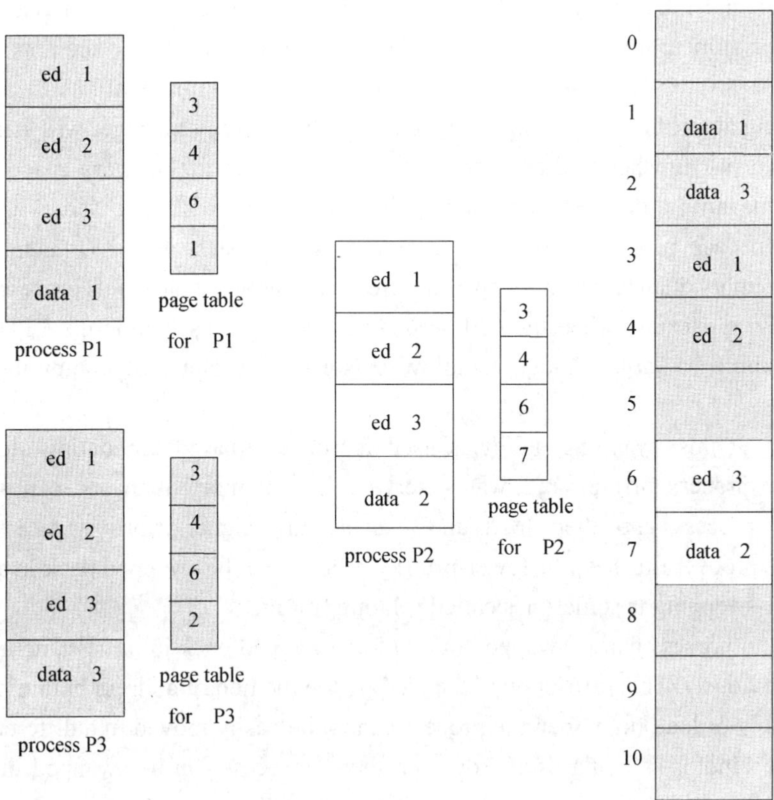

Figure 6.1 Sharing of code in a paging environment

Reentrant code is non-self-modifying code; it never changes during execution. Thus, two or more processes can execute the same code at the same time. Each process has its own copy of registers and data storage to hold the data for the process's execution. The data for two different processes will, of course, be different.

Only one copy of the editor need be kept in physical memory. Each user's page table maps onto the same physical copy of the editor, but data pages are mapped onto different frames. Thus, to support 40 users, we need only one copy of the editor (150 KB), plus 40 copies of the 50 KB of data space per user. The total space required is now 2,150 KB instead of 8,000 KB a significant savings.

Other heavily used programs can also be shared—compilers, window systems, run-time libraries, database systems, and so on. To be sharable, the code must be reentrant. The read-only nature of shared code should not be left to the correctness of the code; the operating system should enforce this property.

Similarly, virtual memory enables processes to share memory. Two or more processes can communicate through the use of shared memory. Virtual memory allows one process to create a region of memory that it can share with another process. Processes sharing this region consider it part of their virtual address space, yet the actual physical pages of memory are shared.

2. Swapping

A process must be in memory to be executed. A process, however, can be swapped temporarily out of memory to a backing store and then brought back into memory for continued execution. For example, assume a multiprogramming environment with a round-robin CPU-scheduling algorithm. When a quantum expires, the memory manager will start to swap out the process that just finished and to swap another process into the memory space that has been freed. In the meantime, the CPU scheduler will allocate a time slice to some other process in memory. When each process finishes its quantum, it will be swapped with another process. Ideally, the memory manager can swap processes fast enough that some processes will be in memory; ready to execute, when the CPU scheduler wants to reschedule the CPU. In addition, the quantum must be large enough to allow reasonable amounts of computing to be done between swaps.

A variant of this swapping policy is used for priority-based scheduling algorithms. If a higher-priority process arrives and wants service, the memory manager can swap out the lower-priority process and then load and execute the higher-priority process. When the higher-priority process finishes, the lower-priority process can be swapped back in and continued. This variant of swapping is sometimes called roll out, roll in.

Normally, a process that is swapped out will be swapped back into the same memory space it occupied previously. This restriction is dictated by the method of address binding. If binding is done at assembly or load time, then the process cannot be easily moved to a different location. If execution-time binding is being used, however, then a process can be swapped into a different memory space, because the physical addresses are computed during execution time.

Swapping requires a backing store. The backing store is commonly a fast disk. It must be large enough to accommodate copies of all memory images for all users, and it must provide direct access to these memory images. The system maintains a ready queue consisting of all processes whose memory images are on the backing store or in memory and are ready to run. Whenever the CPU scheduler decides to execute a process, it calls the dispatcher. The dispatcher checks to see whether the next process in the queue is in memory. If it is not, and if there is no free memory region, the dispatcher swaps out a process currently in memory and swaps in the desired process. It then reloads registers and transfers control to the selected process.

6.2 Demand Paging

6.2.1 Locality Model

The locality model states that, as a process executes, it moves from locality to locality. A locality is a set of pages that are actively used together. A program is generally composed of several different localities, which may overlap.

For example, when a function is called, it defines a new locality. In this locality, memory references are made to the instructions of the function call, its local variables, and a subset of the global variables. When we exit the function, the process leaves this locality, since the local variables and instructions of the function are no longer in active use. We may return to this locality later.

Thus, we see that localities are defined by the program structure and its data structures. The locality model states that all programs will exhibit this basic memory reference structure. Note that the locality model is the unstated principle behind the caching discussions so far in this book. If accesses to any types of data were random rather than patterned, caching would be useless.

Suppose we allocate enough frames to a process to accommodate its current locality. It will fault for the pages in its locality until all these pages are in memory; then, it will not fault again until it changes localities. If we allocate fewer frames than the size of the current locality, the process will thrash, since it cannot keep in memory all the pages that it is actively using.

There are two different types of locality:

(1) Temporal locality (locality in time):, if an item is references, it will tend to be referenced again soon;

(2) Spatial locality (locality in space); if an item is referenced, items whose addresses are close by will tend to be referenced soon.

6.2.2 Working-Set Model

The working-set model is based on the assumption of locality. This model uses a parameter, Δ, to define the working-set window. The idea is to examine the most recent Δ page references. The set of pages in the most recent Δ page references is the working set. If a page is inactive use, it will be in the working set. If it is no longer being used, it will drop from the working set A time units after its last reference. Thus, the working set is an approximation of the program's locality.

The accuracy of the working set depends on the selection of Δ. If Δ is too small, it will not encompass the entire locality; if Δ is too large, it may overlap several localities. In the extreme, if Δ is infinite, the working set is the set of pages touched during the process execution.

The most important property of the working set, then, is its size. If we compute the working-set size, WSS_i, for each process in the system, we can then consider that

$$D = \sum WSS_i,$$

where D is the total demand for frames. Each process is actively using the pages in its working set. Thus, process i needs WSS_i frames. If the total demand is greater than the total number of available frames ($D > m$), thrashing will occur, because some processes will not have enough frames.

Once A has been selected, use of the working-set model is simple. The operating system monitors the working set of each process and allocates to that working set enough frames to provide it with its working-set size. If there are enough extra frames, another process can be initiated. If the sum of the working-set sizes increases, exceeding the total number of available frames, the operating system selects a process to suspend. The process's pages are written out (swapped), and its frames are reallocated to other processes. The suspended process can be restarted later.

This working-set strategy prevents thrashing while keeping the degree of multiprogramming as high as possible. Thus, it optimizes CPU utilization.

6.2.3 Page Faults

Access to a page marked invalid causes a page-fault trap. The paging hardware, in translating the address through the page table, will notice that the invalid bit is set, causing a trap to the operating system. This trap is the result of the operating system's failure to bring the desired page into memory. The procedure for handling this page fault is straightforward.

(1) We check an internal table (usually kept with the process control block) for this process to determine whether the reference was a valid or an invalid memory access.

(2) If the reference was invalid, we terminate the process. If it was valid, but we have not yet brought in that page, we now page it in.

(3) We find a free frame (by taking one from the free-frame list, for example).

(4) We schedule a disk operation to read the desired page into the newly allocated frame.

(5) When the disk read is complete, we modify the internal table kept with the process and the page table to indicate that the page is now in memory.

(6) We restart the instruction that was interrupted by the trap. The process can now access the page as though it had always been in memory.

The hardware to support demand paging is the same as the hardware for paging and swapping:

Page table. This table has the ability to mark an entry invalid through a valid—invalid bit or special value of protection bits.

Secondary memory. This memory holds those pages that are not present in main memory. The secondary memory is usually a high-speed disk. It is known as the swap device, and the section of disk used for this purpose is known as swap space.

A crucial requirement for demand paging is the need to be able to restart any instruction after a page fault. Because we save the state (registers, condition code, instruction counter) of the interrupted process when the page fault occurs, we must be able to restart the process in exactly the same place and state, except that the desired page is now in memory and is accessible. In most cases, this requirement is easy to meet. A page fault may occur at any memory reference. If the page fault occurs on the instruction fetch, we can restart by fetching the instruction again. If a page fault occurs while we are fetching an operand, we must fetch and decode the instruction again and then fetch the operand.

6.2.4 Address Translation

To convert a virtual address into a physical address, the CPU uses the page number as an index into the page table. If the page is resident, the physical frame address in the page table is concatenated in front of the offset to create the physical address as is shown in Figure 6.2.

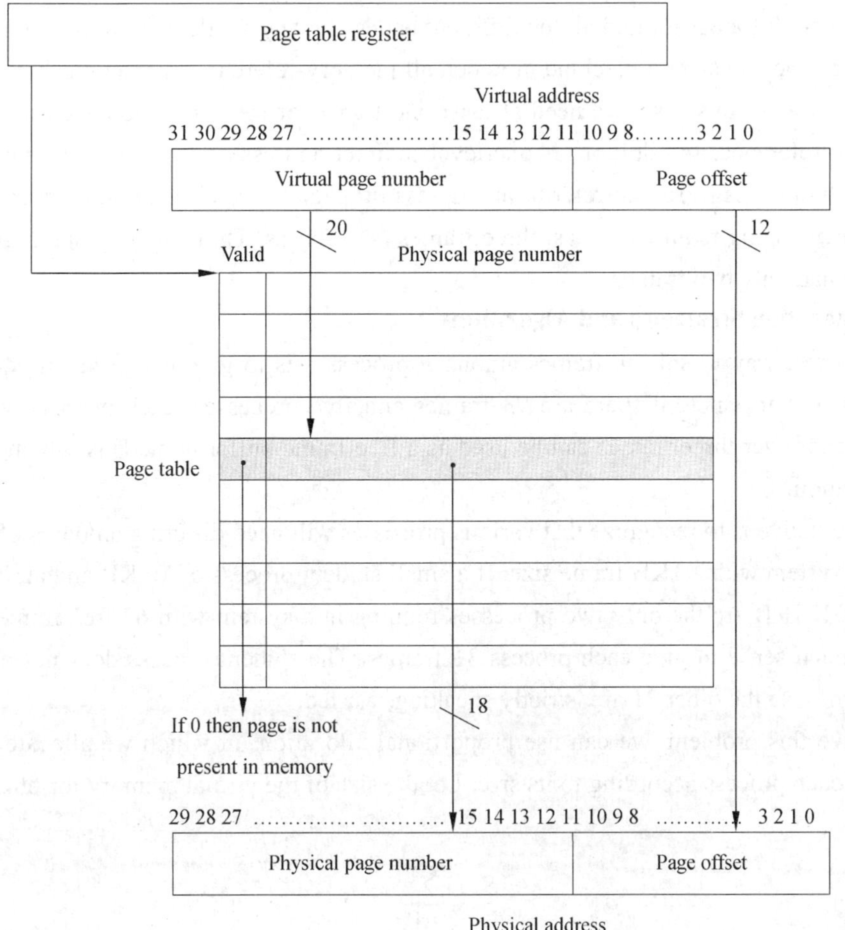

Figure 6.2 Address translation

6.2.5 Allocation of Frames

Consider a single-user system with 128 KB of memory composed of pages 1 KB in size. This system has 128 frames. The operating system may take 35 KB, leaving 93 frames for the user process. Under pure demand paging, all 93 frames would initially be put on the free-frame list. When a user process started execution, it would generate a sequence of page faults. The first 93 page faults would all get free frames from the free-frame list. When the free-frame list was exhausted, a page-replacement algorithm would be used to select one of the 93 in-memory pages to be replaced with the 94th, and so on. When the process terminated, the 93 frames would once again be placed on the free-frame list.

1. Minimum Number of Frames

One reason for allocating at least a minimum number of frames involves performance. Obviously, as the number of frames allocated to each process decreases, the page-fault rate increases, slowing process execution. In addition, remember that, when a page fault occurs before an executing instruction is complete, the instruction must be restarted. Consequently, we must have enough frames to hold all the different pages that any single instruction can reference.

For example, consider a machine in which all memory-reference instructions have only one memory address. In this case, we need at least one frame for the instruction and one frame for the memory reference. In addition, if one-level indirect addressing is allowed (for example, a load instruction on page 16 can refer to an address on page 0, which is an indirect reference to page 23), then paging requires at least three frames per process. Think about what might happen if a process had only two frames.

2. Allocation Strategies and Algorithms

The easiest way to split *m* frames among *n* processes is to give everyone an equal share, *m*/*n* frames. For instance, if there are 93 frames and five processes, each process will get 18 frames. The leftover three frames can be used as a free-frame buffer pool. This scheme is called equal allocation.

An alternative is to recognize that various processes will need differing amounts of memory. Consider a system with a 1KB frame size. If a small student process of 10 KB and an interactive database of 127KB are the only two processes running in a system with 62 free frames, it does not make much sense to give each process 31 frames. The student process does not need more than 10 frames, so the other 21 are, strictly speaking, wasted.

To solve this problem, we can use proportional allocation, in which we allocate available memory to each process according to its size. Let the size of the virtual memory for process pi be and define

$$S = \sum_{i=1}^{n} s_i$$

Then, if the total number of available frames is in, we allocate a frames to process p_i, where

b_i is approximately

$$b_i = \frac{s_i}{S} \times m$$

Of course, we must adjust each b_i to be an integer that is greater than the minimum number of frames required by the instruction set, with a sum not exceeding m.

For proportional allocation, we would split 62 frames between two processes, one of 10 pages and one of 127 pages, by allocating 4 frames and 57 frames, respectively, since

$$10 / 137 \times 62 \approx 4, \quad \text{and} \quad 127 / 137 \times 62 \approx 57$$

In this way, both processes share the available frames according to their "needs," rather than equally.

In both equal and proportional allocation, of course, the allocation may vary according to the multiprogramming level. If the multiprogramming level is increased, each process will lose some frames to provide the memory needed for the new process. Conversely, if the multiprogramming level decreases, the frames that were allocated to the departed process can be spread over the remaining processes.

6.3 Page Replacement Algorithms

Page replacement takes the following approach. If no frame is free, we find one that is not currently being used and free it. We can free a frame by writing its contents to swap space and changing the page table (and all other tables) to indicate that the page is no longer in memory.

There are many different page-replacement algorithms. Every operating system probably has its own replacement scheme. How do we select a particular replacement algorithm? In general, we want the one with the lowest page-fault rate.

6.3.1 First In First Out Page Replacement

The simplest page-replacement algorithm is a first-in, first-out (FIFO) algorithm. A FIFO replacement algorithm associates with each page the time when that page was brought into memory. When a page must be replaced, the oldest page is chosen. Notice that it is not strictly necessary to record the time when a page is brought in. We can create a FIFO queue to hold all pages in memory. We replace the page at the head of the queue. When a page is brought into memory, we insert it at the tail of the queue.

For our example reference string, our three frames are initially empty. The first three references (7, 0, 1) cause page faults and are brought into these empty frames. The next reference (2) replaces page 7, because page 7 was brought in first. Since 0 is the next reference and 0 is already in memory, we have no fault for this reference. The first reference to 3 results in replacement of page 0, since it is now first in line. Because of this replacement, the next reference, to 0, will fault. Page 1 is then replaced by page 0. This process continues as shown in Figure 6.3. Every time a fault occurs, we show which pages are in our three frames. There are 15 faults altogether.

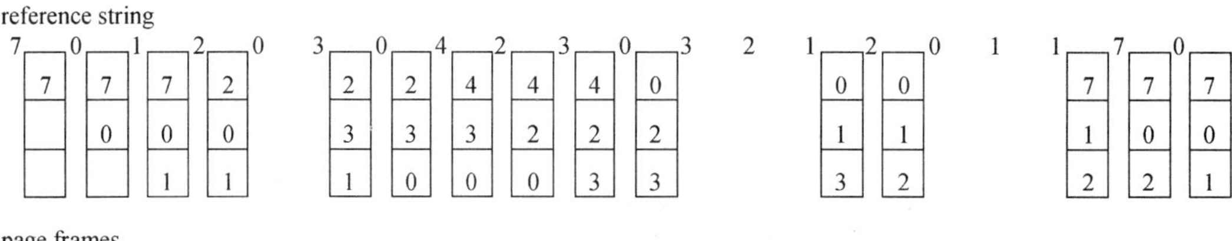

Figure 6.3 FIFO page-replacement algorithm.

The FIFO page-replacement algorithm is easy to understand and program. However, its performance is not always good. On the one hand, the page replaced may be an initialization module that was used a long time ago and is no longer needed. On the other hand, it could contain a heavily used variable that was initialized early and is in constant use.

Notice that, even if we select for replacement a page that is in active use, everything still works correctly. After we replace an active page with a new one, a fault occurs almost immediately to retrieve the active page. Some other page will need to be replaced to bring the active page back into memory. Thus, a bad replacement choice increases the page-fault rate and slows process execution. It does not, however, cause incorrect execution.

To illustrate the problems that are possible with a FIFO page-replacement algorithm, we consider the following reference string:

1, 2, 3, 4, 1, 2, 5, 1, 2, 3, 4, 5

6.3.2 Optimal Page Replacement

One result of the discovery of Belady's anomaly was the search for an optimal page-replacement algorithm. An optimal page-replacement algorithm has the lowest page-fault rate of all algorithms and will never suffer from Belady's anomaly. Such an algorithm does exist and has been called OPT or MIN. It is simply this:

Use of this page-replacement algorithm guarantees the lowest possible page-fault rate for a fixed number of frames.

For example, on our sample reference string, the optimal page-replacement algorithm would yield nine page faults, as shown in Figure 6.4. The first three references cause faults that fill the three empty frames. The reference to page 2 replaces page 7, because 7 will not be used until reference 18, whereas page 0 will be used at 5, and page 1 at 14. The reference to page 3 replaces page 1, as page 1 will be the last of the three pages in memory to be referenced again. With only nine page faults, optimal replacement is much better than a FIFO algorithm, which resulted in fifteen faults. (If we ignore the first three, which all algorithms must suffer, then optimal replacement is twice as good as FIFO replacement.) In fact, no replacement algorithm can process this reference string in three frames with fewer than nine faults.

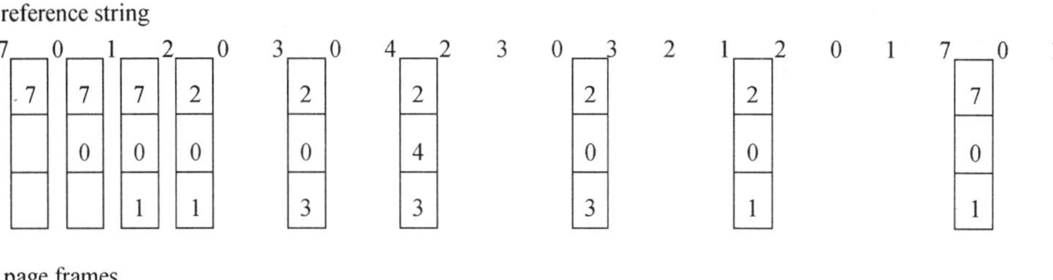

Figure 6.4 Optimal page-replacement algorithm.

Unfortunately, the optimal page-replacement algorithm is difficult to implement, because it requires future knowledge of the reference string.

6.3.3 Least-Recently Used Page Replacement

If the optimal algorithm is not feasible, perhaps an approximation of the optimal algorithm is possible. The key distinction between the FIFO and OPT algorithms (other than looking backward versus forward in time) is that the FIFO algorithm uses the time when a page was brought into memory, whereas the OPT algorithm uses the time when a page is to be used. If we use the recent past as an approximation of the near future, then we can replace the page that has not been used for the longest period of time (Figure 6.5). This approach is the least-recently-used (LRU) algorithm.

LRU replacement associates with each page the time of that page's last use. When a page must be replaced, LRU chooses the page that has not been used for the longest period of time. We can think of this strategy as the optimal page-replacement algorithm looking backward in time, rather than forward. (Strangely, if we let 9 be the reverse of a reference string S, then the page-fault rate for the OPT algorithm on S is the same as the page-fault rate for the OPT algorithm on SR. Similarly, the page-fault rate for the LRU algorithm on S is the same as the page-fault rate for the LRU algorithm on SR.)

The result of applying LRU replacement to our example reference string is shown in Figure 6.5. The LRU algorithm produces 12 faults. Notice that the first 5 faults are the same as those for optimal replacement. When the reference to page 4 occurs, however, LRU replacement sees that, of the three frames in memory, page 2 was used least recently. Thus, the LRU algorithm replaces page 2, not knowing that page 2 is about to be used. When it then faults for page 2, the LRU algorithm replaces page 3, since it is now the least recently used of the three pages in memory. Despite these problems, LRU replacement with 12 faults is much better than FIFO replacement with 15.

The LRU policy is often used as a page-replacement algorithm and is considered to be good. The major problem is how to implement LRU replacement. A LRU page-replacement algorithm may require substantial hardware assistance. The problem is to determine an order for the frames defined by the time of last use.

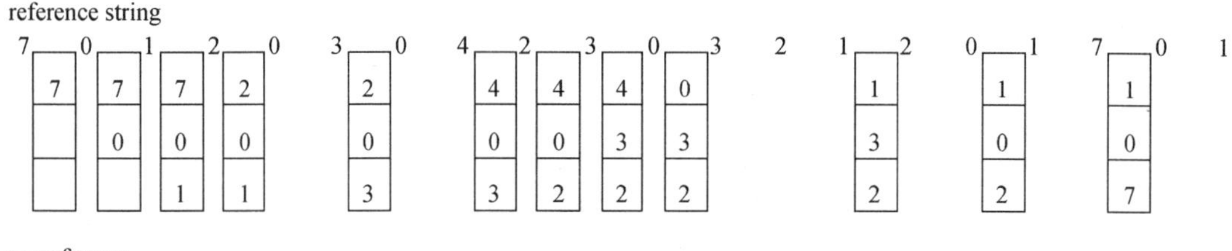

Figure 6.5 LRU page-replacement algorithm

6.3.4 Clock (Second Chance) Algorithm

1. Simple Clock Algorithm

The basic clock algorithm replacement is a FIFO replacement algorithm. When a page has been selected, however, we inspect its reference bit. If the value is 0, we proceed to replace this page; but if the reference bit is set to 1, we give the page a second chance and move on to select the next FIFO page. When a page gets a second chance, its reference bit is cleared, and its arrival time is reset to the current time. Thus, a page that is given a second chance will not be replaced until all other pages have been replaced (or given second chances). In addition, if a page is used often enough to keep its reference bit set, it will never be replaced.

One way to implement the second-chance algorithm (sometimes referred to as the clock algorithm) is as a circular queue. A pointer (that is, a hand on the clock) indicates which page is to be replaced next. When a frame is needed, the pointer advances until it finds a page with a 0 reference bit. As it advances, it clears the reference bits (Figure 6.6). Once a victim page is found, the page is replaced, and the new page is inserted in the circular queue in that position. Notice that, in the worst case, when all bits are set, the pointer cycles through the whole queue, giving each page a second chance. It clears all the reference bits before selecting the next page for replacement. Second-chance replacement degenerates to FIFO replacement if all bits are set.

2. Enhanced Clock Algorithm

We can enhance the clock algorithm by considering the reference bit and the modify bit an ordered pair. With these two bits, we have the following four possible classes:

(1) (0, 0) neither recently used nor modified—best page to replace;

(2) (0, 1) not recently used but modified—not quite as good, because the page will need to be written out before replacement;

(3) (1, 0) recently used but clean—probably will be used again soon;

(4) (1, 1) recently used and modified — probably will be used again soon, and the page will be needed to be written out to disk before it can be replaced;

Each page is in one of these four classes. When page replacement is called for, we use the same scheme as in the clock algorithm; but instead of examining whether the page to which we are pointing has the reference bit set to 1, we examine the class to which that page belongs. We replace the first page encountered in the lowest nonempty class. Notice that we may have to scan the circular queue several times before we find a page to be replaced.

The major difference between this algorithm and the simpler clock algorithm is that here we give preference to those pages that have been modified to reduce the number of I/Os required(Figure6.6).

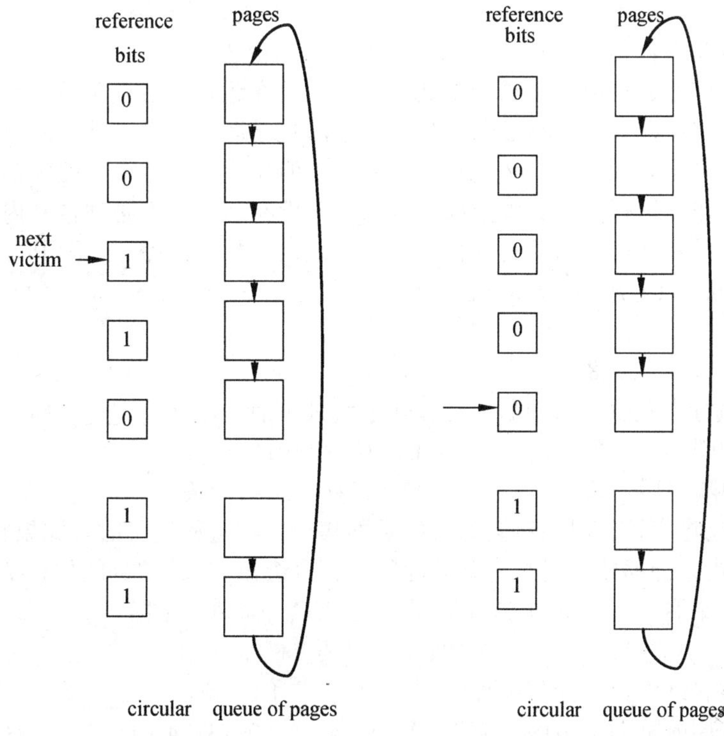

Figure 6.6 Second-chance (clock) page-replacement algorithm

习　　题

一、选择题

1．虚拟存储管理系统的基础是程序的________理论，这个理论的基本含义是指程序执行时往往会不均匀地访问主存储器单元。

A．全局性　　B．局部性　　C．时间全局性　　D．空间全局性

2．虚拟存储器的最大容量________。

A．为内外存容量和　　B．由计算机的地址结构决定

C．是任意的　　D．由作业的地址决定

3．实现虚拟存储器的目的是________。

A．实现存储保护　　B．实现程序浮动

C．扩充辅存容量　　D．扩充主存容量

4．虚拟存储管理系统的基础是程序的________理论。

A. 局部性　　B. 全局性　　C. 动态性　　D. 虚拟性

5. 在请求分页存储管理中，若采用 FIFO 页面淘汰算法，则当分配的页帧数增加时，缺页中断的次数________。

A. 减少　　B. 增加　　C. 无影响　　D. 可能增加也可能减少

6. 系统“抖动”现象的发生是由________引起的。

A. 置换算法选择不当　　B. 交换的信息量过大

C. 内存容量不足　　D. 请求页式管理方案

7. 列关于虚拟存储器的论述中，选出一条正确的论述________。

A. 要求作业运行前，必须全部装入内存，且在运行中必须常驻内存

B. 要求作业运行前，不必全部装入内存，且在运行中不必常驻内存

C. 要求作业运行前，不必全部装入内存，但在运行中必须常驻内存

D. 要求作业运行前，必须全部装入内存，但在运行中不必须常驻内存

8. 在虚拟存储系统中，若进程在内存中占 3 块（开始时为空），采用先进先出页面淘汰算法，当执行访问页号序为 1、2、3、4、1、2、5、1、2、3、4、5、6、时，将产生________次缺页中断。

A. 7　　B. 8　　C. 9　　D. 10

9. 作业在执行中发生缺页中断，经操作系统处理后，应让________执行指令。

A. 被中断的前一条　　B. 被中断的

C. 被中断的后一条　　D. 启动时的第一条

10. 工作集是进程运行时被频繁访问的页面集合。在进程运行时，如果它的工作集页面都在________内，能够使该进程有效地运行，否则会出现频繁的页面调入/调出现象。

A. 主存储器　　B. 虚拟存储器　　C. 辅助存储器　　D. U 盘

二、填空题

1. 请求页式管理中，页面置换算法常用的是___①___和___②___。

2. 假设某程序的页面访问序列为 1、2、3、4、5、2、3、4、5、1、2、3、4 、5 且开始执行时主存中没有页面，则在分配给该程序的物理块数是 3 且采用 FIFO 方式时缺页次数是___①___：在分配给程序的物理块数是 4 且采用 FIFO 方式时，缺页次数是___②___在分配给该程序的物理块数是 3 且采用 LRU 方式时，缺页次数是___③___在分配给该程序的物理块数为 4 且采用 LRU 方式时，缺页次数是___④___。

3. 在虚存管理中，虚拟地址空间是指逻辑地址空间，实地址空间的指___①___；前者的大小只受___②___限制，而后者的大小受___③___。

4. 若选用的算法不合适，可能会出现________抖动现象。

5. 在页式存储管理系统中，常用的页面淘汰算法有：___①___，选择淘汰不再使用或最远的将来才使用的页；___②___，选择淘汰在主存驻留时间最长的页：___③___，选择淘汰离当前时刻最近的一段时间内使用得少的页。

6. 在虚拟段式存储管理中，若逻辑地址的段内地址大于段表中该段的段长，则________发生。

三、综合题

1. 已知页面走向为 1、2、1、3、1、2、4、2、1、3、4，分配给该程序的物理块数是 3，且开始采用 FIFO 页面淘汰算法时缺率为多少？假定现有一种淘汰算法，该算法淘汰页面的策略为当需要淘汰页面时，就把刚使用过的页面作为淘汰对象，试问就相同的页面走

向，其缺页率又为多少？

2. 有一请求分页存储管理系统，页面大小为每页 100 字节。有一个 50×50 的整型数组按行连续存放，每个整数占两个字节，将数组初始化为 0 的程序描述如下：

```
int a[50][50];
int i,j;
for(i=0;i<=49;i++)
for(j=0;j<=49;j++)
a[i][j]=0;
```

若在程序执行时内存中只有一个存储块用来存放数组信息，试问该程序执行时产生多少次缺页中断？

3. 设有一页式存储管理系统，向用户提供的逻辑地址空间最大为 16 页，每页 2048 字节，内存总共有 8 个存储块，试问逻辑地址至少应为多少位？，内存空间有多大？

4. 在一个请求分页存储管理系统中，一个作业的页面走向为 4、3、2、1、4、3、5、4、3、2、1、5，当分配给该作业的物理块数分别为 3、4 时，试计算采用下述页面淘汰算法时的缺页率（假设开始执行时主存中没有页面），并比较所得结果。(1) 最佳置换淘汰算法 (2) 先进先出页面淘汰算法 (3) 最近最久未使用页面淘汰算法。

5. 在一个请求分页系统中，假定系统分配给一个作业的物理块数为 3，并且此作业的页面走向为 2、3、2、1、5、2、4、5、3、2、5、2。试用 FIFO 和 LRU 两种算法分别计算出程序访问过程中所发生的缺页次数。

6. 设系统为某进程分配 3 个物理块数，页引用序列为 1 2 3 4 2 1 5 6 2 1 2 3 7 6 3 2 1 2 3 6，回答下列问题

（1）采用 FIFO 页面替换算法，其缺页次数是多少？（画√表示缺页）

	1	2	3	4	2	1	5	6	2	1	2	3	7	6	3	2	1	2	3	6
页框																				
缺页																				

缺页次数：

（2）采用 LRU 页面替换算法，其缺页次数是多少？（画√表示缺页）

	1	2	3	4	2	1	5	6	2	1	2	3	7	6	3	2	1	2	3	6
页框																				
缺页																				

缺页次数：

7. 某虚拟存储器的地址空间共有 64 个页，每页 1KB，主存 32KB。假定某时刻系统为用户的第 0、1、2、3 页分别分配的物理块号分别为 5、8、4、7。试回答：

（1）逻辑地址的页码需几个二进制位？页内地址几位？

物理地址的块号需要几个二进制位？块内地址几位？

（2）逻辑地址 0B5CH 对应的物理地址是多少？

（3）画图表示该逻辑地址地址映射到物理地址的变换过程。

8. 有一个请求页式系统，整数占 4B，页面大小为 256B，采用 LRU 算法，系统为每个进程分配 3 个页帧。

程序段：

```
int[][]  a=new int[200][200];
  int  i=0,  j=0;
  while (i++<200)
  {
     j=0;
     while (j++<200)
     {
       a[i][j]=0;
     }
  }
```

代码在第 0 页上，问：(1) a 数据占多少页面？(2) 执行这个程序段会产生多少次缺页？

9. 在某页式系统中，有 2^{32}B 的物理内存，2^{12} 页的虚地址空间，且页的大小为 512B。问：

（1）虚地址有多少位？

（2）一个页帧有多少字节？

（3）物理地址中有多少位表示页帧？

（4）页表有多少项（页表有多长）？

（5）页表需要多少位来存入一个页表项（页帧号及一个有效位）？

10. 设某计算机的逻辑地址空间和物理地址空间均为 64 KB，按字节编址。若某进程最多需要 6 页（Page）数据存储空间，页的大小为 1 KB，操作系统采用固定分配局部置换策略为此进程分配 4 个页框（Page Frame）。在时刻 260 前的该进程访问情况如下表所示（访问位即使用位）。

页号	页框号	装入时刻	访问位
0	7	130	1
1	4	230	1
2	2	200	1
3	9	160	1

当该进程执行到时刻 260 时，要访问逻辑地址为 17CAH 的数据。请回答下列问题：

（1）该逻辑地址对应的页号是多少？

（2）若采用先进先出（FIFO）置换算法，该逻辑地址对应的物理地址是多少？

（3）若采用时钟（CLOCK）置换算法，该逻辑地址对应的物理地址是多少？（设搜索下一页的指针沿顺时针方向移动，且当前指向 2 号页框，示意图如图 6-15）。

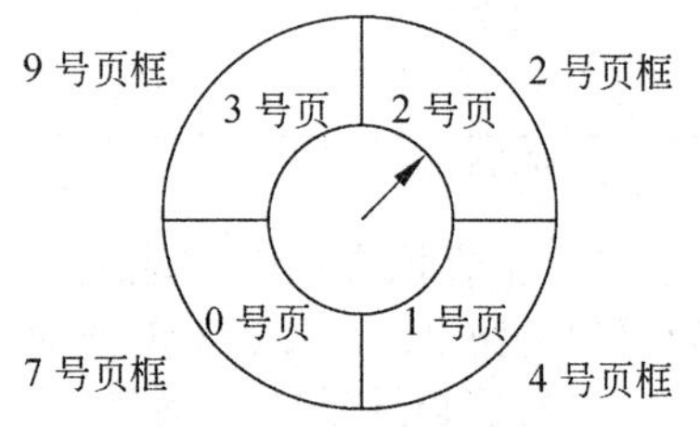

图 6-15　示意图

第7章　文 件 管 理

对大多数用户来说，文件系统是操作系统中最直接可见的部分。计算机的重要作用之一就是能快速处理大量信息，从而，信息的组织、存取和保管就成为一个极为重要的内容。文件系统是计算机组织、存取和保存信息的重要手段。本章主要讨论文件的组织结构、存储空间分配方式、保护以及文件系统空间管理等问题。

7.1　文件和文件系统

文件系统是操作系统用于管理磁盘或分区上的文件的方法和数据结构；即在磁盘上组织文件的方法。操作系统中负责管理和存储文件信息的软件机构称为文件管理系统，简称文件系统。文件系统由三部分组成：与文件管理有关软件、被管理文件以及实施文件管理所需数据结构。从系统角度来看，文件系统是对文件存储器空间进行组织和分配，负责文件存储并对存入的文件进行保护和检索的系统。具体地说，它负责为用户建立文件，存入、读出、修改、转储文件，控制文件的存取，当用户不再使用时撤销文件等。

7.1.1　文件

1．文件和文件名

操作系统将所要处理的信息组织成文件来进行管理，这些信息既包括通常的程序和数据，也包括设备资源。每个文件都有一个文件名，用户通过文件名来存取文件。换句话说，文件就是存储在磁盘上的一组相关信息的集合，具有唯一的标识。

文件名通常由若干 ASCII 码和汉字组成。文件名的格式和长度因系统而异，但大多采用文件名和扩展名组成，前者用于标识文件；后者用于标识文件类型，通常可以有 1～3 个字符，两者之间用一个圆点分隔。文件名是在文件建立时，由用户按规定自行定义的，但为了便于系统管理，每个操作系统都有一些约定的扩展名。例如，MS-DOS 约定的扩展名有：

（1）EXE 表示可执行的目标文件；

（2）COM 表示可执行的二进制代码文件；

（3）LIB 表示库程序文件；

（4）OBJ 表示目标文件；

（5）C 表示 C 语言源程序文件等等。

2．文件的属性和类型

文件具有类型、长度、物理位置和建立时间等属性。从不同的管理角度可将文件划分成不同的文件类型；文件长度单位可以是字节、字或块；文件的物理位置指示了存储该文件的设备和设备上的具体位置；文件的建立时间通常指最后一次修改文件的时间。

为了方便高效地管理文件，不同系统对文件的分类方式各有不同，常见有以下几种分类方式 。

1）按性质和用途分

（1）系统文件。有关操作系统及其他系统软件的信息所组成的文件，这类文件对用户不直接开放，只能通过操作系统调用为用户服务。

（2）库文件。由标准子程序和常用的应用软件组成的文件。这类文件允许用户调用，但不允许用户修改。

（3）用户文件。由用户委托给系统保存的文件。如源程序、目标程序、原始数据、计算结果等文件。

2）按保护级别分

（1）执行文件。用户可将文件当作程序执行，但不能阅读，也不能修改。

（2）只读文件。允许用户文件所有者或授权者读出或执行，但不准写入。

（3）读写文件。限定用户文件所有者或授权者读写，但禁止未核准的用户读写。

（4）不保护文件。所有用户均可存取。

7.1.2 文件系统

操作系统中负责管理和存储文件信息的软件机构称为文件管理系统，简称文件系统。文件系统由被管理文件、管理文件所需的数据结构（如目录表 、文件控制块、存储分配表）和相应的管理软件以及访问文件的一组操作所组成。文件系统模型如图 7-1 所示。

从系统角度来看，文件系统的功能是管理文件存储器空间，包括进行组织、分配和回收，负责文件存储并对存入的文件进行保护和检索。具体地说，它负责为用户管理文件，包括建立文件，当用户不再使用时撤销文件；对指定文件进行打开、关闭、读出、写入、截断、修改、转储、执行等操作；多用户系统中使一个用户能共享其他用户的文件；提供安全性和保密措施防止未被授权的文件访问，能恢复被破坏的文件。

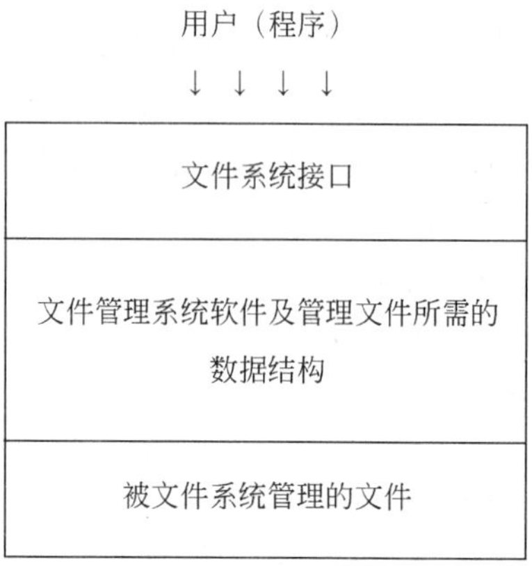

用户（程序）

↓ ↓ ↓ ↓

文件系统接口
文件管理系统软件及管理文件所需的数据结构
被文件系统管理的文件

图 7-1　文件系统模型

从用户角度看，文件系统的功能主要是实现了“按名存取”。即当用户要求保存一个已命名的文件时，文件系统能够根据一定的格式把该文件存放到存储器适当的地方；当用户要使用文件时，系统能够根据用户给出的文件名从文件存储器中找到相应的文件，或文件中某个记录。因此，文件系统的用户（包括操作系统本身及一般用户），只要知道文件名字就可存取文件中信息，而不必关心文件究竟存放在什么地方以及实现过程等细节。

7.2 文件的逻辑组织和物理存储

用户和文件系统往往从不同的角度来对待同一个文件。用户是从使用的角度来组织文件，用户把能观察到的且可以处理的信息根据使用要求按照一定形式构造成的文件，这种用户可见的文件外部形式称为文件的逻辑组织，也称逻辑结构。而文件系统要从文件的存储和检索等管理的角度来组织文件，文件系统根据存储设备的特性、文件的存取方式来决定以怎样的形式把文件存放到存储介质上，即内部的物理存储形式，称为文件的物理结构。实现文件系统的一个重要任务就是选择适当的逻辑组织形式和物理存储方式，在用户的逻辑文件和设备的物理文件之间建立映像关系，实现两者之间的相互转换。

7.2.1 文件的逻辑结构

文件的逻辑结构分成两种形式：无结构的流式文件和有结构的记录式文件。

1．流式文件

流式文件是由相关的一串字符流或字节流构成，字符数就是文件长度，字符是流式文件信息的基本单位。

对这种字符流式文件来说，查找文件中的基本信息单位，例如某个单词，是比较困难的。可以用插入特殊字符作为分界（例如回车换行符、文件结束符），通常按字符数或特殊字符来读取所需信息。但反过来，字符流的无结构文件管理简单，用户可以方便地对其进行操作。所以，那些对基本信息单位操作不多的文件较适于采用字符流的无结构方式。

2．记录式文件

记录式文件是由相关的一组连续有序的记录构成。把文件内的信息按逻辑上独立的含义划分信息单位，每个单位称为一个逻辑记录（简称记录）。记录通常都是描述一个实体集的，例如一个公司的职工档案中，每个职工的基本信息是包含工号、姓名、部门、出生日期、性别等若干数据项组成的一个逻辑记录，该公司的所有职工档案、即全部逻辑记录便组成了该公司的档案信息文件，记录式文件如表 7-1 所示。

表 7-1 记录式文件

	数据项 1	数据项 2	数据项 3	数据项 4	数据项 5	数据项 6
	序号	工号	姓名	部门	出生日期	性别
记录 1	1	11201	李丽	研发	1980.1.2	女
记录 2	2	11401	王洪	销售	1977.5.5	男
记录 3	3	11402	孟庆华	销售	1983.9.18	男
…	…	…	…	…	…	…
记录 *n*	*n*	10501	田馨	人事	1972.11.8	女

记录的长度可分为定长和不定长记录两类。定长记录由相同数目的数据项构成，所有记录的长度是相同的；不定长记录由不同数目或不定长度的数据项构成。定长记录和变长记录文件的逻辑结构如图 7-2 所示。

对于定长记录文件，如果要查找第 i 个记录， 可直接根据下式计算来获得第 i 个记录相对于第一个记录首址的地址：

$$A_i=i\times L$$

然而，对于可变长度记录的文件，要查找其第 i 个记录时，须首先计算出该记录的首地址。为此，须顺序地查找每个记录，从中获得相应记录的长度 L_i，然后才能按下式计算出第 i 个记录的首址。假定在每个记录前用一个字节指明该记录的长度，则

$$A_i=\sum_{i=0}^{i-1}L_i+i$$

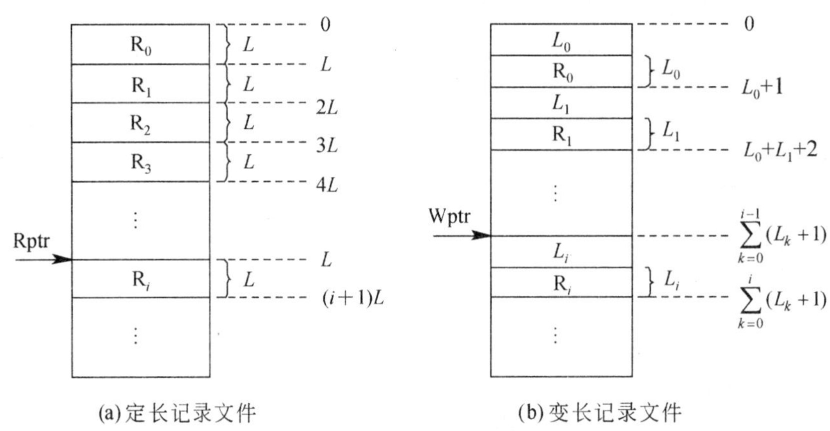

图 7-2　定长和变长记录文件

为便于对记录式文件中的逻辑记录进行存取、检索或更新等操作，可以将记录按不同的方式排列，构成不同的逻辑结构。用户根据需要选择逻辑结构，可组织成三种不同形式的逻辑文件。

（1）顺序文件。顺序文件是最常用的文件组织形式。在这类文件中，每个记录都使用一种固定的格式，所有记录都具有相同的长度，并且由相同数目、长度固定的数据项按特定的顺序组成。

（2）索引文件。由索引表和主文件两部分构成。索引表是一张指示逻辑记录和物理记录之间对应关系的表。索引表中为每一条记录建立索引项。索引项通常是按记录键顺序排列，索引表本身是定长记录文件。索引文件方式方便地实现了不定长记录文件的存取。

（3）索引顺序文件。主文件按主关键字有序的文件称为索引顺序文件。在索引顺序文件中，一组记录建立一个索引项。在索引表中为每组记录中的第一个记录建立一个索引项。

3．文件的存取访问方式

1）顺序存取

顺序存取是按照文件的逻辑地址顺序存取。在记录式文件中，即为按记录的排列顺序

来存取。顺序存取主要用于磁带文件，但也适用于磁盘上的顺序文件。

2）直接存取

存储介质上连续的存储区域划分为物理块（也称物理记录），块的长度通常是固定的，例如 512 字节。在直接存取（随机存取）文件中，记录的关键字和物理块之间通过某种方式建立对应关系，利用这种关系实现来实现文件记录存取。

很多应用场合要求以任意次序直接读写某个记录。例如，航空订票系统把特定航班的所有信息用航班号作标识，存放在某物理块中，用户预订某航班时，需要直接将该航班的信息取出。直接存取方法便适合于这类应用，它通常用于磁盘文件。

3）索引存取

索引存取（按键存取）是基于索引文件的存取方法，文件的存取是按照给定的记录键或记录名进行的。由于文件中的记录不按它在文件中的位置，而按它的记录键来编址，所以，用户提供给操作系统记录键后就可查找到所需记录。

实现按键存取包含对键的搜索和对记录的搜索两个步骤。对键的搜索是在用户给定所要搜索的键名和记录之后，确定该键名在文件中的位置；记录的搜索是在搜索到所要查找的键之后，在含有该键的所有记录中找出所需要的记录。通常记录按记录键的某种顺序存放，例如，按代表健的字母先后次序来排序。对于这种文件，除可采用按键存取外，也可以采用顺序存取或直接存取的方法。实际的系统中，大都采用多级索引，以加速记录查找过程。

7.2.2 文件的物理存储

文件在设备上的物理存储形式决定了对文件的存取方式，因而，对文件系统的性能有很大影响。文件系统往往根据存储设备类型、存取要求、记录使用频度和存储空间容量等因素提供多种不同的外存分配方式，基本的三种：连续分配、链接分配和索引分配。使用这三种分配方式形成的物理结构分别称为连续存储结构、链接存储结构和索引存储结构，对应地形成的物理文件类型为连续顺序文件、链接文件和索引文件。

1. 连续分配

连续分配就是将一个逻辑文件中的信息依次存储在一组物理地址相邻的块上，又称顺序存储分配。在这种存储结构中文件的逻辑记录顺序和物理记录顺序完全一致，为使系统找到文件，应该在文件目录项的“文件物理地址”项中记录文件的第一记录的块号和文件长度。例如，定长记录文件 Count1 长度为 2000B，存放在连续分块的磁带上，每物理块大小为 512B，则需要为其分配 4 个块，假设为其分配第一个块号为 10，Count 在磁带上的存放方式如图 7-3 所示。

磁带机文件、卡片机、打印机、纸带机介质上的文件都是顺序文件，存储在磁盘上的文件也可以是顺序文件。

连续文件的优点是简单、支持顺序存取和随机存取，所需的磁盘寻道次数和寻道时间最少。一旦知道了文件在文件存储设备上的起址和文件长度能很快地进行存取。这是因为从文件的逻辑块号到物理块号的变换可以很简单地完成。但是连续存储文件必须在建立文件时要预先确定文件长度，以便分配连续的存储空间，并且以后不能动态增长；在删除文件的某些部分又会留下无法使用的零头空间。因此，连续分配适用于很少更新的文件，如批处理文件、系统文件用得最多。不适合用来存放经常被修改的文件，如用户文件、数据库文件等。

目录		
文件名	始址	长度
Count	10	4
…	…	…
…	…	…

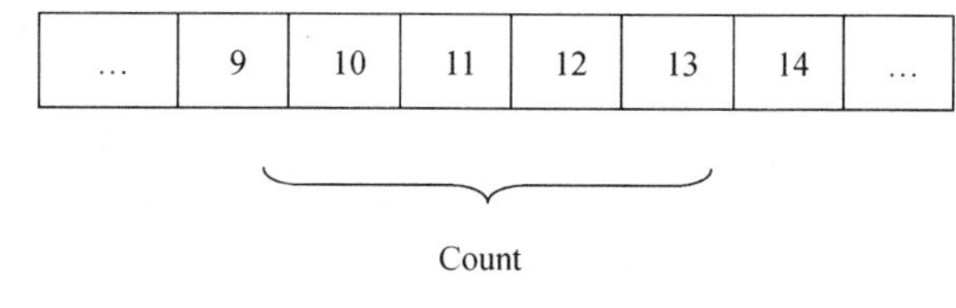

图 7-3　连续文件结构

2．链接分配

链接分配是采用非连续的物理块来存放文件信息。第一块文件信息的物理地址由文件目录给出，其中每个物理块设有一个指针指向其后续连接的另一个物理块，指针的内容为“0”时表示文件至本块结束，如图 7-4 所示。由此，链接分配使得存放同一文件的物理块链接成一个串联队列，这种文件称链接文件，又称串联文件。在系统软件中，常常使用链接存储结构，如输入井、输出井文件等。

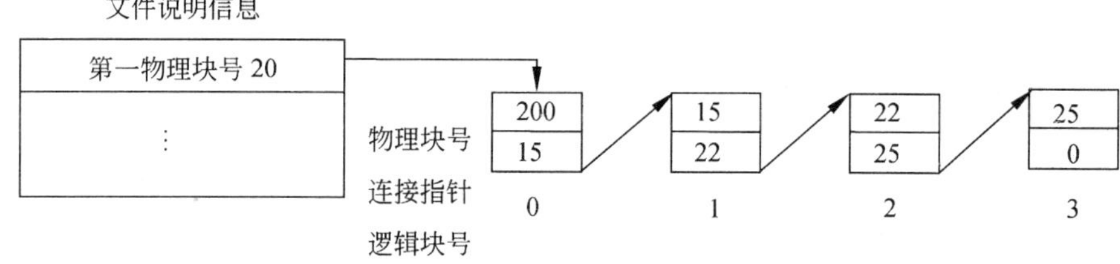

图 7-4　链接文件结构

链接分配采取离散分配方式，不必连续分配存储空间，消除了外部碎片，从而存储空间利用率高；不必预先确定文件的长度，文件长度可以动态地增长；易于对文件记录增、删、改。

链接分配的缺点是存放指针要占额外的存储空间；必须将指针和数据信息存放在一起，而破坏了物理块的完整性；由于存取需通过缓冲区，获得指针后，才能找到下一块的地址，因此仅适用于顺序存取。

3．索引分配

索引分配是实现非连续分配的另一种方法，适用于数据记录存放在随机存取存储设备上的文件。

在索引分配方式中，一个逻辑文件的信息存放在若干不连续物理块中，系统为每个文件建立一个索引块（表），块中记录了分配给该文件的所有物理块号并与文件的逻辑块号对应。通常，索引块的物理地址可由文件目录给出，

如图 7-5 给出了一个索引文件的示例，这种只有一个索引块的索引文件为一级索引。

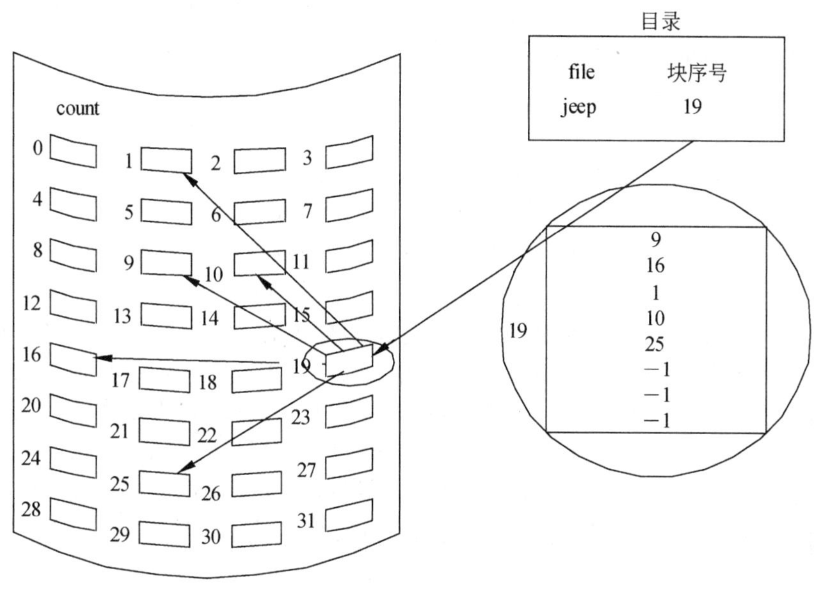

图 7-5　索引文件结构

索引文件在文件存储器上分两个区：索引区和数据区。进行文件存取时先搜索索引块找到相应记录的物理块号，然后根据物理块号获得相应数据记录。索引文件结构既可以满足文件动态增长的需要又可以方便地实现随机存取，但索引块需要占用额外的存储空间。

4．多级索引分配

有时太大，一个文件的索引块能够描述的物理块有限。一个很大的文件要占用许多物理块。可以设计成二级索引。二级索引机构如图 7-6 所示。

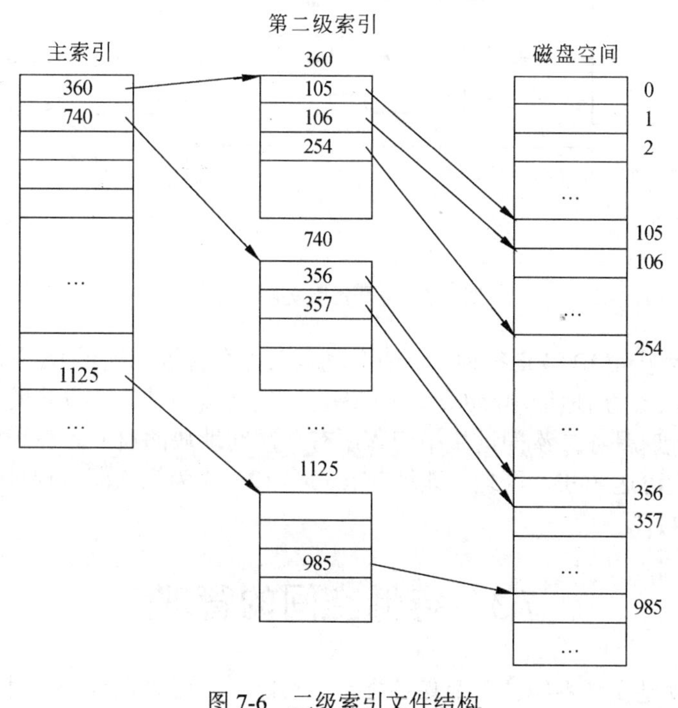

图 7-6　二级索引文件结构

二级索引的表项列出一级索引块的最大键值及该索引表区地址。查找时先查二级索引表找到一级索引表地址，再查一级索引表找出数据记录。同样，当文件相当大时，还可做三级索引等。

5. 混合索引结构

多级索引适合较大的文件的存储分配方式，但是对于较小的文件，又因为太多的索引块，导致存储效率不高。设计一种混合索引结构，当文件较小时可以直接存储，当文件较大时，可以采用多级索引的方式，实现大文件存取。混合索引结构如图 7-7 所示。

在混合索引结构中可设置 10 个直接地址项，即用 iaddr(0)～iaddr(9)来存放直接地址。换言之，在这里的每项中所存放的是该文件数据的盘块的盘块号。较小的文件可以直接存储访问。假如每个盘块的大小为 4KB，当文件不大于 40KB 时，便可直接从索引节点中读出该文件的全部盘块号。对于大、中型文件，只采用直接地址是不现实的。为此，可再利用索引节点中的地址项 iaddr(10)来提供一次间接地址。这种方式的实质就是一级索引分配方式。图 7-7 中的一次间址块也就是索引块，系统将分配给文件的多个盘块号记入其中。在一次间址块中可存放 1KB 个盘块号，因而允许文件长达 4MB。

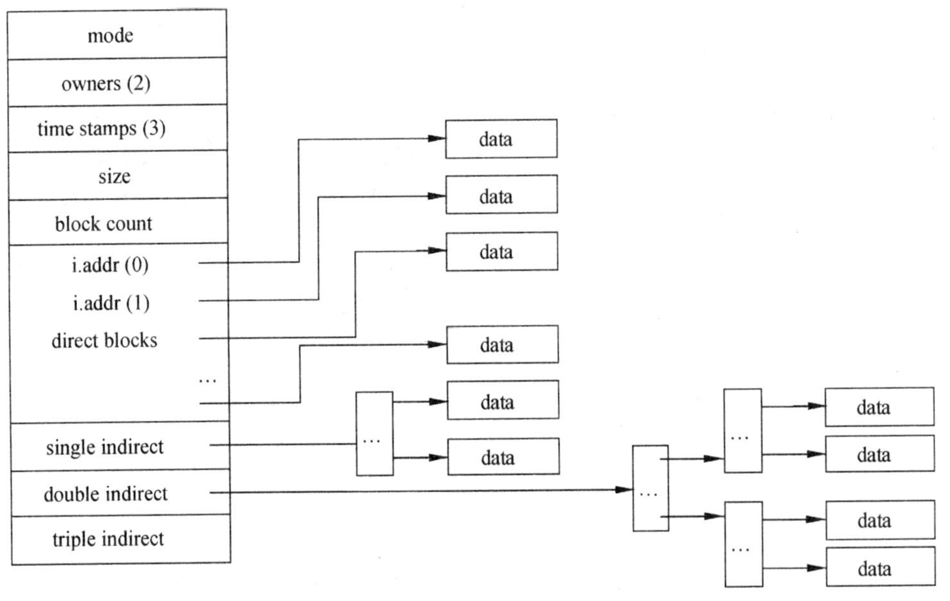

图 7-7 混合索引文件结构

当文件长度大于 4MB+40KB 时（一次间址与 10 个直接地址项），系统还须采用二次间址分配方式。这时，用地址项 iaddr（11）提供二次间接地址。该方式的实质是两级索引分配方式。系统此时是在二次间址块中记入所有一次间址块的盘号。在采用二次间址方式时，文件最大长度可达 4GB。同理，地址项 iaddr（12）作为三次间接地址，其所允许的文件最大长度可达 4TB。

7.3 存储空间的管理

存储空间管理是文件系统的重要任务之一。实际上，磁盘至少分成目录区和数据区，目录区存放一定数量的文件目录项；数据区存放文件的有效信息。磁盘存储空间的管理主要是指对数据区的管理。

用户作业在执行期间经常要求建立一个新文件或撤销一个不再需要的文件，因此，文件系统必须要为它们分配存储空间或回收它所占的存储空间。如何实现存储空间的分配和回收，取决于对空闲块的管理方法，有几种不同的空闲块管理方法：空闲表法、空闲块链、位示图法、成组链接法。

7.3.1 空闲表法

系统为外存上的所有空闲区建立一个空闲表，空闲表中每个表项对应由一个或多个空闲块构成的空闲区，它包括表项序号、第一个空闲块号、空闲块个数等，见表 7-2。

表 7-2 空闲表

序号	第一个空闲块号	空闲块个数
1	5	3
2	12	6
3	20	8
4	…	…

系统为文件分配空闲块与内存的动态分配类似，可采用首次适应算法、循环首次适应算法等。首先扫描空闲表项，直至找到第一个其大小能满足要求的空闲区，则将该盘区分配给申请者，并修改该表项。如果一个空闲区项不能满足申请者要求，则把目录中另一项分配给申请者（连续文件结构除外）。

系统在对用户所释放的存储空间进行回收时，也采取类似于内存回收的方法， 即要考虑回收区是否与空闲表中插入点的前区和后区相邻接，对相邻接者应予以合并。当一个文件被删除释放存储物理块时，系统则把被释放空间的第一空闲块号、空闲块个数置入空闲表的新表项中。

空闲表法在系空闲区较少时有较好的效果，当系统中有大量较小的空闲区导致空闲表变得很大时，效率大为降低。空闲表法适用于连续文件结构的文件存储区的分配与回收。

7.3.2 空闲块链表法

空闲块链表法是把所有的空闲块用指针链接起来，每个空闲块中都设置一个指向另一块的指针，形成了空闲块链。系统设置一个链首指针，指向链中第一个空闲块，最后一个空闲块中指针为“0”。

为文件分配空间时，根据链首指针把链头的一块或几块分配给申请者，并修改链首指针。当文件删除需要回收存储空间时，把释放的空闲块逐个插入链尾上。这种方法的优点是分配和回收过程简单，但是分配和回收通常需要重复操作多次，效率较低。

将空闲块链表法中的空闲块改为将空闲区链接，便形成空闲区链表法。

7.3.3 位示图法

磁盘被分块后，可用一张位示图（简称位图）来指示磁盘存储空间的使用情况。系统首先从内存中划出若干个字节，为每个文件存储设备建立一张位示图。在位示图中，每个文件存储设备的物理块都对应一个比特位。如果该位为“0”，表示所对应的块是空闲块；反之，如果该位为“1”则表示所对应的块已被分配出去。

利用位示图来进行空闲块分配时，只需查找图中的“0”位，并将其置为“1”位；反之，利用位示图回收空闲块时只需把相应的比特位由“1”改为“0”即可。

从位示图中很容易找到一个或一组相邻接的空闲盘块。位示图很小，占用空间少，可将它保存在内存中，进而使在每次进行盘区分配时，无须首先把盘区分配表读入内存，从而节省了许多磁盘的启动操作，位示图常用于微型机和小型机系统中。

例：某个磁盘组共有 100 个柱面,每个柱面上有 8 个磁道，每个盘面分成 4 个扇区。那么,整个磁盘空间共有 4×8×100=3200 个存储块。如果用字长 32 位的字来构造位示图，共需 100 个字，如表 7-3 所示。

表 7-3　构造的位示图

	0 位	1 位	2 位	…	30 位	31 位
第 0 字	0/1	0/1	0/1	…	0/1	0/1
第 1 字	0/1	0/1	0/1	…	0/1	0/1
…	…	…	…	…	…	…
第 99 字	0/1	0/1	0/1	…	0/1	0/1

若磁盘存储空间按柱面编号，则第一个柱面上的盘块号应该为 0～31，第二个柱面上的存储块号为 31～63，…，依次计算，位示图中第 i 个字的第 j 位（i=0，1，2，…，99；j=0，1，2，…，31)对应的块号为：

$$块号 = i\times32+j$$

当有文件要存放到磁盘上时，根据需要的块数查位图中为“0”的位，表示对应的存储块空闲可供使用。一方面在位图中查到的位上置占用标志“1”，另一方面根据查到的位计算出对应的块号，然后确定这些可用的存储块在哪个柱面，哪个磁头，哪个扇区：

柱面号 = [块号/32]

磁头号 = [(块号 mod 32)/4]

扇区号 = [块号 mod 32]mod 4

文件信息就可按确定的地址（柱面号，磁头号，扇区号）存放到磁盘上。

当要删去一个文件，归还磁盘空间时，可根据归还块的物理地址计算出相应的块号，由块号推算出它在位图中的对应位，把这一位的占用标志“ 1 ”清成“ 0 ”，表示该块已成为空闲块。仍以表 7-3 为例，根据归还块所在的柱面号、磁头号、扇区号，计算对应位图中的字号和位号：

块号 = [柱面号×32+磁头号×4+扇区号]

字号 = [块号/32]

位号 = [块号 mod 32]

7.3.4　成组链接法

空闲表法和空闲块链表法都不适用于大型文件系统，因为会使空闲表或空闲链表太长。将上述两种方法相结合而形成的一种空闲盘块管理方法，即为成组链接法，它兼容了上述两种方法的优点，且克服了上述两种方法的缺点。UNIX 操作系统采用的就是成组链接法。

1．空闲盘块的组织

（1）空闲盘块号栈用来存放当前可用的一组空闲盘块的盘块号（最多含 100 个号），以及栈中尚有的空闲盘块号数 *N*。*N* 还兼做栈顶指针用。

例如，当 *N*=100 时，它指向 S.free(99)。由于栈是临界资源，每次只允许一个进程去访问，故系统为栈设置了一把锁。如图 7-8 所示。其中，S.free(0)是栈底，栈满时的栈顶为 S.free(99)。

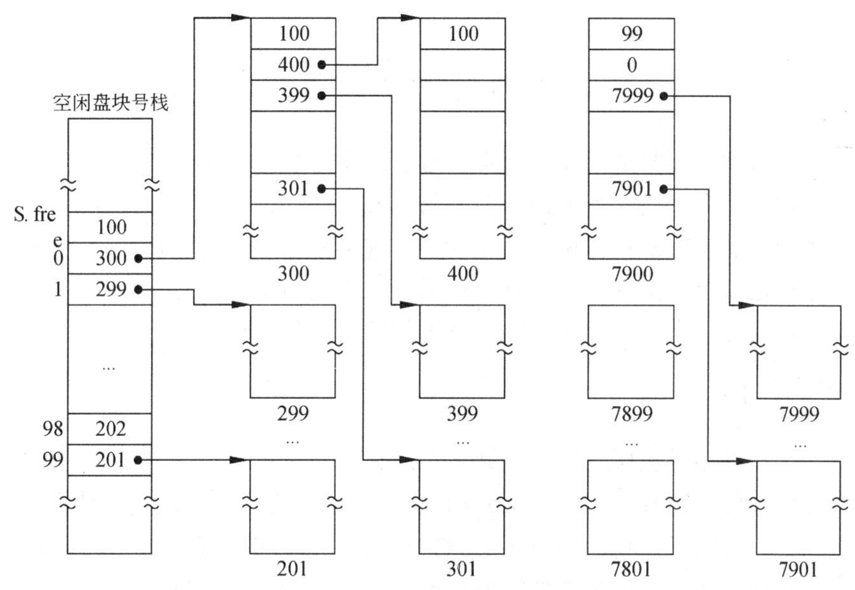

图 7-8　空闲盘块的成组链接法

（2）文件区中的所有空闲盘块，被分成若干个组，例如，将每 100 个盘块作为一组。假定盘上共有 10000 个盘块，每块大小为 1KB，其中第 201～7999 号盘块用于存放文件，即作为文件区，这样，该区的最后一组盘块号应为 7901～7999；次末组为 7801～7900，…，倒数第二级的盘块号为 301～400，第一组为 201～300。

（3）将每一组含有的盘块总数 *N* 和该组所有的盘块号，记入其前一组的第一个盘块的 S.free(0)～S.free(99)中。这样，由各组的第一个盘块可链成一条链。

（4）将第一组的盘块总数的所有盘块号，记入空闲盘块号栈中，作为当前可供分配的空闲盘块号。

（5）最末一组只有 99 个盘块，其盘块号记入其前一组的 S.free(1)～S.free(99)中，而在 S.free(0)中则存放“0”，作为空闲盘块链的结束标志。

2．空闲盘块的分配和回收

1）空闲盘块的分配

首先检查空闲盘块号栈是否上锁，如未上锁，便从栈顶取出一空闲盘块号，将之与对应的盘块分配给用户，然后将栈顶指针下移一格。若该该盘块号已经是栈底，即 S.free(0)，这是当前栈中最后一个可分配的盘块号。由于在该盘块号所对应盘块中记录有下一组可用的盘块号，因此，须调用磁盘读过程，将栈底盘块号所对应盘块的内容读入栈中，作为新的盘块号栈的内容，并把原栈底对应的盘块分配出去（其中的有用数据已经读入栈中）。然后再分配一个相应的缓冲区（作为该盘块的缓冲区）。最后把栈中的空闲盘块数减 1 并返回。

2）磁盘块的回收

在系统回收空闲盘块时，须调用盘块回收过程进行回收。它是将回收盘块的盘块号记入空闲盘块号栈的顶部，并执行空闲盘块数加 1 操作。当栈中空闲盘块号数目已经达到 100 时，表示栈已经满，便将现有栈中的 100 个盘块号，记入新回收的盘块中，再将其盘块号作为新栈底。

7.4 文 件 目 录

系统对文件的管理主要是通过文件目录实现的。文件目录将每个文件的符号名和它们在外存空间的物理地址和有关文件的情况说明信息联系起来，用户只需向系统提供一个文件的符号名，系统就能准确地找出所要的文件。实现符号名和具有物理地址之间的转换，其主要环节就是检索目录。文件系统的基本功能之一就是负责文件目录的建立、维护和检索，要求编排的目录便于查找、防止冲突，目录的检索方便迅速。

7.4.1 文件控制块

为了能对一个文件进行正确的存取，必须为之设置用于描述和控制文件的数据结构，这个数据结构称为文件控制块（File Control Block，FCB）。文件控制块是文件存在的标志，当建立一个文件时，系统就为其建立相应的 FCB。文件管理程序可借助于文件控制块中的信息对文件实施各种操作。人们把文件控制块的有序集合称为文件目录，即一个文件控制块就是一个文件目录项。通常，一个文件目录也被看作一个文件，称为目录文件。

文件控制块中一般包含如下内容。

（1）文件的存取控制信息。例如用户名、文件名、文件类型、属性。

（2）文件的结构信息。例如逻辑结构、物理结构、物理位置、记录数等。

（3）文件的使用信息。如建立日期和时间、修改的日期和时间、当前使用信息等。

7.4.2 索引节点

文件目录通常是存放在磁盘上的，当文件很多时，文件目录可能要占用大量的盘块。在查找目录的过程中，先将存放目录文件的第一个盘块中的目录调入内存，然后把用户所给定的文件名与目录项中的文件名逐一比较。若未找到指定文件，便再将下一个盘块中的目录项调入内存。

在检索目录文件的过程中，只用到了文件名，仅当找到一个目录项（即其中的文件名与指定要查找的文件名相匹配）时，才需从该目录项中读出该文件的物理地址。而其他一些对该文件进行描述的信息，在检索目录时一概不用，显然，这些信息在检索目录时，不需调入内存。为此在有的系统中，如 UNIX 系统便采用了把文件名与文件描述信息分开的办法，使文件描述信息单独形成一个称为索引节点的数据结构，简称为 i 节点。在文件目录中的每个目录项，仅由文件名和指向该文件所对应的 i 节点的指针所构成。

在 UNIX 系统中一个目录仅占 16 个字节，其中 14 个字节是文件名，2 个字节为 i 节点指针。在 1KB 的盘块中可作 64 个目录项，这样，为找到一个文件，可使平均启动磁盘次数减少到原来的 1/4，大大节省了系统开销。

7.4.3 目录结构

目录结构的组织，关系到文件的存取速度，也关系到文件的共享性和安全性。目前常用的目录结构有单级目录、二级目录和多级目录。

1．单级文件目录

为了实现“按名存取”的功能，在整个文件系统中只建立一张目录表，用以标识和描述用户和系统进程可以存取的全部文件 。其中，每个文件占一个目录项，主要由文件名、文件长度、文件类型、文件物理地址以及其他文件属性、状态位等组成。这样的表称为单级文件目录。状态位标识每个目录项是否空闲。单级目录表结构如表 7-4 所示。

表 7-4　单击目录表结构

文件名	物理地址	文件说明	状态位
文件名 1			
文件名 2			
…			

该目录表存放在存储设备的某固定区域，在系统初启时或需要时，将其调入内存（或部分调入内存）。文件系统通过对该表提供的信息对文件进行创建、搜索、删除等操作。

单级文件目录的优点是比较简单，但是对于稍具规模的文件系统，会拥有数目可观的目录项，致使为找到一个指定的目录项要花费较多的时间。单级目录不允许两个文件有相同的名字。对于这一点，只有在单用户情况下才能做得到。在多道程序设计系统中，特别是在有许多用户的分时系统中，“重名”问题是难以避免的。一个灵活的文件系统应该允许文件重名而又能正确地区分它们。另外，还应允许“别名”的存在，即允许用户用不同的文件名访问同一个文件。这种情况在多用户共享文件时可能发生，因为每个用户都喜欢以自己习惯的助记符来调用一个共享文件。

为了解决命名冲突及提供更灵活的命名能力，文件系统采用简单的单级文件目录结构是不行的。为此，必须采用二级文件目录、多级文件目录。

2．二级文件目录

在二级文件目录结构中，将文件目录分成两级，第一级为主文件目录（Master File Directory，MFD），它用于管理所有用户文件目录，它的目录项登记了系统接受的用户的名字及该用户文件目录表的地址；第二级为用户目录（User File Directory，UFD），它为该用户的每个文件保存一登记栏,其内容和一级文件目录的目录项相同。图 7-9 是二级文件目录结构示意。

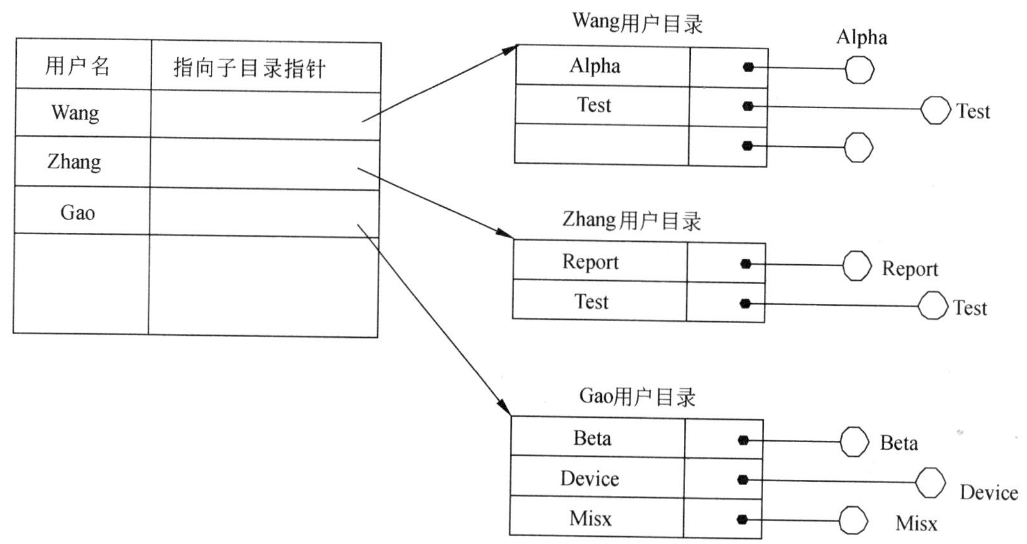

图 7-9　二级文件目录结构

每一用户只允许查看自己的文件目录。当一个新用户作业进入系统时，系统为其在主文件目录中开辟一栏，登记其用户名，并准备一个存放这个用户文件目录的区域，这个区域的首地址填入主文件目录中该用户名所在项。当用户需要访问某个文件时，系统根据用户名从主文件目录中找出该用户的文件目录表的物理位置，其余操作与单级文件目录类似。

使用二级目录可以解决文件重名和文件共享问题，并可获得较高的搜索速度。采用二级文件目录管理文件时，因为任何文件的存取都要通过主文件目录，于是可以检查访问者的存取权限，避免一个用户未经授权就存取另一个用户的文件，使用户文件的私有性得到保证，实现了文件的保护和保密。特别是不同用户具有同名文件时，由于各自有自己的文件目录表而不会导致混乱。对于文件的共享，原则上只要把对应目录项指向同一个物理位置的文件即可。

该结构虽然能有效地将多个用户隔开，在各用户之间完全无关时，这种隔离是一个优点，但当多个用户之间要相互合作去完成一个大任务，且一个用户又需去访问其他用户的文件时，这种隔离便成为一个缺点。

3．多级树状文件目录

很多系统都利用二级目录结构。二级目录虽比较简单实用，但却缺乏灵活性，特别是不容易反映真实世界复杂的文件结构形式。为了使用灵活和管理方便，对二级目录进行扩充成为多级目录。

在多级树状文件目录系统中（除最末一级外），任何一级目录的登记项可以对应一个目录文件，也可以对应一个非目录文件，形成了层次结构。这样，就形成一棵倒向的有根树，树根是根目录；从根向下，每一个树枝是一个子目录；而树叶是文件。树状多级目录有许多优点：可以很好反映现实世界复杂层次结构的数据集合；可以重名，只要这些文件不在同一个子目录中；易于实现子树中文件保护、保密和共享。UNIX 系统和 MS-DOS 系统都采用了树状目录结构。树状目录结构如图 7-10 所示。

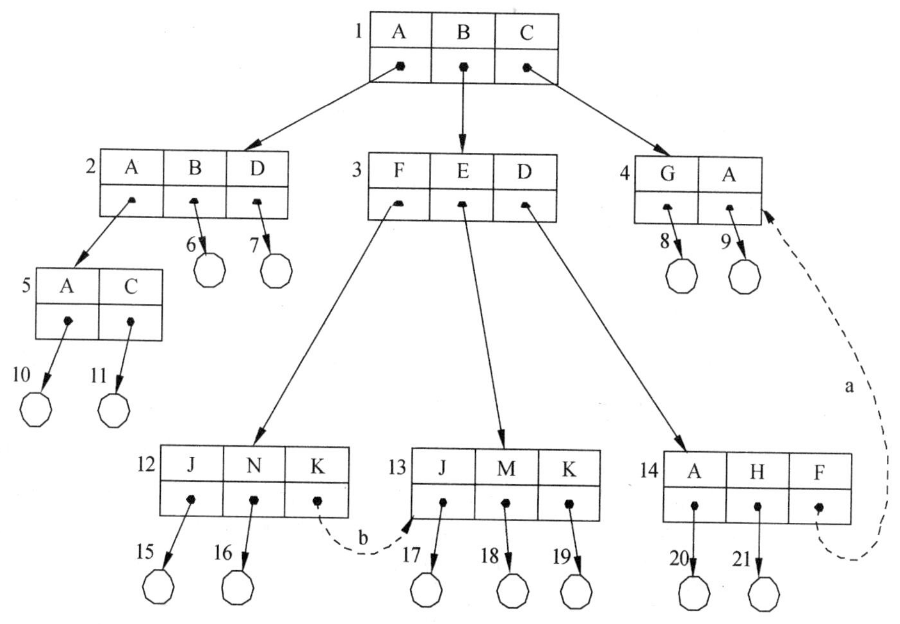

图 7-10　多级树状目录结构

在该树状目录结构中，主（根）目录中有三个用户的总目录项 A、B 和 C。在 B 项所指出的 B 用户的总目录 B 中，又包括三个子目录 F、E 和 D，其中每个子目录又包含多个文件。如 B 目录中的 F 子目录中，包含 J 和 N 两个文件。为了提高文件系统的灵活性，应允许在一个目录文件中的目录项既是作为目录文件的 FCB，又是数据文件的 FCB，这一信息可用目录项中的一位来指示。例如用户 A 的总目录中，目录项 A 是目录文件的 FCB，而目录项 B 和 D 则是数据文件的 FCB。

多级目录文件中，一个文件的路径名是由根目录到该文件的通路上所有目录文件名和该文件名组成的。当用户进程使用路径名访问该件时，文件系统就根据这个路径名的顺序来查访各级目录，从而确定该文件的位置。UNIX 的文件系统就采用这种方式。

树型目录结构中，文件在系统中的搜索路径是从根开始到文件名为止的各文件名组成，因此解决了重名问题；由于对多级目录的查找每次只查找目录的一个子集，因此其搜索速度较单级、二级目录时更快；层次清楚便于管理；不同层次、不同用户的文件可以被赋予不同的存取权限，有利于文件的保护。树状目录结构的缺点是查找一个文件需按路径名逐层检查，由于每个文件都放在外存，多次访盘影响速度。

7.5　文件的共享与保护

文件的共享和文件的安全性是文件系统中的一个重要问题，为了系统的可靠性和数据的安全性，文件的共享必须是有限制的，也就是说，要有对文件的保护和保密措施。文件共享是指不同用户共同使用某些文件；文件保护是指防止文件主或其他用户破坏文件；文件保密则是指文件不能被未经文件主授权的任何用户访问。

7.5.1 文件的共享

在现代计算机操作系统中，必须提供文件共享手段，即指系统应允许多个用户（进程）共享同一个文件。这样，在系统中只需保留该共享文件的一份副本。如果系统不提供文件共享功能，就意味着凡是需要该文件的用户，都须各自备有此文件的副本，会造成对存储空间的极大浪费。此外，相互协同的进程通过文件共享可以相互交换信息，以达到相互通信的目的。

文件共享有下列两种模式。

（1）多个进程不同时使用同一文件。任意时刻最多只有一个进程访问，是比较简单的情形，操作系统只需核对使用者对文件的可访问性。

（2）多个进程同时使用同一文件。此时又存在两种情况：①所有进程都不修改共享文件，同时进行只读操作，系统允许读；②某些进程要求修改共享文件，如果有进程要进行写操作，同时另一进程要进行读或写操作，则操作系统要进行同步控制，以保证数据完整性。对此有两种处理方法，其一是不允许读者与写者，或者写者与写者同时打开同一文件，即实现互斥，这种方式不仅降低系统的并发性，也增加死锁的可能性；其二是允许读者与写者，或者写者与写者同时打开同一文件，但操作系统需要为用户提供相应的互斥手段，文件使用者借助这种互斥手段保证文件共享时不发生冲突。

从系统管理的观点看，有三种方法可以实现文件共享。

1）绕道法

绕道法要求每个用户处在当前目录下工作，用户对所有文件的访问都是相对于当前目录进行的。用户文件的固有名（为了访问某个文件而必须访问的各个目录和文件的目录名与文件名的顺序连接称为固有名）是由当前目录到信息文件通路上所有各级目录的目录名加上该信息文件的符号名组成。使用绕道法进行文件共享时，用户从当前目录出发向上返回到与所要共享文件所在路径的交叉点，再顺序下访到共享文件。绕道法需要用户指定所要共享文件的逻辑位置或到达被共享文件的路径。

2）链接法

通过链接一个文件拥有多个名字，不同用户使用不同名称访问同一文件从而实现共享。

3）基本文件目录表

该方法把所有文件目录的内容分成两部分，一部分包括文件的结构信息、物理块号、存取控制和管理信息等，并由系统赋予唯一的内部标识符来标识；另一部分则由用户给出的符号名和系统赋给文件说明信息的内部标识符组成。这两部分分别称为符号文件目录表（SFD）和基本文件目录表（BFD）。SFD 中存放文件名和文件内部标识符，BFD 中存放除了文件名之外的文件说明信息和文件的内部标识符。只要在不同文件符号中使用相同的文件内部标识符，就可实现文件共享。

7.5.2 文件的保护

文件保护则指文件本身需要防止文件的拥有者本人或其他用户破坏文件内容。

造成文件可能被破坏的原因有时是硬件故障或软件失误引起的；其次，文件信息可能被窃取、破坏，或他人对文件进行不正确的访问。前者是系统可靠性问题，后者则是文件保护的问题。所以，操作系统中文件系统的设计必须考虑这些问题。

为了能在软、硬件失效的意外情况下恢复文件，保证数据的连续可利用性，文件系统应当提供适当的机构，以便复制备份。即系统必须保存所有文件的双份拷贝，以便在任何不幸的偶然事件后，能够重新恢复所有的文件。

文件系统可以采用建立副本和定时转储的办法来保护文件。

1．建立副本

建立副本是指同一文件保存在多个存储介质上，当某个存储介质上的文件被破坏时，可以用其他存储介质上的备用副本来替换或恢复。这种方法简单，但开销大，仅适用于特别重要且容量较小的文件，例如 MS -DOS 的目录文件就是双备份文件。

2．定时转储

定时转储是即定时地把文件转储到其他存储介质上。当文件发生故障时，就用转储的文件来复原，把有故障的文件恢复到某一时刻的状态，仅丢失从上次转储以来新修改或增加的信息。定时转储有两种方式：全量转储和增量转储。

全量转储：把文件存储器中全部文件定期（如每天一次）复制到磁带上（可装卸），视磁带为硬盘的备份。卸下的磁带可保存到可靠的地方，当系统文件故障时，用备份的磁带来恢复。如银行系统账本文件就可用这方法，虽麻烦费时，但相当可靠。

增量转储：对要求快速复原和恢复到故障当时状态的系统，定期将整个文件系统转储是不够的。一种更为适用的技术称作增量转储，即每隔一段更短的时间（例如每隔 2 小时）便把上次转储以来改变过的文件和新文件用磁带转储，关键性的文件也可再次转储。这能克服全量转储的缺点,但磁带上的转储信息不紧凑。为此，可采用定期（例如一周一次）归档方法将属于同一文件主的文件搜集到一条磁带上。

在实际工作中，两种转储方式配合使用。一旦系统发生故障，文件系统的恢复过程大致如下：

（1）从最近一次全量转储盘中装入全部系统文件，使系统得以重新启动，并在其控制下进行后续恢复操作；

（2）从近到远从增量转储盘上恢复文件。可能同一文件曾被转储过若干次，但只要恢复最后一次转储的副本，其他则被略去；

（3）从最近一次全量转储盘中，恢复没有恢复过的文件。

7.5.3 文件的保密

文件保密指未经文件拥有者许可，任何用户不得访问该文件。欲防止系统中的文件被他人窃取、破坏，就必须对文件采取有效的保密措施。

1．口令

分文件口令和终端口令两种。用户在建立文件时即提供一个文件口令，系统为其建立文件目录时附上相应口令（并隐蔽起来），同时告诉允许共享该文件的其他用户。用户请求访问文件时，必须提供相应的口令，仅当口令正确才能打开文件进行访问。另一种是终端口令，由系统分配或用户预先设定一个口令，仅当回答的口令相符时才能使用终端。此法简单易用，但口令直接存在系统内，不诚实的系统程序员可能得到全部口令。另外，对文件的存取权限不能控制，得知文件口令的用户均具有与文件主相同的存取权限。因此须和其他方法配合使用，即系统用口令识别访问文件的用户，而用其他方法实现对文件存取权限的控制。

2．密码

“密码”能有效地防止文件被人窃取而导致涉密。在文件写入时进行保密编码，读出时进行译码。这一编译码工作可由系统替用户完成，但用户请求读/写文件时须提供密钥，以供系统进行加密和解密。由于密钥不直接存入系统，由用户请求读/写文件时动态提供，故可防止不诚实的系统程序员窃取或破坏他人的文件。

密码技术发展迅速，保密性很强，实现时也很节省存储空间，但需花费较大的编译码时间。目前许多系统上的系统文件均采用密码方法实施保护的。

3．隐蔽文件目录

用户将需要保密的文件的文件目录隐蔽起来，在显示时因其他用户不知道文件名而无法使用。MS-DOS 操作系统就可使用这种保密方法。其方法是将该文件的文件属性改为“隐含”方式，于是在列目录时不会将该文件名列在屏幕上。

4．访问控制

“口令”和“密码”都是防止用户文件被他人冒充存取或被窃取，从外部实施文件保密的方法。这就需要通过检查用户拥有的访问权限与本次存取要求是否一致，来防止未授权用户访问文件和被授权用户的越权访问。

Chapter 7 File Management

For most users, the file system is the most visible aspect of an operating system. It provides the mechanism for on-line storage of and access to both data and programs of the operating system and all the users of the computer system. The file system consists of two distinct parts: a collection of files, each storing related data, and a directory structure, which organizes and provides information about all the files in the system.

7.1 File and File System

7.1.1 File Concept

Computers can store information on various storage media, such as magnetic disks, magnetic tapes, and optical disks. So that the computer system will be convenient to use, the operating system provides a uniform logical view of information storage. The operating system abstracts from the physical properties of its storage devices to define a logical storage unit, the file. Files are mapped by the operating system onto physical devices. These storage devices are usually nonvolatile, so the contents are persistent through power failures and system reboots.

A file is a named collection of related information that is recorded on secondary storage. From a user's perspective, a file is the smallest allotment of logical secondary storage; that is, data cannot be written to secondary storage unless they are within a file. Commonly files represent programs (both source and object forms) and data. Data files may be numeric, alphabetic, alphanumeric, or binary. Files may be free form, such as text files, or may be formatted rigidly. In general, a file is a sequence of bits, bytes, lines, or records, the meaning of which is defined by the file's creator and user. The concept of a file is thus extremely general.

The information in a file is defined by its creator. Many different types of information may be stored in a file source programs, object programs, executable programs, numeric data, text, payroll records, graphic images, sound recordings, and so on. A file has a certain defined structure, which depends on its type. A text file is a sequence of characters organized into lines (and possibly pages). A source file is a sequence of subroutines and functions, each of which is further organized as declarations followed by executable statements. An object file is a sequence of bytes organized into blocks understandable by the system's linker. An executable file is a series of code sections that the loader can bring into memory and execute.

1. File Name

Part of the file name. The name is split into two parts—a name and an extension, usually separated by a period character. In this way, the user and the operating system can tell from the name alone what the type of a file is. For example, most operating systems allow users to specify file names as a sequence of characters followed by a period and terminated by an extension of additional characters. File name examples include resume.doc, Server.java, and ReaderThread.c. The system uses the extension to indicate the type of the file and the type of operations that can be done on that file. Only a file with a .com, .exe, or .bat extension can be executed, for instance. The .com and .exe files are two forms of binary executable files, whereas a .bat file is a batch file containing, in ASCII format, commands to the operating system. MS-DOS recognizes only a few extensions, but application programs also use extensions to indicate file types in which they are interested. For example, assemblers expect source files to have an .asm extension, and the Microsoft Word processor expects its files to end with a .doc extension. These extensions are not required, so a user may specify a file without the extension (to save typing), and the application will look for a file with the given name and the extension it expects. Because these extensions are not supported by the operating system, they can be considered as "hints" to the applications that operate on them.

2. File Attribute

A file is named, for the convenience of its human users, and is referred to by its name. A name is usually a string of characters, such as example.c. Some systems differentiate between uppercase and lowercase characters in names, whereas other systems do not. When a file is named, it becomes independent of the process, the user, and even the system that created it. For instance, one user might create the file example.c, and another user might edit that file by specifying its name. The file's owner might write the file to a floppy disk, send it in an e-mail, or copy it across a network, and it could still be called example.c on the destination system.

A file's attributes vary from one operating system to another but typically consist of these following.

(1) Name. The symbolic file name is the only information kept in human-readable form,

(2) Identifier. This unique tag, usually a number, identifies the file within the file system; it is the non-human-readable name for the file.

(3) Type. This information is needed for systems that support different types of files.

(4) Location. This information is a pointer to a device and to the location of the file on that device.

(5) Size. The current size of the file (in bytes, words, or blocks) and possibly the maximum allowed size are included in this attribute.

(6) Protection. Access-control information determines who can do reading, writing, executing, and so on.

(7) Time, date, and user identification. This information may be kept for creation, last modification, and last use. These data can be useful for protection, security, and usage monitoring.

The information about all files is kept in the directory structure, which also resides on secondary storage. Typically, a directory entry consists of the file's name and its unique identifier. The identifier in turn locates the other file attributes. It may take more than a kilobyte to record this information for each file. In a system with many files, the size of the directory itself may be megabytes. Because directories, like files, must be nonvolatile, they must be stored on the device and brought into memory piecemeal, as needed.

7.1.2 File-System

Disks provide the bulk of secondary storage on which a file system is maintained. They have two characteristics that make them a convenient medium for storing multiple files.

(1) A disk can be rewritten in place; it is possible to read a block from the disk, modify the block, and write it back into the same place.

(2) A disk can access directly any given block of information it contains. Thus, it is simple to access any file either sequentially or randomly, and switching from one file to another requires only moving the read—write heads and waiting for the disk to rotate.

To provide efficient and convenient access to the disk, the operating system imposes one or more file systems to allow the data to be stored, located, and retrieved easily. A file system poses two quite different design problems. The first problem is defining how the file system should look to the user. This task involves defining a file and its attributes, the operations allowed on a file, and the directory structure for organizing files. The second problem is creating algorithms and data structures to map the logical file system onto the physical secondary-storage devices.

The file system itself is generally composed of many different levels. The lowest level, the I/O control, consists of device drivers and interrupt handlers to transfer information between the main memory and the disk system. A device driver can be thought of as a translator. Its input consists of high-level commands such as "retrieve block 123." Its output consists of low-level, hardware-specific instructions that are used by the hardware controller, which interfaces the I/O device to the rest of the system. The device driver usually writes specific bit patterns to special locations in the I/O controller's memory to tell the controller which device location to act on and what actions to take.

The basic file system needs only to issue generic commands to the appropriate device driver to read and write physical blocks on the disk.

The file-organization module knows about files and their logical blocks as well as physical blocks. By knowing the type of file allocation used and the location of the file, the file-organization module can translate logical block addresses to physical block addresses for the basic file system to transfer. Each file's logical blocks are numbered from 0 (or 1) through N. Since the physical blocks containing the data usually do not match the logical numbers, a translation is needed to locate each block. The file-organization module also includes the free-space manager, which tracks unallocated blocks and provides these blocks to the file-organization module when requested.

Finally, the logical file system manages metadata information. Metadata includes all of the file-system structure except the actual data (or contents of the files). The logical file system manages the directory structure to provide the file-organization module with the information the latter needs, given a symbolic file name. It maintains file structure via file-control blocks. A file-control block (FCB) contains information about the file, including ownership, permissions, and location of the file contents.

7.2 File Structure and Storage

7.2.1 File Structure

1. Internal Structure

File types also can be used to indicate the internal structure of the file. Further, certain files must conform to a required structure that is understood by the operating system. For example, the operating system requires that an executable file have a specific structure so that it can determine where in memory to load the file and what the location of the first instruction is. Some operating systems extend this idea into a set of system-supported file structures, with sets of special operations for manipulating files with those structures.

This point brings us to one of the disadvantages of having the operating system support multiple file structures: The resulting size of the operating system is cumbersome. If the operating system defines five different file structures, it needs to contain the code to support these file structures. In addition, every file may need to be definable as one of the file types supported by the operating system. When new applications require information structured in ways not supported by the operating system, severe problems may result.

For example, assume that a system supports two types of files: text files (composed of ASCII characters separated by a carriage return and line feed) and executable binary files. Now, if we (as users) want to define an encrypted file to protect the contents from being read by unauthorized people, we may find neither file type to be appropriate. The encrypted file is not ASCII text lines but rather is (apparently) random bits. Although it may appear to be a binary file, it is not executable. As a result, we may have to circumvent or misuse the operating system's file-types mechanism or abandon our encryption scheme.

Some operating systems impose (and support) a minimal number of file structures. This approach has been adopted in UNIX, MS-DOS, and others. UNIX considers each file to be a sequence of 8-bit bytes; no interpretation of these bits is made by the operating system. This scheme provides maximum flexibility but little support. Each application program must include its own code to interpret an input file as to the appropriate structure. However, all operating systems must support at least one structure—that of an executable file so that the system is able to load and run programs.

The Macintosh operating system also supports a minimal number of file structures. It expects files to contain two parts: a resource fork and a data fork. The resource fork contains information of interest to the user. For instance, it holds the labels of any buttons displayed by the program. A foreign user may want to re-label these buttons in his own language, and the Macintosh operating system provides tools to allow modification of the data in the resource fork. The data fork contains program code or data—the traditional file contents. To accomplish the same task on a UNIX or MS-DOS system, the programmer would need to change and recompile the source code, unless she created her own user-changeable data file. Clearly, it is useful for an operating system to support structures that will be used frequently and that will save the programmer substantial effort. Too few structures make programming inconvenient, whereas too many cause operating-system bloat and programmer contusion.

2. Access Methods

1) Sequential Access

The simplest access method is sequential access. Information in the file is processed in order, one record after the other. This mode of access is by far the most common; for example, editors and compilers usually access files in this fashion.

Reads and writes make up the bulk of the operations on a file. A read operation—reed next—reads the next portion of the file and automatically advances a file pointer, which tracks the I/O location. Similarly, the write operation—write next—appends to the end of the file and advances to the end of the newly written material (the new end of file). Such a file can he reset to the beginning; and on some systems, a program may be able to skip forward or backward n records for some integer n —perhaps only for $n = 1$.

2) Direct Access

Another method is direct access (or relative access). A file is made up of fixed-length logical records that allow programs to read and write records rapidly in no particular order. The direct-access method is based on a disk model of a file, since disks allow random access to any file block. For direct access, the file is viewed as a numbered sequence of blocks or records. Thus, we may read block 14, then read block 53, and then write block 7. There are no restrictions on the order of reading or writing for a direct-access file.

Direct-access files are of great use for immediate access to large amounts of information. Databases are often of this type. When a query concerning a particular subject arrives, we compute which block contains the answer and then read that block directly to provide the desired information.

3) Indexing

Other access methods can be built on top of a direct-access method. These methods generally involve the construction of an index for the file. The index, like an index in the back of a book, contains pointers to the various blocks. To find a record in the file, we first search the index and then use thc pointer to access the file directly and to find the desired record.

With large files, the index file itself may become too large to be kept in memory. One solution is to create an index for the index file. The primary index file would contain pointers to secondary index files, which would point to the actual data items.

7.2.2 Allocation Methods

The direct-access nature of disks allows us flexibility in the implementation of files. In almost every case, many files are stored on the same disk. The main problem is how to allocate space to these files so that disk space is utilized effectively and files can be accessed quickly. Three major methods of allocating disk space are in wide use: contiguous, linked, and indexed. Each method has advantages and disadvantages. Some systems (such as Data General's RDOS for its Nova line of computers) support all three. More commonly, a system uses one method for all files within a file system type.

1. Contiguous Allocation

Contiguous allocation requires that each file occupy a set of contiguous blocks on the disk. Disk addresses define a linear ordering on the disk. With this ordering, assuming that only one job is accessing the disk, accessing block $b + 1$ after block b normally requires no head movement. When head movement is needed (from the last sector of one cylinder to the first sector of the next cylinder), the head needs only move from one track to the next. Thus, the number of disk seeks required for accessing contiguously allocated files is minimal, as is seek time when a seek is finally needed.

Contiguous allocation of a file is defined by the disk address and length (in block units) of the first block. If the file is a block long and starts at location b, then it occupies blocks b, $b + 1$, $b + 2$, ..., $b + n - 1$. The directory entry for each file indicates the address of the starting block and the length of the area allocated for this file (Figure 7.1).

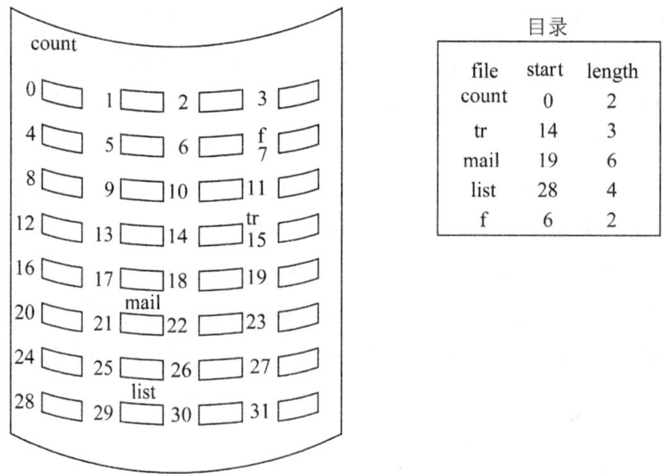

Figure 7.1 Contiguous allocation of disk space

Accessing a file that has been allocated contiguously is easy. For sequential access, the file system remembers the disk address of the last block referenced and, when necessary, reads the next block. For direct access to block i of a file that starts at block b, we can immediately access block $b + i$. Thus, both sequential and direct access can be supported by contiguous allocation.

Contiguous allocation has some problems, however. One difficulty is finding space for a new file. The system chosen to manage free space determines how this task is accomplished. Any management system can be used, but some are slower than others.

2. Linked Allocation

Linked allocation solves all problems of contiguous allocation. With linked allocation, each file is a linked list of disk blocks; the disk blocks may be scattered anywhere on the disk. The directory contains a pointer to the first and last blocks of the file. For example, a file of five blocks might start at block 9 and continue at block 16, then block 1, then block 10, and finally block 25. Each block contains a pointer to the next block. These pointers are not made available to the user. Thus, if each block is 512 bytes in size, and a disk address (the pointer) requires 4 bytes, then the user sees blocks of 508 bytes (Figure 7.2).

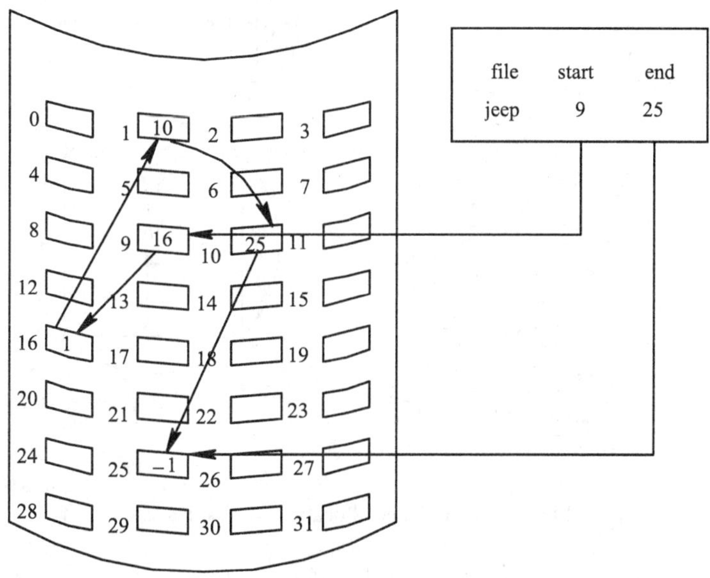

Figure 7.2 Linked allocation of disk space

To create a new file, we simply create a new entry in the directory. With linked allocation, each directory entry has a pointer to the first disk block of the file. This pointer is initialized to nil (the end-of-list pointer value) to signify an empty file. The size field is also set to 0. A write to the file causes the free-space management system to find a free block, and this new block is written to and is linked to the end of the file. To read a file, we simply read blocks by following the pointers from block to block. There is no external fragmentation with linked allocation, and any free block on the free-space list can be used to satisfy a request. The size of a file need not be declared when that file is created. A file can continue to grow as long as free blocks are available. Consequently, it is never necessary to compact disk space.

3. Indexed Allocation

Linked allocation solves the external-fragmentation and size-declaration problems of contiguous allocation. However, in the absence of a FAT, linked allocation cannot support efficient direct access, since the pointers to the blocks are scattered with, the blocks themselves all over the disk and must be retrieved in order. Indexed allocation solves this problem by bringing all the pointers together into one location: the index block.

Each file has its own index block, which is an array of disk-block addresses. The next entry in the index block points to the next block of the file. The directory contains the address of the index. To find and read the next block, we use the pointer in the next index-block entry.

When the file is created, all pointers in the index block are set to nil. When the next block is first written, a block is obtained from the free-space manager, and its address is put in the next index-block entry (Figure 7.3).

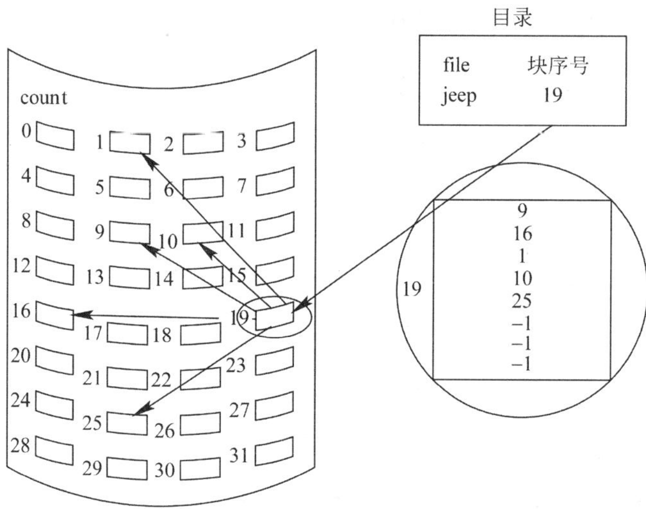

Figure 7.3 Indexed allocation of disk space

Indexed allocation supports direct access, without suffering from external fragmentation, because any free block on the disk can satisfy a request for more space. Indexed allocation does suffer from wasted space, however. The pointer overhead of the index block is generally greater than the pointer overhead of linked allocation. Consider a common case in which we have a file of only one or two blocks. With linked allocation, we lose the space of only one pointer per block. With indexed allocation, an entire index block must be allocated, even if only one or two pointers will be non-nil.

This point raises the question of how large the index block should he. Every file must have an index block, so we want the index block to be as small as possible. If the index block is too small, however, it will not be able to hold enough pointers for a large file, and a mechanism will have to be available to deal with this issue. Mechanisms for this purpose include the following.

(1) Linked scheme. An index block is normally one disk block. Thus, it can be read and written directly by itself. To allow for large files, we can link together several index blocks. For example, an index block might contain a small header giving the name of the file and a set of the

first 100 disk-block addresses. The next address (the last word in the index block) is nil (for a small file) or is a pointer to another index block (for a large file).

(2) Multilevel index. A variant of the linked representation is to use a first-level index block to point to a set of second-level index blocks, which in turn point to the file blocks. To access a block, the operating system uses the first-level index to find a second-level index block and then uses that block to find the desired data block. This approach could be continued to a third or fourth level, depending on the desired maximum file size. With 4,096-byte blocks, we could store 1,024 4-byte pointers in an index block. Two levels of indexes allow 1,048,576 data blocks and a file size of up to 4GB.

(3) Combined scheme. Another alternative, used in the UPS, is to keep the first, say, 15 pointers of the index block in the file's mode. The first 12 of these pointers point to direct blocks; that is, they contain addresses of blocks that contain data of the file. Thus, the data for small files (of no more than 12 blocks) do not need a separate index block. If the block size is 4 KB, then up to 48 KB of data can be accessed directly. The next three pointers point to indirect blocks. The first points to a single indirect block, which is an index block containing not data but the addresses of blocks that do contain data. The second points to a double indirect block, which contains the address of a block that contains the addresses of blocks that contain pointers to the actual data blocks. The last pointer contains the address of a triple indirect block. Under this method, the number of blocks that can be allocated to a file exceeds the amount of space addressable by the 4-byte file pointers used by many operating systems. A 32-bit file pointer reaches only 232 bytes, or 4 GB. Many UNIX implementations, including Solaris and IBM's AIX, now support up to 64-bit file pointers. Pointers of this size allow files and file systems to be terabytes in size.

Indexed-allocation schemes suffer from some of the same performance problems as does linked allocation. Specifically, the index blocks can be cached in memory, but the data blocks may be spread all over a volume.

7.3 Free-Space Management

Since disk space is limited, we need to reuse the space from deleted files for new files, if possible. (Write-once optical disks only allow one write to any given sector, and thus such reuse is not physically possible.) To keep track of free disk space, the system maintains a free-space list. The free-space list records all free disk blocks—those not allocated to some file or directory. To create a file, we search the free-space list for the required amount of space and allocate that space to the new file. This space is then removed from the free-space list. When a file is deleted, its disk space is added to the free-space list. The free-space list, despite its name, might not be implemented as a list, as we discuss next.

7.3.1 Bit Vector

Frequently, the free-space list is implemented as a bit map or bit vector. Each block is represented by 1 bit. If the block is free, the bit is 1; if the block is allocated, the bit is 0.

For example, consider a disk where blocks 2, 3, 4, 5, 8, 9, 10, 11, 12, 13, 17, 18, 25, 26, and 27 are free and the rest of the blocks are allocated. The free-space bit map would be

001111001111110001100000011100000 ...

The main advantage of this approach is its relative simplicity and its efficiency in finding the first free block or *n* consecutive free blocks on the disk. Indeed, many computers supply bit-manipulation instructions that can be used effectively for that purpose. One technique for finding the first free block on a system that uses a bit-vector to allocate disk space is to sequentially check each word in the bit map to see whether that value is not 0, since a 0-valued word has all 0 bits and represents a set of allocated blocks. The first non-0 word is scanned for the first 1 bit, which is the location of the first free block. The calculation of the block number is

(number of bits per word) x (number of 0-value words) + offset of first 1 bit.

Again, we see hardware features driving software functionality. Unfortunately, bit vectors are inefficient unless the entire vector is kept in main memory (and is written to disk occasionally for recovery needs). Keeping it in main memory is possible for smaller disks but not necessarily for larger ones. A 1.3-GB disk with 512-byte blocks would need a bit map of over 332 KB to track its free blocks, although clustering the blocks in groups of four reduces this number to over 83 KB per disk. A 40-GB disk with 1-KB blocks requires over 5 MB to store its bit map.

7.3.2 Linked List

Another approach to free-space management is to link together all the free disk blocks, keeping a pointer to the first free block in a special location on the disk and caching it in memory. This first block contains a pointer to the next free disk block, and so on. In our earlier example, we would keep a pointer to block 2 as the first free block. Block 2 would contain a pointer to block 3, which would point to block 4, which would point to block 5, which would point to block 8, and so on. However, this scheme is not efficient to traverse the list, we must read each block, which requires substantial I/O time. Fortunately, traversing the free list is not a frequent action. Usually, the operating system simply needs a free block so that it can allocate that block to a file, so the first block in the free list is used. The FAT method incorporates free-block accounting into the allocation data structure. No separate method is needed (Figure 7.4).

7.3.3 Grouping

A modification of the free-list approach is to store the addresses of n free blocks in the first free block. The first n-1 of these blocks are actually free. The last block contains the addresses of another free blocks, and so on. The addresses of a large number of free blocks can now be found quickly, unlike the situation when the standard linked-list approach is used.

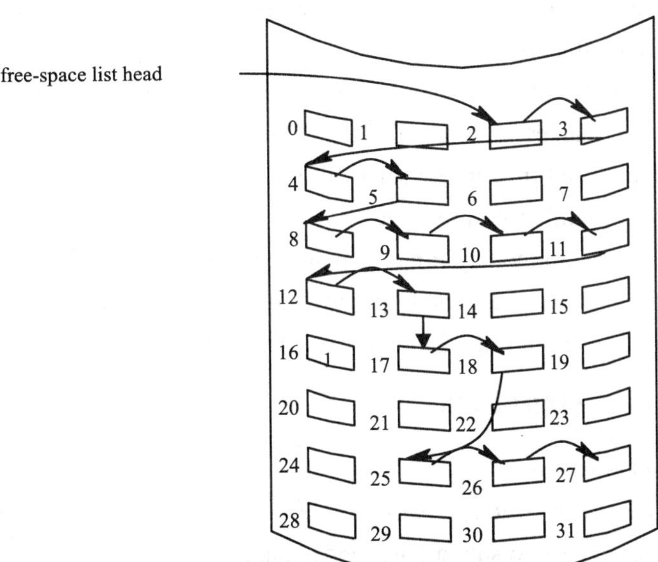

Figure 7.4 Linked free-space list on disk

7.3.4 Counting

Another approach is to take advantage of the fact that, generally, several contiguous blocks may be allocated or freed simultaneously, particularly when space is allocated with the contiguous-allocation algorithm or through clustering. Thus, rather than keeping a list of a free disk addresses, we can keep the address of the first free block and the number of free contiguous blocks that follow the first block. Each entry in the free-space list then consists of a disk address and a count. Although each entry requires more space than would a simple disk address, the overall list will be shorter, as long as the count is generally greater than 1.

7.4 Directory

The directory can be viewed as a symbol table that translates file names into their directory entries. If we take such a view, we see that the directory itself can be organized in many ways. We want to be able to insert entries, to delete entries, to search for a named entry, and to list all the entries in the directory.

7.4.1 Inodes

Unix directories and files don't really have names. They are numbered, using node numbers called inodes, vnodes, or even gnodes (depending on the version of Unix). You won't find the name of a particular file or directory in or near the file or directory itself. All the name-to-number mappings of files and directories are stored in the parent directories. For each file or directory, a link count keeps track of how many parent directories contain a name-number mapping for each node. When a link count goes to zero, no directory points to the node and UNIX is free to reclaim the disk space.

UNIX permits all files to have many name-to-number mappings. So, a file may appear to have several different "names" (Unix calls them "links"), that is, several names that all map to the same node number (and thus to the same file). Or, the file may have the same "name", but, that name may appear in different directories.

Anyone can create a link to any file to which they have access. They don't need to be able to read or write the file itself to make the link; they only need write permission on the directory in which the name-to-number map (the name or "link") is being created.

7.4.2 Directory Structure

1. Single-Level Directory

The simplest directory structure is the single-level directory. All files are contained in the same directory, which is easy to support and understand.

A single-level directory has significant limitations, however, when the number of files increases or when the system has more than one user. Since all files are in the same directory, they must have unique names. If two users call their data file test, then the unique-name rule is violated. For example, in one programming class, 23 students called the program for their second assignment prog2; another 11. called it assign2. Although file names are generally selected to reflect the content of the file, they are often limited in length, complicating the task of making file names unique.

2. Two-Level Directory

In the two-level directory structure, each user has his own user file directory (UFD). The UFDs have similar structures, but and each lists only the files of a single user. When a user job starts or a user logs in, the system's master file directory (MFD) is searched. The MFD is indexed by user name or account number, and each entry points to the UFD for that user.

When a user refers to a particular file, only his own UFD is searched. Thus, different users may have files with the same name, as long as all the file names within each UFD are unique. To create a file for a user, the operating system searches only that user's UFD to ascertain whether another file of that name exists. To delete a file, the operating system confines its search to the local UFD; thus, it cannot accidentally delete another user's file that has the same name.

习　　题

一、选择题

1．操作系统中对数据进行管理的部分叫做________。

A．数据库系统　　B．文件系统　　C．检索系统　　D．数据存储系统

2．文件系统是指________。

A．文件的集合　　B．文件的目录

C．实现文件管理的一组软件　　D．文件、管理文件的软件及数据结构的总体

3．从用户角度看，引入文件系统的主要目的是________。

A．实现虚拟存储　　B．保存系统文档

C．保存用户和系统文档　　D．实现对文件的按名存取

4．文件的逻辑组织将文件分为记录式文件和________文件。

A．索引文件　　B．流式文件　　C．字符文件　　D．读写文件

5．文件系统中用________管理文件。

A．作业控制块　　B．外页表　　C．目录　　D．软硬件结合的方法

6．为了对文件系统中的文件进行安全管理，任何一个用户在进入系统时都必须进行注册，这一级安全管理是________安全管理。

A．系统级　　B．目录级　　C．用户级　　D．文件级

7．为了解决不同用户文件的"命名冲突"问题，通常在文件系统中采用________。

A．约定的方法　　B．多级目录　　C．路径　　D．索引

8．一个文件的绝对路径是从________开始，逐步沿着每一级子目录向下追溯，最后到指定文件的整个通路上所有子目录名组成的一个字符串。

A．当前目录　　B．根目录　　C．多级目录　　D．二级目录

9．对一个文件的访问，常由________共同限制。

A．用户访问权限和文件属性　　B．用户访问权限和用户优先级

C．优先级和文件属性　　D．文件属性和口令

10．磁盘上的文件以________单位读写。

A．块　　B．记录　　C．柱面　　D．磁道

11．磁带上的文件一般只能________。

A．顺序存取　　B．随机存取　　C．以字节为单位存取　　D．直接存取

12．使用文件前必须先________文件。

A．命名　　B．建立　　C．打开　　D．备份

13．文件使用完毕后应该________。

A．释放　　B．关闭　　C．卸下　　D．备份

14．位示图可用于________。

A．文件目录的查找　　B．磁盘空间的管理

C．主存空间的共享　　D．实现文件的保护和保密

15．一般来说，文件名及属性可以收纳在________中以便查找。

A．目录　　B．索引　　C．字典　　D．作业控制块

16．最常用的流式文件是字符流文件，它可看成是________的集合。

A．字符序列　　B．数据　　C．记录　　D．页面

17．按物理结构划分，文件主要有三类：___①___．___②___和___③___。

A．索引文件　　B．读写文件　　C．顺序文件　　D．链接文件

18．在文件系统中，文件的不同物理结构有不同的优点。在下列文件的物理结构，________不具有直接读写文件任意一个记录的能力。

A．顺序结构　　B．链接结构　　C．索引结构　　D．Hash 结构

19．在下列文件的物理结构中，________不利于文件长度动态增长。

A．顺序结构　B．链接结构　C．索引结构　D．Hash 结构

20．如果文件采用直接存取方式且文件大小不固定，则宜选择________文件结构。

A．直接　B．顺序　C．随机　D．索引

21．文件系统采用二级目录结构，这样可以________。

A．缩短访问文件存储器时间　B．实现文件共享

C．节省主存空间　D．解决不同用户之间的文件名冲突问题

22．常用的文件存取方法有两种：顺序存取和________存取。

A．流式　B．串联　C．顺序　D．随机

23．下列叙述中正确的五项是________。

A．在磁带上的顺序文件中插入新的记录时，必须复制整个文件

B．由于磁带的价格比磁盘便宜，用磁带实现索引文件更经济

C．在索引顺序文件的最后添加新的记录时，必须复制整个文件

D．在磁带上的顺序文件的最后添加新的记录时，不必须复制整个文件

E．顺序文件是利用磁带的特有性质实现的，因此顺序文件只有存放在磁带上

F．索引顺序文件既能顺序访问，又能随机访问

G．直接访问文件也能顺序访问，但一般效率较差

H．变更磁盘上的顺序文件的记录内容时，不一定要复制整个文件

I．在磁盘上的顺序文件中插入新的记录时，必须复制整个文件

J．索引顺序文件是一种特殊的顺序文件，因此通常存放在磁带上

24．以下叙述中正确的是

A．文件系统要负责文件存储空间的管理，但不能完成文件名到物理地址的转换。

B．多级目录结构中，对文件的访问是通过路径名和用户目录名来进行的

C．文件被划分成大小相等的若干个物理块，一般物理块的大小是不固定的

D．逻辑记录是对文件进行存取操作的基本单位

25．文件系统中，设立打开文件（Open）系统功能调用的基本操作是________。

A．把文件信息从辅存读到内存

B．把文件的控制管理信息从辅存读到内存

C．把磁盘的超级块从辅存读到内存

D．把文件的 FAT 表信息从辅存读到内存

26．用磁带作为文件存储介质时，文件只能组织成________。

A．顺序文件　B．链接文件　C．索引文件　D．目录文件

27．无结构的文件组织也常被叫做________。

A.顺序文件　B.链接文件　C.索引文件　D.流

二、填空题：

1．索引文件大体上由________区和________构成。其中________区一般按关键字的顺序存放。

2．对操作系统而言，打开文件广义指令的主要作用是装入________目录表。

3．磁盘文件目录表的内容至少应包含________和________。

4．操作系统实现按________进行检索等关键在于解决文件名与文件的存储地址的转换。

5．文件的物理组织有________。

6．在文件系统中，若按逻辑结构划分．可将文件划分成________和________两大类。

7．按用户对文件的存取权限将用户分为若干组，同时规定每一组用户对文件的访问权限。这样，所有用户组存取权限的集合称为该文件的________。

8．________是指避免文件拥有者或其他用户因有意或无意的错误操作使文件受到破坏。

9．从文件管理角度看，文件由________和________两部分组成

10．磁盘与主机之间传递数据是以________为单位进行的。

11．在文件系统中，要求物理块必须连续的物理文件是________。

12．文件系统为每个文件另建立一张指示逻辑记录和物理块之间的对应关系，由此表和文件本身构成的文件是________。

13．文件的结构就是文件的组织形式，从用户观点出发所看到的文件组织形式称为文件的________；从实现观点出发，文件在外存上的存放组织形式称为________。

三、综合题

1．简述采用位视图法管理外存存储空间时，怎样分配和回收存储空间？

2．简述文件系统的组成和功能。

3. 某操作系统的磁盘文件空间共有 500 块，若用字长为 32 位的位示图管理盘空间问：

（1）位示图需要多少个字？

（2）第 i 字第 j 位对应的块号是多少？

4．图 7-11 是文件管理中采用显式链式结构的管理方法示意图，回答问题：

（1）解释 FCB 的含义；

（2）解释 FAT 的含义；

（3）写出文件 A 占用的数据块块号；

（4）写出文件 B 占用的数据块块号；

（5）画出删除文件 B 之后的 FAT 表。

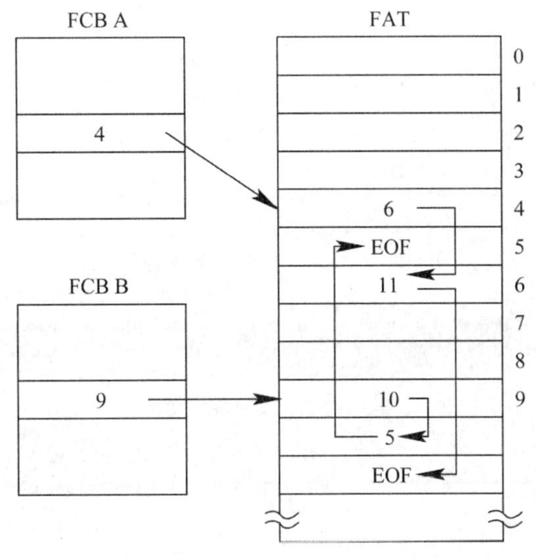

图 7-11　显示链式结构图

5．图 7-12 是空闲盘块成组链接图，请回答相关问题。

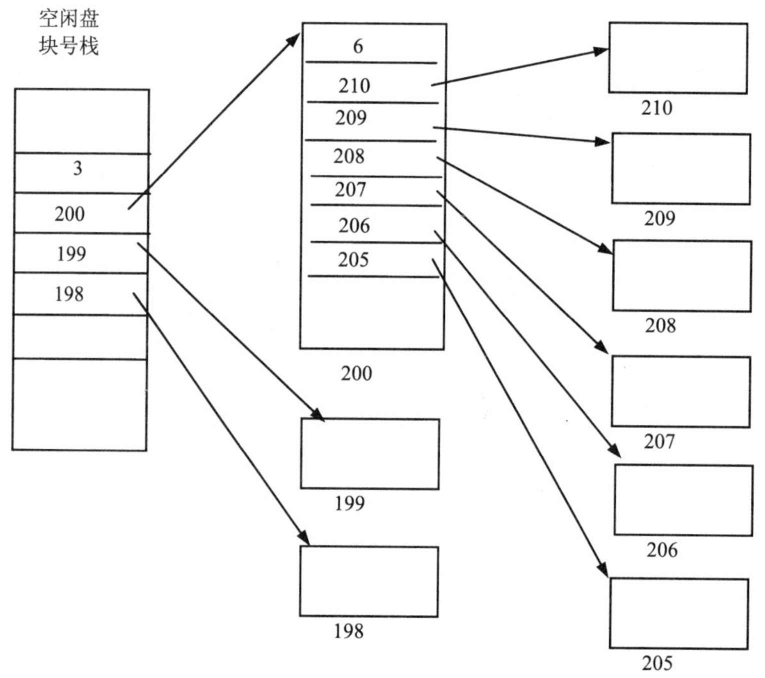

图 7-12　空闲盘块或组链接图

（1）在当前情况下，要创建文件 A，文件 A 需要两个块。把哪两个块分配给文件 A？

（2）接问题（1），请画出创建文件 A 后的空白盘块号栈。

（3）接问题（2），创建文件 B，文件 B 需要 3 个块，把哪三个块分配给文件 B？

（4）接问题（3），请画出创建文件 B 后的空闲盘块成组链接图。

第8章 设 备 管 理

I/O 系统是计算机系统的一个重要组成部分，它包括用于信息输入、输出和存储功能的 I/O 设备和相应的设备控制器。设备管理的主要对象是 I/O 设备和设备控制器，可能还涉及 I/O 通道。设备管理的主要任务是：为用户进程分配其所需的 I/O 设备；完成用户进程提出的 I/O 请求；提高 CPU 和 I/O 设备的利用率；提高 I/O 速度；方便用户使用 I/O 设备。为实现上述任务，设备管理应具有缓冲管理、设备分配、设备处理和虚拟设备等功能。

8.1 I/O 系统的硬件

在 I/O 系统中，除了有直接用于输入、输出和存储信息的 I/O 设备外，还需要有相应的设备控制器。随着计算机技术的发展，在大、中型计算机系统中，又增加了 I/O 通道。I/O 设备、设备控制器、I/O 通道和相应的总线构成了 I/O 系统的硬件。

8.1.1 I/O 设备

I/O 设备即输入输出设备，包括了除主机以外的大部分硬件设备，故也称外部设备或外围设备，简称外设。I/O 设备种类繁多，性能各异。为了便于管理，可以从不同角度对它们进行分类。

1．按工作特性分

（1）输入设备。输入设备是指将程序、数据和操作命令等信息转换成计算机所能接收的电信号并输入计算机的装置。常用的输入设备有键盘、鼠标、触摸屏、扫描仪和字符识别设备等。

（2）输出设备。输出设备是指将计算机处理后的信息转换为数字、文字、字符、声音、图形和图像等形式，并在其信息载体上输出的装置。常用的输出设备有显示器、打印机和绘图仪等。

（3）存储设备。存储设备也称外存储器或辅助存储器，简称外存，是存储信息的主要设备，既可以输入也可以输出。相对内存，它的优点是存储容量大、成本低和具有非易失性，但存取速度较慢。常用的存储设备有磁盘、磁带和光盘等。

2．按传输速率分

（1）低速设备。低速设备是指传输速率为每秒钟几个字节到数百个字节的设备。典型的低速设备有键盘、鼠标等。

（2）中速设备。中速设备是指传输速率在每秒钟数千个字节至数十千个字节的设备。行式打印机、激光打印机都属于中速设备。

（3）高速设备。高速设备是指传输速率在数百千个字节至数兆字节的设备。典型的高速设备有磁带机、磁盘机、光盘机等。

3. 按信息传送单位分

（1）块设备。块设备是指以数据块为单位组织和传送数据信息的设备。这类设备主要用于存储信息，属于有结构设备。典型的块设备是磁盘，每个盘块大小为 512B～4KB，传输速率较高，通常每秒钟几兆位，多采用 DMA 方式进行数据传送。

（2）字符设备。字符设备是指以单个字符为单位传送数据信息的设备。这类设备一般多用于数据的输入和输出，如键盘和打印机，属于无结构设备。字符设备的传输速率较低，常采用中断控制方式进行数据传送。

4. 按设备的共享属性分

（1）独占设备。独占设备是指在一段时间内只允许一个进程访问的设备。系统一旦把此类设备分配给某个进程，便由该进程独占，直至用完释放后，才能分配给其他的进程使用。这类设备属临界资源，如果分配不当，可能会造成死锁。独占设备多为一些慢速设备，如磁卡机、打印机等。

（2）共享设备。共享设备是指在一段时间内允许多个进程同时访问的设备。“同时访问”的含义是指各进程在执行期间内，可以交替、分时地对共享设备进行访问，但某一时刻只允许一个进程访问。与独占设备相比，共享设备不仅可以获得良好的设备利用率，而且是实现文件系统和数据库系统的物质基础。典型的共享设备是磁盘。

（3）虚拟设备。严格地讲，虚拟设备是一种设备管理的技术。采用该技术可以将一台慢速独占设备变换为若干台可供多个进程共享的逻辑设备，可以大大提高独占设备的使用效率。在现代计算机系统中，主要利用假脱机技术（SPOOLing 技术）实现虚拟设备。

由于不同类型的设备之间在工作特性、传输速率、信息传送单位和共享属性等方面存在着较大差异，从而导致对它们的管理的方法也有所不同。例如低速设备通常以字节为单位进行数据传送，传输速率较低，控制方式相对简单，对这类设备的管理相对容易。而存储类设备是以信息块为单位进行数据传送，传输速率高，控制方式复杂，除了需要复杂的设备管理功能，还需要有文件管理子系统和虚拟存储器管理子系统的支持。

8.1.2 设备控制器

设备控制器是在主机与具体设备之间设置的、具有缓冲功能的接口电路。它的主要功能是控制一个或多个 I/O 设备，实现 I/O 设备和主机之间的数据传送。设备控制器是一个可编址的设备，当它控制一个设备时，只有一个设备地址；若控制器连接多个设备时，则有多个设备地址，每一个设备地址对应一个设备。

1. 设备控制器的基本功能

1）实现数据传送

这里所说的数据传送是指 CPU 与控制器之间、控制器与设备之间的数据传送。对于前者，CPU 通过数据总线将数据并行写入控制器，或从控制器中并行地读出数据；对于后者，设备将数据输入到控制器，或从控制器输出数据给设备。为了实现数据传送功能，在设备控制器中必须设置数据寄存器。

2）识别设备地址

一个计算机系统通常有多个I/O设备，每一个设备都有一个唯一的地址。在某一时刻，主机只能与一个I/O设备传送数据。设备控制器中的地址译码电路对地址总线上的设备地址进行译码，以选中指定设备，只有被选中的I/O设备才能与主机进行数据传送。

3）标识和报告I/O设备的状态

主机在与I/O设备进行数据传送的过程中，需要不断地对设备的工作状态进行查询。在设备控制器中设置有状态寄存器，其中的状态位用于反映I/O设备的某些状态。CPU通过读取状态寄存器的内容，就可以很容易了解I/O设备的工作状态。

4）接收和识别CPU命令

为了实现CPU对I/O设备进行控制，在设备控制器中设置有控制寄存器，用来接收和存放CPU发来的各种不同的命令和参数，并对所接收的命令进行译码。

5）实现数据缓冲

为解决主机与I/O设备速度不匹配问题，在设备控制器中，一般设置缓冲器。数据输入时，先将I/O设备输入的数据送入缓冲器暂存起来，待接收一批数据后，再将暂存在缓冲器中的数据高速传送给主机；数据输出时，缓冲器先暂存主机高速传送的数据，然后以适合I/O设备的速度将缓冲器中的数据输出。在设备控制器中设置数据缓冲器，不但使I/O设备有足够时间处理高速系统传送过来的数据，而且也便于CPU处理其他事务。

6）实现差错控制

在数据传送过程中，由于各种原因，可能会导致数据传输错误。设备控制器具有对I/O设备传送来的数据进行差错检测的功能。若发现传送中出现了错误，通常是将差错检测码置位，并向 CPU报告，于是CPU将本次传送来的数据作废，并重新进行一次数据传送，从而保证传输数据的正确性。

2．设备控制器的组成

设备控制器位于处理机与设备之间，它一方面通过系统总线实现与CPU之间的通信，另一方面又通过数据、控制和状态信号线实现与设备之间的通信。设备控制器的组成如图8-1所示，主要由以下三功能部件组成的。

1）设备控制器与处理机的接口

该接口主要用于实现CPU与设备控制器之间的通信，主要包括两类寄存器，这两类寄存器都通过数据线与处理机相连。第一类是数据寄存器，控制器中可以有一个或多个数据寄存器，用于存放从设备送来的数据（输入）或从CPU送来的数据（输出）；第二类是控制和状态寄存器，控制器中可以有一个或多个此类寄存器，其中控制寄存器是用于存放从CPU送来的控制信息，状态寄存器是用于存放设备的状态信息。

2）设备控制器与设备的接口

一个设备控制器上可以连接一个或多个设备，相应地，在控制器中便有一个或多个设备接口，一个接口连接一台设备。控制器中的I/O逻辑根据CPU发来的地址信号，进行设备接口的选择。每个接口中都存在数据、控制和状态三种类型的信号。

3）I/O逻辑

I/O逻辑主要实现对设备的控制。CPU每启动一个设备，都向控制器发送I/O命令和设备地址，I/O逻辑对接收到的地址进行译码，选中所指定的设备，对接收到的I/O命令进行译码，产生相应的控制信号，从而完成对所选设备的控制。

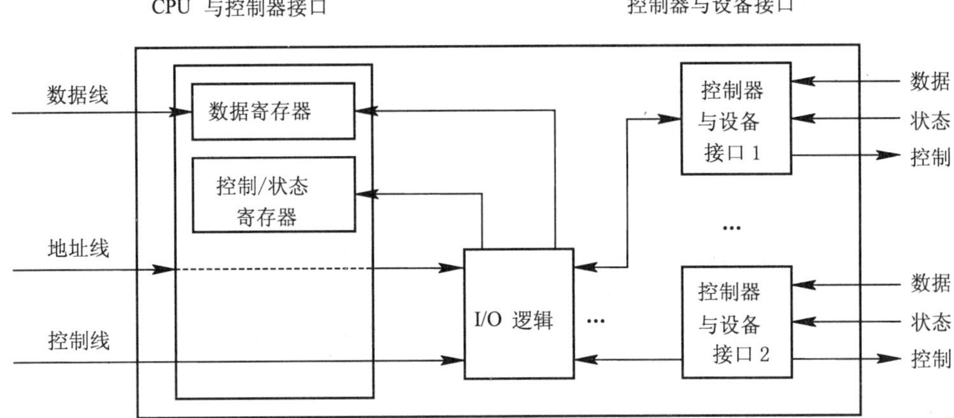

图 8-1　设备控制器的组成

8.1.3　I/O 通道

在大、中型计算机系统中，外设配置多，数据传输频繁，CPU 的负担十分繁重。为此在 CPU 和设备控制器之间增设了 I/O 通道，由 I/O 通道接替 CPU 完成数据传送工作，以保证 CPU 有更多的时间进行数据处理。

I/O 通道也称通道处理器，是一种能执行有限 I/O 指令集合、专用于输入输出控制的 I/O 处理器。一个主机可以连接多个通道，每个通道又可以连接多台不同类型的 I/O 设备。采用 I/O 通道的系统不仅可以灵活地增减 I/O 设备，而且还加强了主机与通道之间、通道与通道之间、设备与设备之间的并行操作能力。

按照数据传输方式和通道工作方式，可把通道分为字节多路通道、选择通道和数组多路通道三种类型。一个机器系统可以同时拥有三种类型通道，也可以只包含其中一种或两种类型。

1．字节多路通道

字节多路通道可以轮流为多个设备服务，这些设备可以同时处于工作状态，交叉进行数据传送。每次只传输一个字节数据。字节多路通道主要用于连接慢速或中速的字符设备。

2．选择通道

选择通道可以连接多个不同的 I/O 设备，但最多只能有一个设备处于工作状态，只有当该设备结束数据传送之后，才能选择其他设备工作。选择通道主要用于连接数据传输速度较高的块设备。

3．数组多路通道

数组多路通道可以轮流为多个设备服务，与字节多路通道不同的是，每次传输一个数据块（数组）。数组多路通道既保留了选择通道高速传输的优点，同时也可以为多个设备提供服务，使通道作用得到充分发挥。磁盘、磁带等一些块设备多采用数组多路通道。

8.2　I/O 控制方式

I/O 设备种类繁多，工作速度千差万别。由于 I/O 设备在工作速度上的差异，使得主机与 I/O 设备之间的数据传送存在多种不同的控制方式。采用合适的 I/O 控制方式，不但有利于提高 CPU 与 I/O 设备并行处理的效率，而且还可以形成多种 I/O 设备之间的并行操作。特别是在多通道程序设计环境下，I/O 操作的控制能力已经成为评价计算机系统综合处理能力的重要因素。

按照 CPU 与 I/O 设备并行处理的程度及数据传输的控制能力，将 I/O 控制方式分为四类：程序 I/O 方式、中断控制方式、直接存取方式和 I/O 通道方式。

8.2.1　程序 I/O 方式

主机与 I/O 设备在进行数据传送之前时，需要先查询设备的工作状态，只有这样，才能保证数据传送的可靠性。一般在设备控制器中都设置有状态寄存器，I/O 设备通过状态线把自己的工作状态写入该寄存器中。CPU 只需通过 I/O 指令读取状态寄存器的内容，就可以很容易获取 I/O 设备的状态信息。

程序 I/O 方式是指通过程序的方式来控制主机与 I/O 设备之间进行数据传送。通常的做法是：在用户程序中安排一段由 I/O 指令和其他指令组成的程序段，用来完成数据传送工作。数据传送时，CPU 首先向相应的设备控制器发出一条 I/O 命令，启动设备工作，同时将状态寄存器设置为数据未准备好状态（输入设备）或设备忙状态（输出设备）；接着 CPU 等待 I/O 设备完成输入或输出数据的工作，在等待的过程中，CPU 通过不断地读取状态寄存器的内容，来了解 I/O 操作的完成情况；I/O 设备在完成输入或输出操作后，会将状态寄存器设置成数据准备好或设备闲的状态，CPU 只有在检测此种状态后，才执行数据的传送。

主机与 I/O 设备之间采用程序 I/O 方式实现数据传送，可以很好地解决 CPU 与 I/O 设备之间工作速度不匹配的矛盾。但在这种控制方式下，CPU 和 I/O 设备之间、I/O 设备和 I/O 设备之间只能串行工作。当 I/O 设备的工作速度很慢时，CPU 长时间处于等待状态，造成 CPU 的极大浪费。因此这种数据传送方式只适合简单、I/O 设备较少的计算机系统。

8.2.2　中断控制方式

中断是指计算机在执行程序期间，系统内发生急需处理的事件，使得 CPU 暂时中断当前正在执行的程序，转去执行相应的事件处理程序，待处理完毕后又继续执行原来的程序。

在中断控制方式下，当进程要启动某个 I/O 设备时，先由 CPU 发出启动命令，启动设备工作，之后 CPU 继续执行其他程序，此时 CPU 与 I/O 设备并行工作。当 I/O 设备完成输入或输出数据的准备工作后，由 I/O 设备主动向 CPU 发出中断请求，要求 CPU 执行数据传送。CPU 接到中断请求后，可以暂时停止正在执行的程序，转去执行输入输出程序，待数据传送工作完成后，仍继续执行被中断的程序。

中断方式下，I/O 设备具有申请 CPU 服务的主动权。所谓的 CPU 服务，实际是执行一个数据输入输出程序，此程序称为中断服务程序或中断处理程序。

引入中断后，CPU 与 I/O 设备之间、I/O 设备和 I/O 设备之间可以并行工作，极大地提高了系统资源的利用率和吞吐量。同时 I/O 设备一旦处于就绪状态，就可以立即得到 CPU 服务，系统实时性较好，特别适合 I/O 设备工作速度很慢或随机传送数据的情况。

8.2.3 直接存取方式

中断控制方式在一定程度上提高了 CPU 和系统的工作效率，但它主要还是通过 CPU 执行程序来完成数据传送，且一次中断处理只能传送一个字节的数据。对于一些数据传输率较高且需要频繁与存储器之间进行批量数据传送的块设备来说，采用中断控制方式显然是极其低效的。为了解决这个问题，人们提出直接内存存取方式，也称为 DMA（Direct Memory Access）方式，实现了以数据块为单位进行传送，即每传送一次至少为一个数据块。

DMA 方式最突出的特点是数据传送的基本单位为数据块。其次，它在存储器与 I/O 设备之间开辟一条高速数据通道，实现了 I/O 设备与存储器之间直接进行高速、批量的数据传送。DMA 方式的第三个特点是：数据的传送主要是在专用硬件设备（DMA 控制器）的控制下完成。

DMA 控制器包括 DMA 控制逻辑、中断控制逻辑和若干个设备寄存器等。常用的设备寄存器有内存地址寄存器、设备地址寄存器、字数计数器、控制与状态寄存器和数据缓冲寄存器。

采用 DMA 方式进行数据传送的过程如下。

（1）CPU 首先通过执行若干条 I/O 指令，向 DMA 控制器发送必要的传送参数，如发送操作方式、设备地址、主存地址和传送的数据个数等，并启动 I/O 设备工作。这个过程称为 DMA 预处理，在完成了 DMA 预处理工作后，CPU 继续执行原来的程序。

（2）I/O 设备在准备好一次数据传送后，向 DMA 控制器发 DMA 请求，DMA 控制器在接收到该请求信号后，又向 CPU 发出总线请求信号，申请总线的使用权。

（3）CPU 响应 DMA 控制器的总线请求，将总线控制权交给 DMA 控制器。DMA 控制器在接管总线控制权后，负责在存储器与 I/O 设备之间直接进行高速的数据传送。

（4）当数据传送完毕，DMA 控制器向发出 CPU 中断请求，同时把总线控制权交还给 CPU。

不难看出，采用 DMA 方式进行数据传送，绝大部分工作都是由 DMA 控制器的控制完成，只有在数据块传送的开始和结束时需要 CPU 少量干预。与中断控制方式相比，DMA 方式极大地提高了 CPU 与 I/O 设备的并行操作程度。

8.2.4 I/O 通道方式

在大、中型计算机系统中，I/O 设备配置多，数据传送频繁。如仍采用 DMA 方式，则存在以下两个问题。第一，众多的 I/O 设备都需要配置专用的 DMA 控制器，将大幅度增加系统的硬件成本，同时也使控制复杂化。第二，所有的 I/O 设备都需要 CPU 进行 DMA 预处理工作，势必会占用 CPU 较多的时间，从而降低 CPU 执行程序的效率。

为避免上述弊病，在大、中型计算机系统中多采用 I/O 通道方式。I/O 通道是计算机系统中代替 CPU 管理和控制外设的独立部件，是一种能执行有限 I/O 指令集合的 I/O 处理机。通道接受 CPU 的委托，通过独立执行自己的通道程序来实现 I/O 设备与存储器之间的数据传送。在通道机构中含有通道指令（或称通道命令），每一条通道指令规定了 I/O 设备的一种操作，由一系列通道指令构成了指挥 I/O 设备工作的通道程序。在不同的计算机系统中，通道的指令格式和指令码可能不同，因此在通道程序的编制方式上也会有所不同。

I/O 通道方式与 DMA 方式的区别在于：DMA 方式要求 CPU 执行设备驱动程序启动 I/O 设备，并要求给出存放数据的内存起始地址、设备地址、传送的数据个数和操作方式等信息；若采用通道方式，CPU 只需发出 I/O 启动命令，其他所有工作均由通道程序来完成。不难看出，通道作为专用于 I/O 控制的处理器，分担了 CPU 大部分的 I/O 处理工作，使 CPU 从繁琐的 I/O 操作中解脱出来，真正发挥其“计算”的能力。另外，通道技术把以一个数据块为单位的读写干预，减少到以一组数据块为单位，实现 CPU、通道和 I/O 设备三者之间的并行工作，更加有效地提高了整个系统的资源利用率和系统吞吐量。

8.3 I/O 软件

I/O 软件的总体设计目标是高效性和通用性。为了达到这一目标，通常将 I/O 软件组织成一种层次结构，可以极大地提高了设备管理的效率。

8.3.1 I/O 软件的设计目标和层次结构

1. I/O 软件的目标

随着 CPU 速度的不断提高，I/O 设备低速性和 CPU 运行高速性之间的矛盾越来越突出，严重地影响了整个系统效率。同时 I/O 设备种类繁多，硬件构造复杂，物理特性各异，这不仅增加了设备管理和操作的复杂性，也给用户的使用带来了极大的困难。因此 I/O 软件的总体设计目标主要体现在以下两点。

1）高效性

通过采用多种技术和措施，尽可能提高 CPU 与 I/O 设备之间的并行操作程度，提高系统效率。主要用到的技术有：中断技术、DMA 技术、通道技术和缓冲技术。

2）通用性

通用性是指为用户提供方便、统一的界面。所谓方便，是指用户无须了解具体设备复杂的物理特性，就可以安全、方便地使用各类设备。所谓统一，是尽可能采用统一标准的方法来管理所有的设备及所需的 I/O 操作。为达到这一目标，引入设备的独立性，即用户操作的是逻辑设备，由操作系统与具体的 I/O 物理设备打交道。

2. I/O 软件的层次结构

I/O 软件涉及的范围很广，向上与用户直接交互，向下与硬件密切相关，和进程管理、存储器管理和文件管理有着密切的联系。为了使复杂 I/O 软件具有清晰的结构、较好的移植性和可适应性，在 I/O 软件中普遍采用层次结构，即把系统中所有用于完成设备操作和管理的软件组织成四个层次，如图 8-2 所示。其中最低层与硬件有关，它把硬件与较高层次的软件隔离开来；而最高层次的软件负责向用户提供一个统一、友好的 I/O 设备接口。每个层次都具有定义明确的功能和与邻近层次的交互接口。只要层次间的接口不变，对任何层次软件

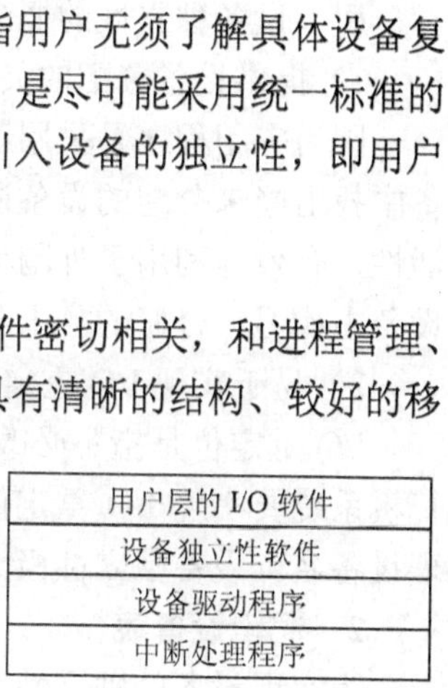

用户层的 I/O 软件
设备独立性软件
设备驱动程序
中断处理程序

图 8-2　I/O 软件的层次结构图

进行修改都不会引起它的上层和下层代码的变更。下面按照从高到低的顺序对每一层次软件进行讨论。

8.3.2 用户层的 I/O 软件

尽管大部分 I/O 软件都在操作系统内部，但是仍然有一小部分在用户层，包括与用户程序连接在一起的库函数和完全运行于内核之外的程序。

在用户层，用户程序是通过内核提供的一组系统调用来获取操作系统服务。在 C 语言等一些现代高级语言中，系统调用通常由库函数实现，用户程序通过调用库函数就可以十分方便进行系统调用。

用户层的 I/O 软件除了包含库函数外，还有一个重要的类别就是 SPOOLing 系统。SPOOLing 系统是多道程序设计系统中处理独占 I/O 设备的一种方法，需要创建一个特殊的进程和一个特殊的目录，分别称为守护进程和假脱机目录。以典型 SPOOLing 设备打印机为例，当一个进程要打印一个文件时，首先生成要打印的整个文件，并将其放在假脱机目录下。由守护进程打印假脱机目录下的文件。由于用户不能直接使用打印机，守护进程是唯一允许使用打印机的进程，从而避免某些进程长期空占打印机的问题。虽然守护进程是运行于内核之外的程序，但仍属于 I/O 系统。

8.3.3 设备独立性软件

1．设备独立性

设备独立性也称设备无关性，其基本含义是：应用程序独立于具体使用的物理设备。为实现设备独立性，引入了物理设备和逻辑设备的概念。物理设备是一个具体设备。物理设备名是为了识别外部设备，系统为每台设备所起的、具有标识作用的符号名。而逻辑设备是对实际物理设备属性的抽象，并不限于某个具体设备。逻辑设备名是由用户命名且可以更改。设备独立性的基本思想是：用户程序不直接使用物理设备名，而以逻辑设备名来请求使用某类设备。系统在实际执行时，先将逻辑设备名转换为某个具体的物理设备名，然后才能实施 I/O 操作。设备管理的功能之一就是实现逻辑设备名转换成物理设备名。

引入设备独立性的概念后，可带来两方面的好处。

1）提供设备分配的灵活性

因为应用程序是采用逻辑设备名而不是物理设备名提出设备申请，系统可以从该类设备中找出尚未分配的设备进行分配，而不是仅仅局限一个设备，大大增加了设备分配的灵活性，有效地利用了外围设备。同时，当用户使用的设备发生故障时，系统也可以从同类设备中找另一台好的且未分配的设备来替换，从而提高了系统的可靠性。

2）易于实现 I/O 重定位

I/O 重定位是指不必改变应用程序，就可以实现 I/O 设备的更换。具有设备独立性的计算机系统要实现 I/O 重定位很简单，只需在分配设备时，把逻辑设备名转换成另外一个物理设备名。设备独立性极大地提高了用户程序的可适应性。

2．逻辑设备表

为实现设备的独立性，系统必须设置一张逻辑设备表（Logical Unit Table，LUT），如图 8-3 所示，用于完成从逻辑设备名到物理设备名的映射。通常，表中的每一条表目包含逻辑设备名、物理设备名和设备驱动程序的入口地址等信息。当进程用逻辑设备名请求分配 I/O 设备时，设备分配程序会按照一定的分配策略和分配算法为它分配相应的物理设备，当分配成功后，便在 LUT 表中建立一个新表目，填上应用程序中使用的逻辑设备名、系统分配的物理设备名和该设备驱动程序的入口地址。以后该进程再使用相同的逻辑设备名请求 I/O 操作，系统便通过查找 LUT，找到对应的物理设备和对应的驱动程序。

对于单用户、单进程的系统，整个系统中只设置一张 LUT，表中记录了系统所有进程的设备分配情况，表中不允许有相同的逻辑设备名。对于多用户多进程系统，系统为每一个用户设置一张 LUT。每当用户登录时，便为该用户创建一个进程，同时也为之建立一张 LUT，并将该表放入进程的 PCB 中，只需要查询进程 PCB 即可实现逻辑设备名到物理设备名的映射。

逻辑设备名	物理设备名	驱动程序入口地址
/dev/tty	3	1024
/dev/printer	5	2048
…	…	…

图 8-3　逻辑设备表

3．设备独立性软件

设备驱动程序是一个与硬件（或设备）紧密相关的软件，为了实现设备的独立性，必须在驱动程序之上设置一层软件，称为设备独立性软件。至于设备驱动程序与设备独立性软件之间的确切界限，主要取决于操作系统、设备独立性和设备驱动程序的运行效率等多方面因素的权衡。有时出于效率和其他原因的考虑，一些本来应由设备独立性软件实现的功能，实际上是由设备驱动程序来完成。

设备独立性软件的主要功能分为以下两个方面。

1）执行所有设备的公有操作

（1）负责将逻辑设备名映射为物理设备名，并进一步找到相应物理设备的驱动程序。

（2）实现对缓冲区进行有效管理。

（3）负责独占设备的分配和回收。

（4）提供独立于设备的逻辑数据块。不同类型的设备，它们的信息交换单位、处理速度和传输速率也各不相同。设备独立性软件应隐藏着这些差异，向高层软件提供大小统一的逻辑数据块，保证高层软件只与抽象的设备打交道，不必考虑实际设备的数据块大小。

（5）负责必要的出错处理。因为绝大多数错误都与设备无关，所以一般是由设备驱动程序来处理，设备独立性软件只处理那些驱动程序无法处理的错误。

2）向用户层软件提供统一接口

无论何种设备，它们向用户提供的接口都应该是相同的。独立性软件负责提供统一的接口，从而保证了用户更方便地访问系统资源。

8.3.4 设备驱动程序

设备驱动程序也称设备处理程序，它是请求 I/O 的进程与设备控制器之间的通信和转换程序。其主要任务是接收来自上层软件的抽象 I/O 请求，如 Read 或 Write 命令，并将抽象请求转化为具体要求，发送给设备控制器，由设备控制器启动设备执行；此外，它也将设备控制器发来的信号如设备状态、I/O 操作完成情况及时地传递给上层软件。

设备驱动程序和一般的应用程序及系统程序有着明显的区别，主要体现在它与 I/O 设备的控制方式及硬件结构紧密相关，常由设备的制造商编写并随同设备一起交付。通常不同类型的设备配置不同的设备驱动程序。这些设备驱动程序大体上可以分成两部分，分别是启动设备的设备驱动程序和用于负责处理 I/O 完成工作的设备中断处理程序。

设备驱动程序主要任务是启动指定设备。但在启动之前需要进行必要的准备工作，只有完成所有的准备工作后，才向设备控制器发出启动命令。设备驱动程序的处理过程如下。

1．将接收到的抽象请求转化为具体要求

通常在每个设备控制器中都包含若干个寄存器，分别用于暂存数据、命令和设备的状态等信息。不同的设备在设备寄存器的数量、命令的性质等方面有着根本性的不同。因为用户层的 I/O 软件和上层设备独立性软件对这些设备控制器的具体情况不了解，只能发出抽象的 I/O 请求，需要设备驱动程序将这些抽象请求转化为具体要求后，再发送给设备控制器。例如当进行磁盘操作时，磁盘驱动程序要将抽象请求中的线性的盘块号转换为磁盘的柱面号、盘面号和扇区号。

2．检查 I/O 请求的合法性

在真正启动某个设备进行 I/O 操作之前，设备驱动程序要对 I/O 请求的合法性进行验证。通常每类设备只能完成一组特定的功能，对于设备不能支持的 I/O 请求，则认为是非法的。如键盘驱动程序会拒绝向键盘输出数据的请求。

3．了解 I/O 设备的状态

设备驱动程序需要读取设备的工作状态，只有当测试到这个设备处于接收就绪状态，才启动设备控制器工作，否则会将请求挂在对应的设备队列上等待。

4．传递有关参数

对于有些设备如块设备，还需要传送一些必要的参数。例如，在启动磁盘进行读写之前，应将本次传送的字节数、数据在主存的首址等信息送入控制器的相应寄存器中。

5．设置设备工作方式

有些设备具有多种工作方式，可满足不同系统的需求。对于这类设备，设备驱动程序应在完成工作方式的设置之后，再向设备控制器发出启动命令。

6．启动分配到的 I/O 设备

只有在完成上述各项准备工作之后，设备驱动程序才向设备控制器发出启动命令，然后把自己阻塞起来，由设备控制器控制执行后续的 I/O 操作。

8.3.5 中断处理程序

I/O 设备完成 I/O 操作后，设备控制器便向 CPU 发出一个中断请求，CPU 响应中断后，根据中断类型号调用相应的中断处理程序进行中断处理。中断处理时要唤醒被阻塞的驱动程序。

由于中断处理与硬件紧密相关，对用户和用户程序而言，应该尽可能加以屏蔽，故中断处理程序放在 I/O 软件的最底层，系统的其余部分也应尽量少与它发生联系。

中断处理程序的处理过程分为以下几个步骤。

1．唤醒被阻塞的进程

当中断处理程序开始执行时，要唤醒被阻塞的驱动进程。

2．中断响应

中断响应是指从 CPU 获取引起中断的中断源类型码，到最后找到该中断源的中断处理程序的全过程。在这个过程中，CPU 完成了以下工作：

① 首先 CPU 发出中断响应信号，通知 I/O 设备 CPU 已响应中断，应准备好中断类型码并将其放到数据总线上，然后 CPU 从数据总线上获取中断类型码并暂存；

② 由硬件自动将处理机的状态字（PSW）内容压入堆栈保护起来；

③ 将 PSW 中的单步跟踪标志位和中断允许标志位清 0；

④ 由硬件自动将断点地址即程序计数器的内容压入堆栈保护起来，以便中断结束后能返回断点处继续执行；

⑤ 根据 I/O 设备提供的中断类型码，查找中断向量表，找到中断处理程序入口地址；

⑥ 将找到的中断处理程序入口地址送入程序计数器，转去执行中断处理程序。

3．中断处理

中断处理过程实际就是执行中断处理程序。对于不同的设备，有不同的中断处理程序。

4．中断结束

当中断处理完成后，便将存放在堆栈中保护的信息再重新恢复，同时将断点地址送入程序计数器，继续执行被中断的程序。

8.4 缓 冲 管 理

为了缓解 CPU 与 I/O 设备之间速度不匹配的矛盾，提高 I/O 速度和资源利用率，在所有的 I/O 设备与处理机（内存）之间，都采用了缓冲区实现数据传送。因此设备管理的功能之一就是组织和管理缓冲区，并提供建立、分配和释放缓冲区的手段。

8.4.1 缓冲技术概述

随着计算机技术和电子技术的发展，各种新的 I/O 设备不断推出，I/O 设备的工作速度也在不断提高，但与 CPU 的处理速度相比，仍然存在较大的差异，从而造成 CPU 的工作效率低下。同时由于 I/O 通道数量不足而产生“瓶颈”现象，使得 CPU、通道和 I/O 设备之间的并行能力并未得到充分发挥。

为了缓解 CPU 与 I/O 设备之间速度不匹配的矛盾，提高的 I/O 速度和设备利用率，在设备管理中引入用来暂存数据的缓冲技术。缓冲是为了协调速度相差很大的设备之间进行数据传送工作而常采用的一种手段。

缓冲技术的实现原理是：系统首先建立缓冲区，当一个进程进行输入操作时，先将输入设备上的数据送入缓冲区，此时 CPU 可以处理其他任务，直到缓冲区满，才以中断方式通知 CPU 处理缓冲区中的数据；反之，当某个进程进行输出操作时，也先将数据送入缓冲区，当缓冲区满时再将缓冲区的内容送到输出设备上。这样原本是 CPU 与 I/O 设备之间进行的输入输出操作变成了 CPU 与缓冲区之间的读写操作，因为缓冲区的速度远高于 I/O 设备，所以大大提高了输入输出的速度。

在设备管理中引入缓冲技术有许多优点，主要归结为以下几点。

（1）缓解 CPU 与 I/O 设备间速度不匹配的矛盾。通常情况下，大多数程序都是输入、计算和输出操作交替出现。如有一个程序，它时而进行长时间的计算而没有输出，时而又阵发性把输出送到打印机。如果没有设置缓冲区，在 CPU 进行计算时，打印机处于空闲状态；而在打印机输出时，由于打印的速度很慢，又使得 CPU 长时间等待。设置了缓冲区后，程序输出的数据先送到缓冲区暂存，然后由打印机从缓冲区取数据慢慢地输出，CPU 不必等待，可以继续执行程序，从而大大提高了 CPU 的工作效率。

（2）提高 CPU 和 I/O 设备之间的并行性。从上面例子可以看出，CPU 和打印机可以并行工作。因此缓冲的引入可显著提高 CPU 和 I/O 设备的并行操作程度，提高设备的利用率和系统的吞吐量。

（3）减少对 CPU 的中断频率。原来 I/O 操作每传送一个字节就要产生一次中断，在设置了 n 个字节的缓冲区后，则可以等到缓冲区满后才产生中断，传送同样数量的数据，中断次数可以减少到 $1/n$，同时中断响应的时间也相应的放宽。

缓冲区的设定有两种方式：采用专门的硬件方法来实现，这在一定程度上增加了系统的硬件成本。除了在关键的地方采用少量、必要的硬件缓冲器外，许多系统都采用另外一种称为软件缓冲的方式，即从主存空间中划定出一个特殊的内存区域作为缓冲区。本节仅对软件缓冲技术进行介绍。

根据系统设置的缓冲区个数，可以将缓冲技术分为：单缓冲、双缓冲、循环缓冲和缓冲池。目前广泛采用公用缓冲池。

8.4.2 单缓冲和双缓冲

1. 单缓冲

单缓冲是操作系统提供的一种最简单的缓冲形式。当用户进程发出一个 I/O 请求时，系统便在主存中为之分配一个缓冲区，用来临时存放输入输出数据。单缓冲工作示意图如图 8-4 所示。

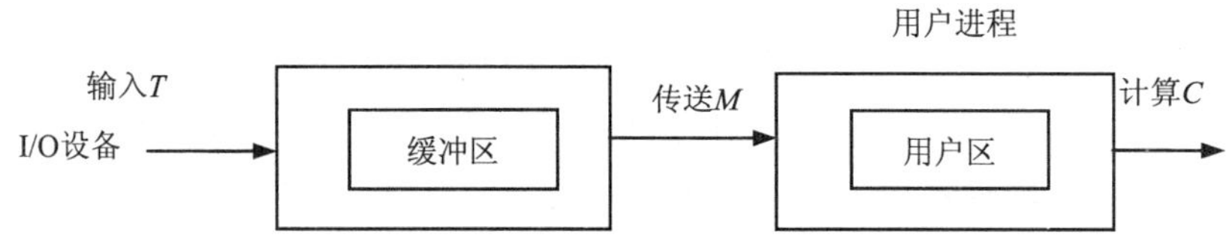

图 8-4　单缓冲工作示意图

在单缓冲方式下，块设备进行输入的过程是：输入的数据块先写入缓冲区，当处理机需要数据时，系统再把此数据块从缓冲区移送到用户区，与此同时请求设备输入下一个数据块。和输入过程类似，块设备进行输出时，用户进程也先把数据块从用户区移送到缓冲区，然后进程继续执行，最终由系统在适当时间将缓冲区中的数据块输出到设备上。对于字符设备，在单缓冲方式下，缓冲区主要暂存用户输入或输出的一行数据。

下面以读取磁盘数据进行计算为例，对单缓冲的工作效率进行分析。首先从磁盘读取一个数据块写入缓冲区，这个过程所需时间为 *T*；然后由系统负责将缓冲区的数据块移送到用户区，所需时间为 *M*，当数据块被移出缓冲区后，立即请求从磁盘读入下一个数据块；最后 CPU 对数据块进行计算，计算时间为 *C*。不难看出，整个过程分为三步，其中对上一个数据块进行计算和把下一个数据块读入缓冲区的这两个操作是可以并行执行，故系统对每个数据块的处理时间为 max（*C*，*T*）+M，其中 *M* 远小于 *T* 或 *C*。而没有设置缓冲区时，系统对每个数据块的处理时间为 *T*+*C*。很显然，CPU 和外设的利用率都有一定的提高。

2．双缓冲

单缓冲区方式下，当 CPU 和外设的工作速度相差较为悬殊时，系统效率的提升十分有限，另外因为只有一个缓冲区，数据输入和数据提取这两个操作不能同步进行，为此引入双缓冲技术。双缓冲指系统为某一设备设置两个缓冲区，如图 8-5 所示。当执行输入操作时，I/O 设备先将数据送入第一个缓冲区，当第一个缓冲区满后，I/O 设备转向对第二个缓冲区操作，此时操作系统可以从第一个缓冲区中提取数据传送到用户区，并释放第一个缓冲区，当第二个缓冲区满后，I/O 设备又可转过来使用被释放的第一个缓冲区。通过交替使用缓冲区方式来提高 CPU 和 I/O 设备的并行程度。

双缓冲技术可以实现对缓冲区中数据的输入、数据的提取和 CPU 的计算三者并行工作。仍以上述 CPU 从磁盘上读一个数据块进行计算为例，如果采用双缓冲技术，系统处理一个数据块的处理时间则可粗略地认为：max（*C*，*T*）。若 *C*＜*T*，可使块设备连续输入；若 *C*＞*T*，可使 CPU 不必等待设备输入。

在计算机系统中，由于 CPU 的速度总是比 I/O 设备快得多，准确地说，双缓冲技术还不能真正实现 CPU 与 I/O 设备的并行操作。同时由于计算机系统中配备有多种外围设备，CPU 与各种 I/O 设备的速度匹配全部由双缓冲来承担也是不现实的。

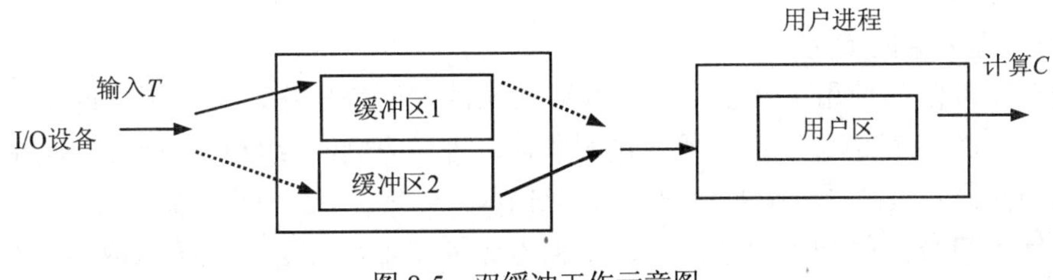

图 8-5　双缓冲工作示意图

8.4.3　循环缓冲

当数据输入速度和数据提取速度基本相匹配时，采用双缓冲可使两者并行工作，效果较好。如果两者的速度相差甚远，双缓冲的效果就不够理想。可以通过增加缓冲区数量的方式加强缓冲效果，常将多个大小相等的缓冲区组织成循环队列的形式，称为循环缓冲。其中一些队列专门用于输入，另一些队列专门用于输出。输入缓冲区循环队列通常提供给输入进程和计算进程使用，输入进程向该队列中空缓冲区输入数据，计算进程则从队列中装满数据的缓冲区提取数据；而输出缓冲区循环队列主要提供给计算进程和输出进程使用，计算进程向队列中空缓冲区输入运算结果，输出进程则从队列中装满数据的缓冲区提取这些结果。

相对单缓冲和双缓冲，循环缓冲的管理相对复杂。下面以输入循环缓冲区为例，对循环缓冲的组成和使用进行介绍。

1．循环缓冲的组成

循环缓冲的组成如图 8-6 所示，主要有以下两部分构成。

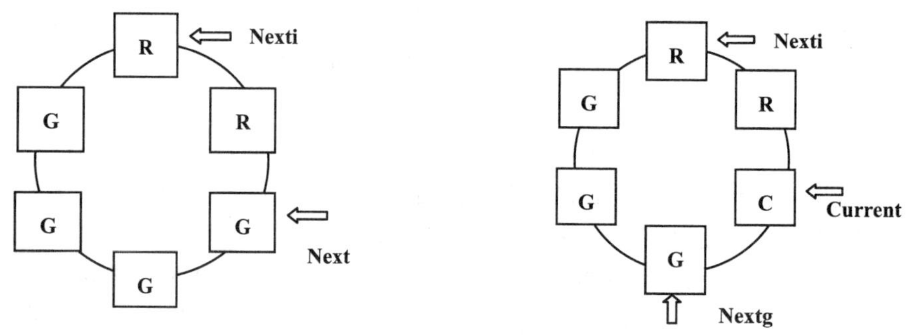

图 8-6　循环缓冲

（1）多个缓冲区。在循环缓冲中包括多个缓冲区，为便于管理，把输入的多缓冲区分为三种类型：用于装入输入数据的空缓冲区 R，已装满数据的满缓冲区 G，计算进程正在使用的工作缓冲区 C。

（2）多个指针。在循环缓冲中设置多个指针，分别指向不同类型的缓冲区。作为输入缓冲区可设置三个指针：用于指示计算进程下一个可用缓冲区的 Nextg 指针，用于指示输入进程下一个可用空缓冲区的 Nexti 指针，用于指示目前计算进程正在工作的缓冲区首地址 Current 指针。使用输入缓冲区时，依据对缓冲区的操作，这三个指针也将沿着顺时针方向进行相应的移动。

2．循环缓冲区的使用

对循环缓冲区的使用主要包括分配和释放缓冲区两个操作，可分别通过调用 Getbuf 和 Releasebuf 过程来实现。下面以输入进程和计算进程为例，介绍利用 Getbuf 和 Releasebuf 过程实现输入循环缓冲区分配和释放的具体过程。

（1）输入进程分配和释放缓冲区。当输入进程要使用空缓冲区装入数据时，可调用 Getbuf 过程，该过程将 Nexti 指针所指示的 R 缓冲区分配给输入进程使用，同时将 Nexti 指针移向下一个 R 空缓冲区。当输入进程把缓冲区装满数据时，应调用 Releasebuf 过程，将该缓冲区释放，并改为满缓冲区 G。

（2）计算进程分配和释放缓冲区。当计算进程要使用缓冲区中的数据时，可调用 Getbuf 过程，该过程将 Nextg 指针所指示的 G 缓冲区提供给计算进程使用，相应地将其改为现行工作缓冲区 C，令 Current 指针指向该缓冲区的第一个单元，同时将 Nextg 指针移向下一个 G 缓冲区。当计算进程把现行工作缓冲区 C 的数据提取完毕时，便调用 Releasebuf 过程，将缓冲区 C 释放，并把现行工作缓冲区 C 改为空缓冲区 R。

使用输入循环缓冲时，由于设置了多个缓冲区，可使输入进程和计算进程并行执行。由于缓冲区为临界资源，所以要特别注意输入进程和计算进程之间的同步问题。当输入进程和计算进程的速度不一致时，可能会出现以下两种情况。

（1）当 Nexti 指针追赶上 Nextg 指针，表明输入进程速度大于计算进程，没有可用的空缓冲区，这时输入进程阻塞（系统受计算限制）。

（2）当 Nextg 指针追赶上 Nexti 指针，表明计算进程速度大于输入进程，已没有装满数据的缓冲区，这时计算进程阻塞（系统受 I/O 限制）。

8.4.4 缓冲池

上述介绍的三种缓冲区都只能由某个特定的 I/O 进程和计算进程使用，属于专用缓冲区。当系统配置较多的设备时，使用专用缓冲区不仅要消耗大量的内存空间，而且缓冲区的利用率也不高。为了提高缓冲区的利用率，目前广泛采用公用缓冲池，在池中设置了多个可供若干个进程共享的缓冲区。

缓冲池中的缓冲区既可以作为输入使用又可以作为输出使用。按照缓冲区中存储的信息，把缓冲池中所有缓冲区分成空缓冲区、装满输入数据的缓冲区和装满输出数据的缓冲区。为了管理方便，将相同类型的缓冲区链成一个队列，于是便形成三个队列：空缓冲区队列、输入缓冲区队列和输出缓冲区队列。

除了有以上三个队列外，缓冲池还应具有四种工作缓冲区，分别是：用于收容输入数据的工作缓冲区 hin；用于收容输出数据的工作缓冲区 hout；用于提取输入数据的工作缓冲区 sin；用于提取输出数据的工作缓冲区 sout。

缓冲池可以工作在收容输入、提取输入、收容输出和提取输出四种方式下，如图 8-7 所示。

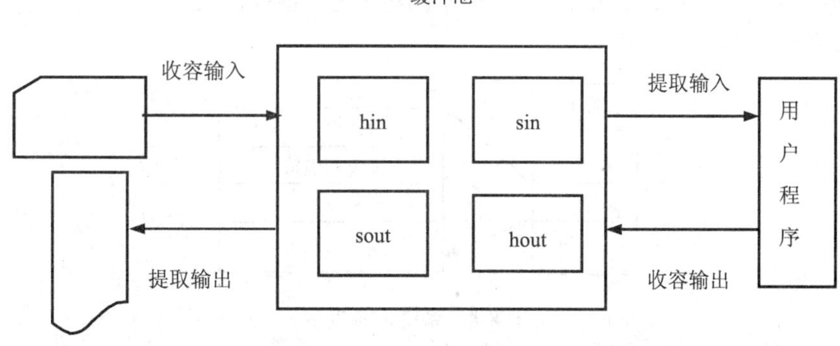

图 8-7 缓冲池的工作方式

下面仅以输入为例说明收容输入和提取输入的实现过程。当输入进程需要从输入设备输入数据时，缓冲池工作在收容输入方式下，实现过程分为三步：首先从空缓冲区队列的队首申请一个空缓冲区，把它作为收容输入数据的工作缓冲区 hin；然后把输入设备的数据输入到工作缓冲区 hin 中；当工作缓冲区 hin 满时，再把这个装满数据的缓冲区插入输入缓冲区队列的队尾。当计算进程需要输入数据时，缓冲池将工作在提取输入方式下，实现过程也分为三步：首先从输入缓冲区队列的队首取一个装满数据的缓冲区，把它作为提取输入数据的工作缓冲区 sin；然后计算进程从工作缓冲区 sin 中提取数据；当计算进程提取完该缓冲区的所有数据后，再将它插入到空缓冲区队尾。

8.5 设 备 分 配

在多道程序环境下，系统中的设备被所有进程共享，对这些设备的分配必须由操作系统参与完成。设备分配是设备管理的重要功能之一。当进程向系统提出 I/O 请求时，设备分配程序应按照一定的策略，将请求的设备及有关资源（如缓冲区、控制器和通道）分配给它。在进行设备分配的同时，还必须考虑系统的安全性，避免发生死锁。

在多通路的 I/O 系统中，为了满足一个用户的 I/O 请求，不仅仅要分配一个 I/O 设备，还应分配相应的设备控制器和 I/O 通道，建立起 CPU 与 I/O 设备之间进行数据传送的通路。

8.5.1 设备分配中的数据结构

为了实现设备分配，系统设置了一些表格形式的数据结构。这些数据结构除了用于记录系统中的每台 I/O 设备、设备控制器和 I/O 通道的状态信息，还提供控制设备所需的信息。主要包括系统设备表、设备控制表、控制器控制表和通道控制表。

1. 系统设备表（SDT）

在整个系统中，只设置一张系统设备表，如图 8-8 所示，用于记录系统中所有 I/O 设备的相关信息。其中，每个设备占一个表目，每个表目包括若干个表项，主要包括设备类型、设备标识符、设备控制表指针及设备驱动程序的入口地址等信息。

（1）设备类型。设备类型用于反映设备的特性，如：块设备或字符设备。

（2）设备标识符。设备标识符主要用于识别设备。

（3）设备控制表指针。该指针用于指向本设备的设备控制表。

（4）设备驱动程序的入口地址。

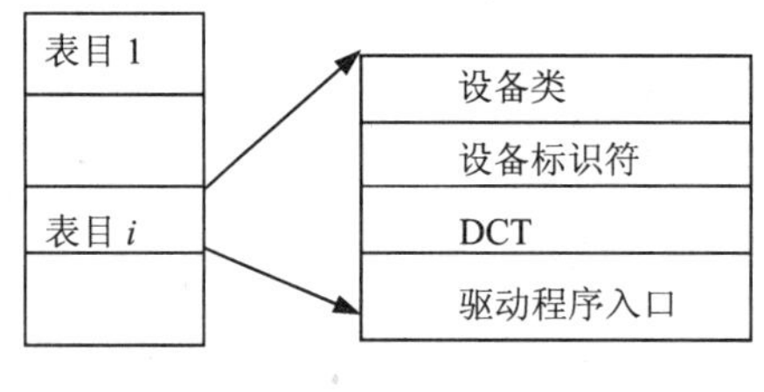

图 8-8 系统设备表

2. 设备控制表（DCT）

系统为每一个设备都配置了一张设备控制表，如图 8-9（a）所示，用于记录该设备的特性、与设备控制器的连接情况。设备控制表中除了有设备标识符、设备类型等信息，还包括以下内容。

（1）设备状态。当设备自身处于“忙”状态时，应将它的忙标志置为“1”，反之，应将它的忙标志置为“0”；若由于与该设备相连接的控制器或通道处于“忙”状态而不能启动该设备时，则将设备的等待标志置为“1”。

（2）设备队列的队首指针。设备队列也称为设备请求队列，是所有因请求该设备而未得到满足的进程的 PCB 按照一定的分配策略排成的队列。队首指针指向该设备队列的队首。

（3）与设备相连的控制器表指针。它指向与该设备相连接的控制器的控制表。在具有多条通路的情况下，一个设备可能与多个控制器相连接，此时，在 DCT 中应设置多个控制器表指针。

3．控制器控制表（COCT）

每个控制器配有一张控制器控制表，如图 8-9（b）所示。它反映设备控制器的使用状态、控制器与通道之间的连接情况等信息。

4．通道控制表（CHCT）

每个通道都配有一张通道控制表（CHCT），如图 8-9（c）所示。它描述通道工作情况，主要包括通道标识符、通道状态、等待获得该通道的进程等待队列指针等信息。

不难看出，上面这四张表在设备分配过程中形成了一个有机整体，可以有效地记录了 I/O 系统硬件资源的使用情况，是设备管理程序进行设备分配时所参考的主要数据结构。设备分配时，先通过系统设备表中指向设备控制表的指针，找到申请设备的设备控制表；然后通过该设备控制表中指向控制器控制表的指针，找到与该设备相连的控制器控制表；最后通过控制器控制表中指向通道控制表的指针，找到与该控制器相连的通道控制表。

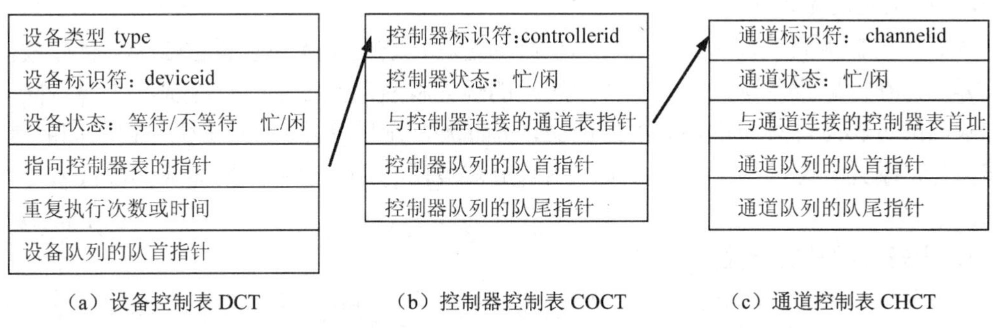

设备类型 type
设备标识符：deviceid
设备状态：等待/不等待　忙/闲
指向控制器表的指针
重复执行次数或时间
设备队列的队首指针
控制器标识符:controllerid
控制器状态：忙/闲
与控制器连接的通道表指针
控制器队列的队首指针
控制器队列的队尾指针
通道标识符：channelid
通道状态：忙/闲
与通道连接的控制器表首址
通道队列的队首指针
通道队列的队尾指针
（a）设备控制表 DCT
（b）控制器控制表 COCT
（c）通道控制表 CHCT

图 8-9　设备控制表、控制器控制表和通道控制表

8.5.2　设备分配应考虑的因素

设备作为一种十分重要的系统资源，是由操作系统统一管理和分配。设备分配的原则是既要充分发挥设备的使用效率，又要避免由于设备分配不合理造成进程死锁，另外设备分配还应具有一定的灵活性。

基于上述原则，进行设备分配时应综合考虑以下几个因素。

1．I/O 设备的固有属性

按照设备的共享属性，I/O 设备可分为三种类型：独占设备、共享设备和虚拟设备，对于这三种不同类型的设备，系统所采取的分配策略也有所不同。

1）独享分配策略

独享分配策略主要思想是：把一个设备分配给某进程后，这个设备便一直由该进程独占，直至进程释放这个设备后，系统才可以对它进行重新分配。这种分配策略的缺点是设备利用率低，而且还会引起系统死锁。独占设备分配时多采用此策略。

2）共享分配策略

对于共享设备如磁盘，多采用共享分配策略，即将共享设备同时分配给多个进程使用。采用共享分配策略时，因为有可能多个进程同时访问共享设备，所以要特别注意合理调度这些进程访问设备的先后顺序，使平均服务时间越短越好。

虽然虚拟设备本身是独占设备，但采用虚拟技术后，可以认为对应多台逻辑设备，这些逻辑设备可同时分配给多个进程使用。因此虚拟设备也被看成共享设备，采用共享分配策略进行设备分配。

2．设备的分配算法

设备的分配机制，除了与I/O设备的固有属性有关，还与系统所采用的分配算法有关。I/O 设备的分配算法与进程调度算法十分相似，主要采用的两种算法是：先请求先服务和优先权最高者优先。

1）先请求先服务

当多个进程对同一个设备提出 I/O 请求时，该算法要求把这些 I/O 请求的进程，按照发出请求的先后顺序排成一个等待队列，设备分配程序优先把设备分配给排在队首的进程。

2）优先权最高者优先

按照优先权最高者优先算法，优先权高的进程，它的 I/O 请求也被赋予高优先权。采用此算法，在形成设备队列时，是按照进程的优先权从高到低的顺序排成一个等待队列，优先级高的进程总是排在设备队列的前面，从而优先分配设备。而对于优先权相同的进程，则按照先请求先分配的原则排队分配。这种分配算法有助于进程尽快完成并释放所占有的资源。

3．设备分配的安全性

设备分配时要特别注意是否会产生死锁，尽量避免各进程循环等待资源的现象发生。基于设备分配的安全性考虑，通常有以下两种方式。

1）安全分配方式

在这种分配方式下，每当进程在发出 I/O 请求后，便进入阻塞状态，直至其 I/O 操作完成后才被唤醒。安全分配方式使得运行过程中的进程不保持任何资源，而处于阻塞状态的进程则没有机会和可能再请求其他新的资源，这就摒弃了造成死锁的四个必要条件之一的“请求和保持”条件，因此分配是安全的。但这种分配方式的缺点是 CPU 与 I/O 设备是串行工作，导致进程进展缓慢。

2）不安全分配方式

在这种分配方式下，进程在发出 I/O 请求后仍继续执行，执行过程中还可以发第二个I/O 请求、第三个 I/O 请求或更多个 I/O 请求，只有当该进程所请求的设备无法满足时，才进入阻塞状态。不安全分配方式的优点是一个进程可同时操作多个设备，进程推进速度快。它的缺点是因为具有“请求和保持”的特点，极易造成死锁。可以通过在设备分配程序中增加安全性计算的方式，避免死锁发生，但这在一定程度上增加了系统的开销。

4．设备的独立性

为了提高 OS 的可适应性和可扩展性，目前几乎所有的 OS 都实现了设备的独立性，也称为设备无关性。因为实施 I/O 操作的逻辑设备并不限于某个具体设备，是实际物理设备的抽象，所以分配设备时适应性好，灵活性强。在设备分配时，系统应提供逻辑设备名转换为某个具体的物理设备名的功能。

8.5.3 独占设备分配程序

在一个具有通道的计算机系统中进行设备分配，不仅要分配设备，而且还要分配与设备相连的设备控制器及通道。考虑到设备的独立性，应按如下步骤进行独占设备的分配。

1. 分配设备

（1）进程以逻辑设备名提出 I/O 请求。

（2）查找逻辑设备表 LUT，获得逻辑设备对应的物理设备在系统设备表 SDT 中的指针。

（3）按照指针所指位置开始顺序检索 SDT，直到找到一个与请求设备同类型、空闲且可安全分配的设备的 DCT，并将这个设备分配给请求进程；如果未找到安全可用的空闲设备，则把请求进程的 PCB 挂到相应类型设备队列上，等待唤醒和重新分配。

2. 分配设备控制器

系统把设备分配给 I/O 请求进程后，通过访问该设备的 DCT 表找到与它相连接的控制器的 COCT，根据 COCT 表中的状态字段判断该控制器是否忙碌，若忙，则把请求进程的 PCB 挂到该控制器的等待队列上，否则，将该控制器分配给进程。

3. 分配通道

系统把控制器分配给 I/O 请求进程后，再到该控制器的 COCT 中找出与其相连接的通道的 CHCT，根据 CHCT 表中的状态字段判断该通道是否忙碌，若忙，则把请求进程的 PCB 挂到该通道的等待队列上；否则将该通道分配给进程。

只有当设备、控制器和通道三者都分配成功后，才算真正完成设备分配工作。这之后便可启动设备，进行数据传送。

8.5.4 SPOOLing 技术

在计算机发展早期，为了缓解高速的 CPU 与慢速的外部设备之间速度不匹配的矛盾，引入了脱机输入输出技术。所谓脱机输入是指在一台专用于输入输出的外围处理机的控制下，事先将低速 I/O 设备上的数据传送到高速磁盘上，这样当 CPU 需要这些数据时，就不必将它们从低速 I/O 设备调入，而是直接从磁盘高速调入内存。同理，脱机输出是指在数据输出时，CPU 也是先将用户的结果数据输出到高速磁盘而不直接输出到 I/O 设备上，然后在适当的时候，再由外围处理机控制将这些数据从磁盘输出到低速 I/O 设备上。

多道程序技术出现后，完全可以通过程序模拟完成外围处理机的功能。用一道程序模拟脱机输入时外围处理机的功能，把低速 I/O 设备上的数据传送到高速的磁盘；再用另一道程序模拟脱机输出时外围处理机的功能，把数据从磁盘传送到低速 I/O 设备上。这样在主机的直接控制下，就可以实现脱机输入输出操作。因为此时的外围操作与 CPU 对数据的处理同时进行，故把这种在联机情况下实现的同时外围操作称为 SPOOLing（Simultaneous Peripheral Operation On Line），或称为假脱机操作。

1. SPOOLing 系统的组成

SPOOLing 技术是利用程序来模拟实现脱机输入、输出时的外围控制机的功能。因此

SPOOLing 系统不仅必须建立在具有多道程序功能的操作系统上，而且还应有高速外存（硬盘）的支持。如图 8-10 所示，SPOOLing 系统主要有以下三部分组成：

（1）输入井和输出井。在磁盘上开辟的两个大的存储空间。输入井模拟脱机输入时的磁盘，用于收容 I/O 设备输入的数据。输出井模拟脱机输出时的磁盘，用于收容用户程序的输出数据。

（2）输入缓冲区和输出缓冲区。在将数据存入输入井或输出井的过程中，为了缓解 CPU 与磁盘之间速度不匹配的矛盾，需要在内存中开辟输入缓冲区和输出缓冲区。输入缓冲区用于暂存由输入设备送来的数据，当输入缓冲区满或数据输入完成后，才将缓冲区内的数据成批传送到输入井。同理，输出缓冲区用于暂存从输出井送来的数据，以后再传送给输出设备。

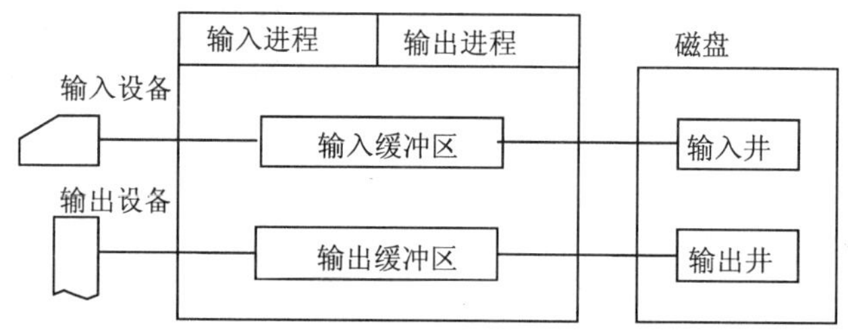

图 8-10　SPOOLing 系统的组成

（3）输入进程 SPi 和输出进程 SPo。进程 SPi 模拟脱机输入时的外围控制机，将用户要求的数据从输入设备通过输入缓冲区送入输入井中，当 CPU 需要输入数据时，直接从输入井读出送入内存。SPo 进程模拟脱机输出时的外围控制机，把用户要求输出的数据，先从内存送到输出井，当输出设备空闲时，再将输出井中的数据，经过输出缓冲区送到输出设备上。

可把输入进程 SPi 的操作分为两个部分：

① 存输入部分，完成将数据从输入设备读入并存放在输入井中；

② 取输入部分，完成将输入井的数据送入内存。

同理，把输出进程 SPo 的操作也分为两个部分：

① 存输出部分，完成将 CPU 处理的结果从内存送到输出井中；

② 取输出部分，完成将输出井的数据送到外部设备。

2．SPOOLing 系统的特点

SPOOLing 系统具有如下特点。

（1）提高了 I/O 速度。采用 SPOOLing 技术可以把对低速设备的 I/O 操作演变成对高速磁盘中输入井或输出井的访问，缓解了 CPU 与外部设备的速度不匹配的矛盾，极大地提高了 I/O 的速度。

（2）实现了虚拟设备功能。宏观上，虽然还是多个进程在同时使用一台独占设备，但对每一个进程而言，它们会认为自己独占了一个设备。当然这个设备只是逻辑上的设备，并不是真正的物理设备。SPOOLing 技术既实现了将独占设备变换为若干台对应的逻辑设备，又实现了将独占设备改造成共享设备。

（3）增加了系统的复杂性。在 SPOOLing 系统中，不仅输入井和输出井占用大量磁盘空间，输入缓冲区和输出缓冲区也占用大量内存空间，同时输入和输出进程分别增加了对输入井和输出井的存取操作，这在一定程度上增加了系统的复杂性。

3．共享打印机

打印机虽然是独占设备，但是通过 SPOOLing 技术，可以将它改造为一台可供多个用户共享的设备，从而提高设备的利用率，也方便用户使用。

当用户进程请求打印输出时，SPOOLing 系统同意为它打印输出，但并不真正把打印机分配给该用户进程，而是由输出进程为它在输出井中申请一个空闲盘块区，并将要打印的数据以文件的形式存放其中。输出进程同时为用户进程申请一张空白的用户打印请求表，把用户的打印要求填写在表中，并将该表挂到请求打印队列上。当打印机空闲时，输出进程便从请求打印队列的队首取出一张请求打印表，根据表中的要求将要打印的数据从输出井传送到内存缓冲区，由打印机进行打印。打印完后，输出进程会查看请求打印队列中是否还有等待打印的请求表，如果有，则按照先进先出顺序依次将打印队列中的文件送入打印机，完成实际的打印工作。

8.6 磁盘存储器管理

磁盘存储器作为一种大容量、高速度的存储设备在现代计算机系统中发挥着非常重要的作用。它不仅是程序、数据和其他信息文件最主要的联机存储器，也是实现虚拟存储系统所必需的存储设备。磁盘系统的存取速度和可靠性，直接影响整个计算机系统的性能。因此，如何改善磁盘系统的性能，有效地管理磁盘存储器，已成为操作系统非常重要的任务之一。

8.6.1 磁盘存储器简述

1．磁盘存储器的信息记录格式

磁盘存储器是利用磁记录技术在涂有磁记录介质的旋转圆盘上进行数据存储的辅助存储器。它通常由磁盘、磁盘驱动器（或称磁盘机）和磁盘控制器构成。其中磁盘是存储介质，是由若干个盘片构成的盘片组。盘片组所有盘片都被安装在主轴电机上，由主轴系统驱动盘片组以额定转速稳定旋转。

盘片组中的每个盘片分为上、下两个盘面。若盘片组中有 S 个盘片，则共有 $2S$ 个盘面。通常情况下，整个盘片组中的最上和最下两个盘面用于磁头定位，因此实际可用的盘面为 $2S-2$ 个。把所有可用的盘面从上至下依次编号，该编号称为盘面号或磁头号。

每个盘面都被划分成数目相等的同心圆，称为磁道。磁道由外向里依次编号，该编号称为磁道号。所有盘面中具有相同编号的磁道构成一个柱面，因此柱面号就是磁道号，柱面数与磁道数相同。

每条磁道在逻辑上又被划分成若干个区域，一个区域称为一个扇区，需要存储的信息存放在扇区中。对每个扇区依次编号，称为扇区号。对于等扇区结构的磁盘，由于每个磁道的扇区数相同，因此内层磁道的扇区存放的信息密度要高于外层。

扇区既是磁盘读取和写入数据的基本单位，同时也是操作系统进行磁盘空间管理和分配的基本单位。磁盘地址是由柱面号、盘面号和扇区号三个参数组成，是一个三维地址，可通过磁盘控制器中的地址翻译器将其转换为一维地址。

2．磁盘的访问时间

磁盘存储器是依据地址存取信息的。进行磁盘访问时，磁盘驱动器首先根据磁盘地址中的柱面（磁道）号，控制所有磁头沿盘面的半径方向一起移动，将磁头定位于该柱面上；其次根据盘面号选定该柱面中某一个盘面上的磁头进行工作；然后再根据磁盘地址中的扇区号，旋转整个磁盘组，将磁头定位到该扇区的起始位置；最后进行磁盘的读写操作。因此执行一次磁盘 I/O 操作所需的访问时间主要包括以下三部分。

1）寻道时间 T_s

寻道时间是指磁头从当前位置移动到指定磁道所花费的时间，该时间包括磁盘的启动时间 s 和磁头移动 n 条磁道所需的时间，可以用如下公式表示：

$$T_s=m\times n+s$$

其中 m 为磁头移动一条磁道所用的时间，是常数，通常与磁盘驱动器的工作速度有关，一般磁盘 m=0.2，高速磁盘 $m \leqslant 0.1$。对于一个磁盘而言，它的启动时间 s 也是相对固定的。因此，只有当磁头移动的距离 n 变小时，寻道时间 T_s 才会相应地缩短。

2）旋转延迟时间 T_r

旋转延迟时间也称为等待时间，是指磁头定位到指定扇区的起始位置所需的时间，与磁盘驱动器的旋转速度有关。若磁盘每秒钟的转数为 r，则平均旋转延迟时间 T_r=1/2r。硬盘的旋转速度大多为 5400 r/min，每转需时 11.1 ms，平均旋转延迟时间 T_r 为 5.55 ms。

3）传输时间 T_t

传输时间是指将扇区上的数据从磁盘读出或向磁盘写入数据所用的时间。T_t 的大小不仅与磁盘的旋转速度有关，而且还与每次读写的字节数 b 有关。

$$T_t=\frac{b}{rN}$$

其中，r 为磁盘每秒钟的转数，N 为一条磁道上的字节数。

因此，磁盘的访问时间 T_a 表示为：

$$T_a=T_s+\frac{1}{2r}+\frac{b}{rN}$$

由上式可以看出，在总的访问时间 T_a 中，寻道时间和旋转延迟时间占据了大部分的访问时间，它们均与读写数据的字节数无关。同时由于磁头移动的动作要比磁头旋转的动作慢，因此寻道时间需要的时间最长。

8.6.2 磁盘调度算法

磁盘是可供多个进程共享的设备，当有多个进程请求访问磁盘时，应采用一种最佳调度算法来决定各进程的执行次序，尽可能地使磁盘的平均访问时间最短，提高磁盘的访问效率。由于在磁盘的访问过程中，寻道是花费时间最长的操作。因此，磁盘调度的主要目标是缩短平均寻道时间。常用的调度算法有先来先服务、最短寻道时间优先、扫描算法和循环扫描算法等。

1．先来先服务

先来先服务（First-Come First Served，FCFS）算法是最简单的磁盘调度算法。该算法只考虑进程提出磁盘访问请求的先后次序，而不考虑所请求访问磁道的物理位置。采用先来先服务磁盘调度算法，每个进程的磁盘访问都会按照请求的先后顺序依次得到处理，不会出现某一个进程的请求长期得不到处理的情况，算法具有一定的公平性。

假如磁盘共有 200 个柱面，其编号为 0～199，当前磁头在 53 号柱面上，现有 8 个进程先后提出磁盘访问请求，按请求时间先后顺序进行排队，请求队列为：98、183、37、122、14、124、65、67。图 8-11（a）给出了按 FCFS 算法进行调度时，各进程调度的次序、每次磁头移动的距离和平均寻道长度。通过这个例子不难看出，该算法存在的问题是当磁盘访问的进程数量较多，并且所访问的磁道相互距离比较分散时，会造成磁头移动距离加大，从而导致平均寻道时间增加。

2．最短寻道时间优先

为了尽可能地减少磁头的移动距离，最短寻道时间优先（Shortest Seek Time First，SSTF）算法的主要思想是优先选择距离当前磁头最近的访问请求进行服务，以保证每次的寻道时间最短，但它无法保证平均寻道时间最短。

图 8-11（b）给出了上面例子中的 8 个进程按 SSTF 算法进行调度时，各进程调度的次序、每次磁头移动的距离和平均寻道长度。与图 8-11（a）比较可以看出：采用 SSTF 算法磁头平均每次移动的距离明显低于 FCFS 算法，有较好的寻道性能。SSTF 算法的缺点是只要有新的进程请求访问距离当前磁头较近的磁道，按照 SSTF 算法就必须优先处理这些访问请求，导致一些与当前磁道距离较远的访问请求长期等待得不到服务，造成进程“饥饿”现象。

被访问的下一个磁道号	移动距离
98	45
183	85
37	146
122	85
14	108
124	110
65	59
67	2
平均寻道长度：80	

（a）FCFS 调度算法

被访问的下一个磁道号	移动距离
65	12
67	2
37	30
14	23
98	84
122	24
124	2
183	59
平均寻道长度：29.5	

（b）SSTF 调度算法

图 8-11　FCFS 和 SSTF 调度算法

3．扫描算法 SCAN

为了避免出现进程“饥饿”的现象，对 SSTF 算法略加进行修改，便形成 SCAN 算法。与 SSTF 算法相比，SCAN 算法不仅考虑所访问的磁道与当前磁道的距离，还优先考虑磁头目前的移动方向。该算法的具体思想是：当没有磁盘访问请求时，磁头不动；当有磁盘访问请求时，磁头按一个方向移动，并优先处理该方向上与当前磁道距离最近的访问请求，直至该方向上没有访问请求时，才改变磁头的移动方向；同理，磁头反向移动时也是优先处理该方向上距离当前磁道最近的磁盘访问。不难看出，因为磁头是沿着一个方向移动，只有当该方向没有访问请求时，才改变移动方向，从而避免了进程“饥饿”现象的产生。此算法与自然界中电梯的工作方式极为相像，也称“电梯调度”算法。图 8-12（a）给出了按 SCAN 算法对 8 个进程进行调度的次序、磁头移动的距离和平均寻道长度等情况。

4．循环扫描算法（CSCAN）

SCAN 算法不仅可以获得较好的寻道性能，而且还防止了进程“饥饿”现象的产生，因此得到较为广泛的应用。但该算法仍然存在一些问题，如当磁头正在向外或向里移动的过程中，恰好刚刚读取过的磁道出现了一个新的磁盘访问请求，按照 SCAN 算法，此时磁头不能改变移动方向，那么这个访问请求只能等到磁头反方向移动到该磁道时，才能进行处理，等待的时间比较长。为了解决这个问题，又引入了循环扫描算法。循环扫描算法又称单向扫描算法，此算法规定磁头移动方向始终保持不变，属于单向移动，或从里向外，或从外向里。假如磁头的移动方向是从里向外，与 SCAN 算法一样，也是优先处理该方向上与当前磁道距离最近的访问请求，直至处理完最外面的磁盘访问请求。与 SCAN 算法不同的是，此时磁头并不改变移动方向，而是立即返回到最里面要访问的磁道，然后仍然按照从里向外的方向依次处理新的磁道访问请求，从而构成循环扫描。图 8-12（b）给出了按 CSCAN 算法对 8 个进程进行调度的次序、磁头移动的距离和平均寻道长度等情况。

被访问的下一个磁道号	移动距离
65	12
67	2
98	31
122	24
124	2
183	59
37	146
14	23
平均寻道长度：37.4	

（a）SCAN 调度算法

被访问的下一个磁道号	移动距离
65	12
67	2
98	31
122	24
124	2
183	59
14	169
37	23
平均寻道长度：40.3	

（b）CSCAN 调度算法

图 8-12　SCAN 和 CSCAN 调度算法

5．N-Step-SCAN 和 FSCAN 算法

在 SSTF、SCAN 和 CSCAN 的算法中，当一个或多个进程对某一磁道有较高的访问频率时，就会不断地产生对这个磁道的 I/O 请求。由于始终是对一个磁道进行访问，磁臂在很长一段时间内都不会移动，把这种现象称为磁臂粘着。磁臂粘着现象会造成对磁盘访问不均衡，为了避免这种粘着现象产生，将磁盘请求队列分成若干个段，一次只有一段被完全处理。这种思想对应的两个算法分别是 N-Step-SCAN 算法和 FSCAN 算法。

1）N-Step-SCAN 算法

该算法主要思想是将磁盘请求队列分成若干个长度为 N 的子队列，按照 FCFS 算法依次处理这些子队列，即对一个子队列处理完后，再处理其他子队列，而每个子队列内部的磁盘请求则是按照 SCAN 算法处理。如果在处理某子队列过程中，出现新的磁盘访问请求，则把新的请求加入其他子队列中。此算法保证了只有处理完一个子队列后，才能处理下一个子队列，从而避免了粘着现象的产生。当 N 很大时，该算法接近于 SCAN；当 N=1 时，该算法蜕化为 FCFS。

2）FSCAN 算法

FSCAN 算法可以看成是 N-Step-SCAN 算法的简化形式。它只将磁盘请求队列分成两个子队列，一个是由当前所有磁盘请求形成的队列，另一个是初始为空的队列。在按 SCAN 算法扫描处理第一个队列时，若有新的磁盘 I/O 请求，则把它加入空队列中，从而形成一个新的等待处理队列。当第一个队列处理完后，再处理这个新的等待处理队列。

8.6.3 独立磁盘冗余阵列

近几年，处理器技术飞速发展，CPU 的速度已提高了几个数量级。与其相比，磁盘技术的发展和性能的提高相对滞后，这就造成了处理器与磁盘之间在速度上的差距变得越来越大，磁盘访问速度慢已成为严重影响计算机系统性能的主要瓶颈。当然，可以采用合理磁盘调度算法来减少磁盘平均服务时间，达到提高磁盘访问速度的目的。但这种方式对于提高计算机系统整体性能，影响十分有限。

随着并行处理技术的广泛应用，人们提出将并行处理的思想应用于磁盘系统。1987 年由美国加州大学伯克利分院提出 RAID 概念，RAID 全称为 Redundant Array of Independent Disks，是“独立磁盘冗余阵列”（最初为“廉价磁盘冗余阵列”）的缩略语，即将 N 台磁盘驱动器组成磁盘阵列，利用一台磁盘阵列控制器实现对整个磁盘阵列统一管理和控制。系统可以把数据分为若干个子盘块数据，分别存储到磁盘阵列中各磁盘的相同位置上。当要将数据传送到内存时，采取并行传输方式，将各个子盘块上的数据同时传输到内存中。由于磁盘阵列采用并行存取方式，可将磁盘访问速度提高 N–1 倍。通过 RAID 阵列技术可以将若干个磁盘组成一个快速、可靠的大容量磁盘系统。

1．RAID 的优点

RAID 技术通过在多个磁盘上同时存储和读取数据的方式，大幅提高存储系统的数据吞吐量。在 RAID 中，可以让很多磁盘驱动器同时传输数据，而这些磁盘驱动器在逻辑上又是一个磁盘驱动器。使用 RAID 可以达到单个磁盘驱动器几倍、几十倍甚至上百倍的速率，解决了由于处理器与磁盘之间的速度不匹配而造成的矛盾。

RAID 技术通过采用冗余存储和数据校验方式，提供了容错功能，极大地提高磁盘存储数据的可靠性。对于普通的磁盘驱动器除了通过在磁盘上保存循环冗余校验码的方式进行数据校验，无法提供其他的容错功能。RAID 技术的容错是建立在每个磁盘驱动器的硬件容错功能之上，具有较高的安全性。在很多 RAID 模式中，都有较为完备的相互校验/恢复的措施，甚至是直接相互的镜像备份，从而大大提高了 RAID 系统的容错度，提高了系统的稳定性。

2．RAID 的分级

按照数据在磁盘上的组织形式和具有的特点，业界已制定了一套工业标准，该标准规定了多种数据存放方法，称为 RAID 级别。最初 RAID 分为 6 级，即 RAID0～5，后来扩充到 RAID 7。需要说明的是，这 8 级只是构造不同 RAID 时的性能体现，不同级别对应不同的数据存放方式，而不是隶属关系，不同级别之间并无继承关系，高级不依赖低级。

（1）RAID 0 级。它仅提供并行交叉存取，虽能有效地提高磁盘 I/O 速度，却没有冗余校验功能。只要阵列中有一个磁盘出现故障，那么数据就会全盘丢失，可靠性很低。

（2）RAID 1 级。具有磁盘镜像功能。每个工作盘都有一个对应的镜像盘，两者保存的数据完全相同。在每次访问磁盘时，可利用并行读、写特性，将数据同时写入工作盘和镜像盘，以保持数据的一致性。在不影响性能情况下，RAID 1 级最大限度地保证系统的可靠性和可修复性，同时它也具有很高的数据冗余，磁盘利用率仅为 50%，成本最高，多用于保存重要数据。

（3）RAID 3 级。采用了奇偶校验技术，将奇偶校验码存放在一个专用的校验盘上，相比 RAID 1 级，减少了所需要的冗余磁盘数。如果某个磁盘出现故障，它上面的正确数据可以通过对其他磁盘上的数据进行异或运算得到。但在写入数据时，需要计算校验位，因而速度受到影响。

（4）RAID 5 级。无独立的校验盘，将用于纠错的校验信息以螺旋方式分布在阵列的所有磁盘上。每个驱动器都有各自独立的数据通路，独立地进行读/写。不论是对大量数据还是少量数据，都具有较好的读写性能。

（5）RAID 6 级和 RAID 7 级。在 RAID6 阵列中，设置了一个专用的、可快速访问的异步校验盘。该盘具有独立的数据访问通路，具有比 RAID3 级及 RAID5 级更好的性能，但其性能改进得很有限，且价格昂贵。RAID7 级是对 RAID6 级的改进，在该阵列中的所有磁盘，都具有较高的传输速率和优异的性能，是目前最高档次的磁盘阵列，但价格也较高。

Chapter 8 I/O Systems

The two main jobs of a computer are I/O and processing. In many cases, the main job is I/O, and the processing is merely incidental. For instance, when we browse a web page or edit a file, our immediate interest is to read or enter some information, not to compute an answer. The control of devices connected to the computer is a major concern of operating-system designers. Because I/O devices vary so widely in their function and speed (consider a mouse, a hard disk, and a CD-ROM jukebox), varied methods are needed to control them. These methods form the I/O subsystem of the kernel, which separates the rest of the kernel from the complexities of managing I/O devices.

8.1 I/O Hardware

Computers operate a great many kinds of devices. Most fit into the general categories of storage devices (disks, tapes), transmission devices (network cards, modems), and human-interface devices (screen, keyboard, mouse). Other devices are more specialized, such as the steering of a military fighter jet or a space shuttle. In these aircraft, a human gives input to the flight computer via a joystick and foot pedals, and the computer sends output commands that cause motors to move rudders, flaps, and thrusters. Despite the incredible variety of I/O devices, though, we need only a few concepts to understand how the devices are attached and how the software can control the hardware.

A device communicates with a computer system by sending signals over a cable or even through the air. The device communicates with the machine via a connection point (or port)—for example, a serial port. If devices use a common set of wires, the connection is called a bus. A bus is a set of wires and a rigidly defined protocol that specifies a set of messages that can be sent on the wires. In terms of the electronics, the messages are conveyed by patterns of electrical voltages applied to the wires with defined timings. When device A has a cable that plugs into device B, and device B has a cable that plugs into device C, and device C plugs into a port on the computer, this arrangement is called a daisy chain. A daisy chain usually operates as a bus.

A controller is a collection of electronics that can operate a port, a bus, or a device. A serial-port controller is a simple device controller. It is a single chip (or portion of a chip) in the computer that controls the signals on the wires of a serial port. By contrast, a SCSI bus controller is not simple. Because the SCSI protocol is complex the SCSI bus controller is often implemented as a separate circuit board (or a host adapter) that plugs into the computer. It typically contains a processor, a microcode, and some private memory to enable it to process the SCSI protocol messages. Some devices have their own built-in controllers. If you look at a disk drive, you will see a circuit board attached to one side. This board is the disk controller. It implements the disk side of the protocol for some kind of connection—SCSI or ATA, for instance. It has microcode and a processor to do many tasks, such as bad-sector mapping, perfecting, buffering, and caching.

8.2 I/O Control

8.2.1 Polling

The complete protocol for interaction between the host and a controller can be intricate, but the basic handshaking notion is simple. We explain handshaking with an example. We assume that 2 bits are used to coordinate the producer —consumer relationship between the controller and the host. The controller indicates its state through the busy bit in the status register. (Recall that to set a bit means to write a 1 into the bit and to clear a bit means to write a 0 into it.) The controller sets the busy bit when it is busy working and clears the busy bit when it is ready to accept the next command. The host signals its wishes via the command-ready bit in the command register. The host sets the command-ready bit when a command is available for the controller to execute. For this example, the host writes output through a port, coordinating with the controller by handshaking as follows:

(1) The host repeatedly reads the busy bit until that bit becomes clear.

(2) The host sets the write bit in the command register and writes a byte into the data-out register.

(3) The host sets the command-ready bit.

(4) When the controller notices that the command-ready bit is set, it sets the busy bit.

(5) The controller reads the command register and sees the write command. It reads the data-out register to get the byte and does the I/O to the device.

(6) The controller clears the command-ready bit, clears the error bit in the status register to indicate that the device 1/O succeeded, and clears the busy bit to indicate that it is finished.

This loop is repeated for each byte.

In step 1, the host is busy-waiting or polling: It is in a loop, reading the status register over and over until the busy bit becomes clear. If the controller and device are fast, this method is a reasonable one. But if the wait may be long, the host should probably switch to another task. Howerver, how does the host know when the controller has become idle? For some devices, the host must service the device quickly, or data will be lost. For instance, when datum are streaming in on a serial port or from a keyboard, the small buffer on the controller will overflow and datum will be lost if the host waits too long before returning to read the bytes.

8.2.2 Interrupts

The basic interrupt mechanism works as follows. The CPU hardware has a wire called the interrupt-request line that the CPU senses after executing every instruction. When the CPU detects that a controller has asserted a signal on the interrupt request line, the CPU performs a state save and jumps to the interrupt-handler routine at a fixed address in memory. The interrupt handler determines the cause of the interrupt, performs the necessary processing, performs a state restore, and executes a return from interrupt instruction to return the CPU to the execution state prior to the interrupt. We say that the device controller raises an interrupt by asserting a signal on the interrupt request line, the CPU catches the interrupt and dispatches it to the interrupt handler, and the handler clears the interrupt by servicing the device.

This basic interrupt mechanism enables the CPU to respond to an asynchronous event, as when a device controller becomes ready for service. In a modern operating system, however, we need more sophisticated interrupt-handling features.

(1) We need the ability to defer interrupt handling during critical processing.

(2) We need an efficient way to dispatch to the proper interrupt handler for a device without first polling all the devices to see which one raised the interrupt.

(3) We need multilevel interrupts, so that the operating system can distinguish between high- and low-priority interrupts and can respond with the appropriate degree of urgency.

In modern computer hardware, these three features are provided by the CPU and by the interrupt-controller hardware.

Most CPUs have two interrupt request lines. One is the nonmaskable interrupt, which is reserved for events such as unrecoverable memory errors. The second interrupt line is maskable: It can be turned off by the CPU before the execution of critical instruction sequences that must not be interrupted. The maskable interrupt is used by device controllers to request service.

The interrupt mechanism accepts an address—a number that selects a specific interrupt-handling routine from a small set. In most architectures, this address is an offset in a table called the interrupt vector. This vector contains the memory addresses of specialized interrupt handlers. The purpose of a vectored interrupt mechanism is to reduce the need for a single interrupt handler to search all possible sources of interrupts to determine which one needs service. In practice, however, computers have more devices (and, hence, interrupt handlers) than they have address elements in the interrupt vector. A common way to solve this problem is to use the technique of interrupt chaining, in which each element in the interrupt vector points to the head of a list of interrupt handlers. When an interrupt is raised, the handlers on the corresponding list are called one by one, until one is found that can service the request. This structure is a compromise between the overhead of a huge interrupt table and the inefficiency of dispatching to a single interrupt handler.

The interrupt mechanism also implements a system of interrupt priority levels. This mechanism enables the CPU to defer the handling of low-priority interrupts without masking off all interrupts and makes it possible for a high-priority interrupt to preempt the execution of a low-priority interrupt.

8.2.3 Direct Memory Access

For a device that does large transfers, such as a disk drive, it seems wasteful to use an expensive general-purpose processor to watch status bits and to feed data into a controller register one byte at a time—a process termed programmed I/O (PIO). Many computers avoid burdening the main CPU with PIO by offloading some of this work to a special-purpose processor called a direct-memory-access (DMA) controller. To initiate a DMA transfer, the host writes a DMA command block into memory. This block contains a pointer to the source of a transfer, a pointer to the destination of the transfer, and a count of the number of bytes to be transferred. The CPU writes the address of this command block to the DMA controller, then goes on with other work. The DMA controller proceeds to operate the memory bus directly, placing addresses on the bus to perform transfers without the help of the main CPU. A simple DMA controller is a standard component in PCs, and bus-mastering I/O boards for the PC usually contain their own high-speed DMA hardware.

Handshaking between the DMA controller and the device controller is performed via a pair of wires called DMA-request and DMA-acknowledge. The device controller places a signal on the DMA-request wire when a word of data is available for transfer. This signal causes the DMA controller to seize the memory bus, to place the desired address on the memory-address wires, and to place a signal on the DMA-acknowledge wire. When the device controller receives the DMA-acknowledge signal, it transfers the word of data to memory and removes the DMA-request signal.

When the entire transfer is finished, the DMA controller interrupts the CPU. When the DMA controller seizes the memory bus, the CPU is momentarily prevented from accessing main memory, although it can still access data items in its primary and secondary caches. Although this cycle stealing can slow down the CPU computation, offloading the data-transfer work to a DMA controller generally improves the total system performance. Some computer architectures use physical memory addresses for DMA, but others perform direct virtual memory access (DVMA), using virtual addresses that undergo translation to physical addresses. DVMA can perform a transfer between two memory-mapped devices without the intervention of the CPU or the use of main memory.

8.3 Principles of I/O Software

Let us now turn away from the I/O hardware and look at the I/O software. First we will look at the goals of the I/O software and then at the different ways I/O can be done from the point of view of the operating system.

8.3.1 Goals of I/O Software

A key concept in the design of I/O software is device independence. What this means is that it should be possible to write programs that can access any I/O device without having to specify the device in advance. For example, a program that reads a file as input should be able to read a file on a floppy disk, on a hard disk, or on a CD-ROM, without having to modify the program for each different device. Similarly, one should be able to type a command such as

sort <input >output

and have it work with input coming from a floppy disk, an IDE disk, a SCSI disk, or the keyboard, and the output going to any kind of disk or the screen. It is up to the operating system to take care of the problems caused by the fact that these devices really are different and require very different command sequences to read or write.

Closely related to device independence is the goal of uniform naming. The name of a file or a device should simply be a string or an integer and not depend on the device in any way. In UNIX and MINIX 3, all disks can be integrated into the file system hierarchy in arbitrary ways so the user need not be aware of which name corresponds to which device. For example, a floppy disk can be mounted on top of the directory /usr/ast/backup so that copying a file to that directory copies the file to the diskette. In this way, all files and devices are addressed the same way: by a path name.

Another important issue for I/O software is error handling. In general, errors should be handled as close to the hardware as possible. If the controller discovers a read error, it should try to correct the error itself if it can. If it cannot, then the device driver should handle it, perhaps by just trying to read the block again. Many errors are transient, such as read errors caused by specks of dust on the read head, and will go away if the operation is repeated. Only if the lower layers are not able to deal with the problem should the upper layers be told about it. In many cases, error recovery can be done transparently at a low level without the upper levels even knowing about the error.

The final concept that we will mention here is sharable versus dedicated devices. Some I/O devices, such as disks, can be used by many users at the same time. No problems are caused by multiple users having open files on the same disk at the same time. Other devices, such as tape drives, have to be dedicated to a single user until that user is finished. Then another user can have the tape drive. Having two or more users writing blocks intermixed at random to the same tape will definitely not work. Introducing dedicated (unshared) devices also introduces a variety of problems, such as deadlocks. Again, the operating system must be able to handle both shared and dedicated devices in a way that avoids problems.

8.3.2 Interrupt Handlers

Interrupts are an unpleasant fact of life; although they cannot be avoided, they should be hidden away, deep in the bowels of the operating system, so that as little of the operating system as possible knows about them. The best way to hide them is to have the driver starting an I/O operation block until the I/O has completed and the interrupt occurs. The driver can block itself by doing a down on a semaphore, a wait on a condition variable, a receive on a message, or something similar, for example.

When the interrupt happens, the interrupt procedure does whatever it has to in order to handle the interrupt. Then it can unblock the driver that started it. In some cases it will just complete up on a semaphore. In others it will do a signal on a condition variable in a monitor. In still others, it will send a message to the blocked driver. In all cases the net effect of the interrupt will be that a driver that was previously blocked will now be able to run. This model works best if drivers are structured as independent processes, with their own states, stacks, and program counters.

8.3.3 Device Drivers

Each device controller has registers used to give it commands or to read out its status or both. The number of registers and the nature of the commands vary radically from device to device. For example, a mouse driver has to accept information from the mouse telling how far it has moved and which buttons are currently depressed. In contrast, a disk driver has to know about sectors, tracks, cylinders, heads, arm motion, motor drives, head settling times, and all the other mechanics of making the disk work properly. Obviously, these drivers will be very different.

Thus, each I/O device attached to a computer needs some device-specific code for controlling it. This code, called the device driver, is generally written by the device's manufacturer and delivered along with the device on a CD-ROM. Since each operating system needs its own drivers, device manufacturers commonly supply drivers for several popular operating systems.

Each device driver normally handles one device type, or one class of closely related devices. For example, it would probably be a good idea to have a single mouse driver, even if the system supports several different brands of mice. As another example, a disk driver can usually handle multiple disks of different sizes and different speeds, and perhaps a CD-ROM as well. On the other hand, a mouse and a disk are so different that different drivers are necessary.

In order to access the device's hardware, meaning the controller's registers, the device driver traditionally has been part of the system kernel. This approach gives the best performance and the worst reliability since a bug in any device driver can crash the entire system. MINIX 3 departs from this model in order to enhance reliability. As we shall see, in MINIX 3 each device driver is now a separate user-mode process.

As we mentioned earlier, operating systems usually classify drivers as block devices, such as disks, or character devices, such as keyboards and printers. Most operating systems define a standard interface that all block drivers must support and a second standard interface that all character drivers must support. These interfaces consist of a number of procedures that the rest of the operating system can call to get the driver to do work for it.

In general terms, the job of a device driver is to accept abstract requests from the device-independent software above it and see to it that the request is executed. A typical request to a disk driver is to read block *n*. If the driver is idle at the time a request comes in, it starts carrying out the request immediately. If, however, it is already busy with a request, it will normally enter the new request into a queue of pending requests to be dealt with as soon as possible.

The first step in actually carrying out an I/O request is to check that the input parameters are valid and to return an error if they are not. If the request is valid the next step is to translate it from abstract to concrete terms. For a disk driver, this means figuring out where on the disk the requested block actually is, checking to see if the drive's motor is running, determining if the arm is positioned on the proper cylinder, and so on. In short, the driver must decide which controller operations are required and in what sequence.

Once the driver has determined which commands to issue to the controller, it starts issuing them by writing into the controller's device registers. Simple controllers can handle only one command at a time. More sophisticated controllers are willing to accept a linked list of commands, which they then carry out by themselves without further help from the operating system.

After the command or commands have been issued, one of two situations will apply. In many cases the device driver must wait until the controller does some work for it, so it blocks itself until the interrupt comes in to unblock it. In other cases, however, the operation finishes without delay, so the driver needs not block. As an example of the latter situation, scrolling the screen on some graphics cards requires just writing a few bytes into the controller's registers. No mechanical motion is needed, so the entire operation can be completed in a few microseconds.

In the former case, the blocked driver will be awakened by the interrupt. In the latter case, it will never go to sleep. Either way, after the operation has been completed, it must check for errors. If everything is all right, the driver may have data to pass to the device-independent software (e.g., a block just read). Finally, it returns some status information for error reporting back to its caller. If any other requests are queued, one of them can now be selected and started. If nothing is queued, the driver blocks waiting for the next request.

Dealing with requests for reading and writing is the main function of a driver, but there may be other requirements. For instance, the driver may need to initialize a device at system startup or the first time it is used. Also, there may be a need to manage power requirements, handle Plug in Play, or log events.

8.4 Buffering and Device Management

8.4.1 I/O Scheduling

To schedule a set of I/O requests means to determine a good order in which to execute them.

The order in which applications issue system calls rarely is the best choice. Scheduling can improve overall system performance, can share device access fairly among processes, and can reduce the average waiting time for I/O to complete. Here is a simple example to illustrate the opportunity. Suppose that a disk arm is near the beginning of a disk and that three applications issue blocking read calls to that disk. Application 1 requests a block near the end of the disk, application 2 requests one near the beginning, and application 3 requests one in the middle of the disk. The operating system can reduce the distance that the disk arm travels by serving the applications in the order 2, 3, 1. Rearranging the order of service in this way is the essence of I/O scheduling.

Operating-system developers implement scheduling by maintaining a wait queue of requests for each device. When an application issues a blocking I/O system call, the request is placed on the queue for that device. The I/O scheduler rearranges the order of the queue to improve the overall system efficiency and the average response time experienced by applications. The operating system may also try to be fair, so that no one application receives especially poor service, or it may give priority service for delay-sensitive requests. For instance, requests from the virtual memory subsystem may take priority over application requests.

When a kernel supports asynchronous I/O, it must be able to keep track of many I/O requests at the same time. For this purpose, the operating system might attach the wait queue to a device-status table. The kernel manages this table, which contains an entry for each I/O device. Each table entry indicates the device's type, address, and state (not functioning, idle, or busy). If the device is busy with a request, the type of request and other parameters will be stored in the table entry for that device.

One way in which the I/O subsystem improves the efficiency of the computer is by scheduling I/O operations. Another way is by using storage space in main memory or on disk via techniques called buffering, caching, and spooling.

8.4.2 Buffering

A buffer is a memory area that stores data while they are transferred between two devices or between a device and an application. Buffering is done for three reasons. One reason is to cope with a speed mismatch between the producer and consumer of a data stream. Suppose, for example, that a file is being received via modem for storage on the hard disk. The modem is about a thousand times slower than the hard disk. So a buffer is created in main memory to accumulate the bytes received from the modern. When an entire buffer of data has arrived, the buffer can be written to disk in a single operation. Since the disk write is not instantaneous and the modem still needs a place to store additional incoming data, two buffers are used. After the modem fills the first buffer, the disk write is requested. The modem then starts to fill the second buffer while the first buffer is written to disk. By the time the modem has filled the second buffer, the disk write from the first one should have completed, so the modem can switch back to the first buffer while the disk writes the second one. This double buffering decouples the producer of data from the consumer, thus relaxing timing requirements between them.

A second use of buffering is to adapt between devices that have different data-transfer sizes. Such disparities are especially common in computer networking, where buffers are used widely for fragmentation and reassembly of messages. At the sending side, a large message is fragmented into small network packets. The packets are sent over the network, and the receiving side places them in a reassembly buffer to form an image of the source data.

A third use of buffering is to support copy semantics for application I/O. An example will clarify the meaning of "copy semantics." Suppose that an application has a buffer of data that it wishes to write to disk. It calls the write() system call, providing a pointer to the buffer and an integer specifying the number of bytes to write. After the system call returns, what happens if the application changes the contents of the buffer? With copy semantics, the version of the data written to disk is guaranteed to be the version at the time of the application system call, independent of any subsequent changes in the application's buffer. A simple way in which the operating system can guarantee copy semantics is for the write() system call to copy the application, data into a kernel buffer before returning control to the application. The disk write is performed from the kernel buffer, so that subsequent changes to the application buffer have no effect. Copying of data between kernel buffers and application data space is common in operating systems, despite the overhead that this operation introduces, because of the clean semantics. The same effect can be obtained more efficiently by clever use of virtual memory mapping and copy-on-write page protection.

8.4.3 Streaming

UNIX System V has an interesting mechanism, called STREAMS, that enables an application to assemble pipelines of driver code dynamically. A stream is a full-duplex connection between a device driver and a user-level process. It consists of a stream head that interfaces with the user process, a driver end that controls the device, and zero or more stream modules between them. The stream head, the driver end, and each module contain a pair of queues a read queue and a write queue. Message passing is used to transfer data between queues.

Modules provide the functionality of STREAMS processing; they are pushed onto a stream by use of the ioctl() system call. For example, a process can open a serial-port device via a stream and can push on a module to handle input editing. Because messages are exchanged between queues in adjacent modules, a queue in one module may overflow an adjacent queue. To prevent this from occurring, a queue may support flow control. Without flow control, a queue accepts all messages and immediately sends them on to the queue in the adjacent module without buffering them. A queue supporting flow control buffers messages and does not accept messages without sufficient buffer space; this process involves exchanges of control messages between queues in adjacent modules.

A user process writes data to a device using either the write() or putmsg () system call. The write() system call writes raw data to the stream, whereas putmsg() allows the user process to specify a message. Regardless of the system call used by the user process, the stream head copies the data into a message and delivers it to the queue for the next module in line. This copying of messages continues until the message is copied to the driver end and hence the device. Similarly, the user process reads data from the stream head using either the read() or getmsg() system call. If read() is used, the stream head gets a message from its adjacent queue and returns ordinary data (an unstructured byte stream) to the process. If getmsg () is used, a message is returned to the process.

STREAMS I/O is asynchronous (or nonblocking) except when the user process communicates with the stream head. When writing to the stream, the user process will block, assuming the next queue uses flow control, until there is room to copy the message. Likewise, the user process will block when reading from the stream until data are available.

The driver end is similar to a stream head or a module in that it has a read and write queue. However, the driver end must respond to interrupts, such as one triggered when a frame is ready to be read from a network. Unlike the stream head, which may block if it is unable to copy a message to the next queue in line, the driver end must handle all incoming data. Drivers must support flow control as well. However, if a device's buffer is full, the device typically resorts to dropping incoming messages. Consider a network card whose input buffer is full. The network card must simply drop further messages until there is ample buffer space to store incoming messages.

8.4.4 Spooling and Device Reservation

A spool is a buffer that holds output for a device, such as a printer, that cannot accept interleaved data streams. Although a printer can serve only one job at a time, several applications may wish to print their output concurrently, without having their output mixed together. The operating system solves this problem by intercepting all output to the printer. Each application's output is spooled to a separate disk file. When an application finishes printing, the spooling system queues the corresponding spool file for output to the printer. The spooling system copies the queued spool files to the printer one at a time. In some operating systems, spooling is managed by a system daemon process. In others, it is handled by an in-kernel thread. In either case, the operating system provides a control interface that enables users and system administrators to display the queue, to remove unwanted jobs before those jobs print, to suspend printing while the printer is serviced, and so on.

Some devices, such as tape drives and printers, cannot usefully multiplex the I/O requests of multiple concurrent applications. Spooling can coordinate concurrent output. Another way to deal with concurrent device access is to provide explicit facilities for coordination. Some operating systems (including VMS) provide support for exclusive device access by enabling a process to allocate an idle device and to deallocate that device when it is no longer needed. Other operating systems enforce a limit of one open file handle to such a device. Many operating systems provide functions that enable processes to coordinate exclusive access among themselves. For instance, Windows NT provides system calls to wait until a device object becomes available. It also has a parameter to the open() system call that declares the types of access to be permitted to other concurrent threads. On these systems, it is up to the applications to avoid deadlock.

习　题

一、选择题

1．设备无关性是指________。

A．系统中的设备必须有一个独立的接口　B．每种设备只能有一个

C．应用程序可以独立于具体的设备　D．系统设备只能由一个进程独占

2．缓冲技术中的缓冲池在________中。

A．主存　B．外存　C．ROM　D．寄存器

3．缓冲技术用于________。

A．扩充相对地址空间　B．提供主．辅存接口

C．提高设备利用率　D．提高主机和设备交换数据的速度

4．引入缓冲的主要目的是________。

A．改善 CPU 和 I/O 设备之间速度不匹配的情况　B．节省内存

C．提高 CPU 的利用率　D．提高 I/O 设备的效率

5．CPU 输出数据的速度远远高于打印速度，为了解决这一矛盾，可采用________。

A．并行技术　B．通道技术　C．缓冲技术　D．虚存技术

6．为了使多个进程能同时处理输入和输出，最好使用________结构的缓冲技术。

A．缓冲池　B．闭缓冲区环　C．单缓冲区　D．双缓冲区

7．通过硬件和软件的功能扩充，把原来独立的设备改造成能为若干用户共享的设备，这种设备称为________。

A．存储设备　B．系统设备　C．用户设备　D．虚拟设备

8．如果 I/O 设备与存储设备进行数据交换不经过 CPU 来完成，这种数据交换方式是________。

A．程序查询　B．中断方式　C．DMA 方式　D．无条件存取方式

9．中断发生后，应保存________。

A．缓冲区指针　B．关键寄存器内容　C．被中断的程序　D．页表

10．在中断处理中，输入输出中断是指________。

Ⅰ．设备出错　Ⅱ．数据传输结束

A．Ⅰ　B．Ⅱ　C．Ⅰ和Ⅱ　D．都不是

11．中断矢量是指________。

A．中断处理程序入口地址

B．中断矢量表起始地址

C．中断处理程序入口地址在中断矢量表中的存放地址

D．中断断点的地址

12．如果有多个中断同时发生，系统将根据中断优先级响应优先级最高的中断请求。若要调整中断事件的响应次序，可以利用________。

A．中断向量　B．中断嵌套　C．中断响应　D．中断屏蔽

13．设备管理程序对设备的管理是借助一些数据结构来进行的，下面的________不属于设备管理数据结构。

A．JCB　　B．DCT　　C．COCT　　D．CHCT

14．多数低速设备都属于________设备。

A．独享　　B．共享　　C．虚拟　　D．Spool

15．________用作连接大量的低速或中速 I/O 设备。

A．数据选择通道　　B．字节多路通道　　C．数据多路通道

16．________是直接存取的存储设备。

A．磁盘　　B．磁带　　C．打印机　　D．键盘显示终端

17．以下叙述中正确的为________。

A．在现代计算机中，只有 I/O 设备才是有效的中断源

B．在中断处理过程中必须屏蔽中断

C．同一用户所使用的 I/O 设备也可能并行工作

D．SPOOLing 是脱机 I/O 系统

18．________是操作系统中采用的以空间换取时间的技术。

A．SPOOLing 技术　　B．虚拟存储技术

C．覆盖与交换技术　　D．通道技术

19．操作系统中的 SPOOLing 技术，实质是将________转化为共享设备的技术。

A．虚拟设备　　B．独占设备　　C．脱机设备　　D．块设备

20．SPOOLing 系统提高了________利用率

A．独占设备　　B．共享设备　　C．文件　　D．主存储器

21．在操作系统中，________是一种硬件机制。

A．通道技术　　B．缓冲池

C．SPOOLing 技术　　D．内存覆盖技术

22．在操作系统中，用户在使用 I/O 设备时，通常采用________。

A．物理设备名　　B．逻辑设备名　　C．虚拟设备名　　D．设备牌号

23．采用假脱机技术，将磁盘的一部分作为公共缓冲区以代替打印机，用户对打印机的操作实际上是对磁盘的存储操作，用以代替打印机的部分是________。

A．独占设备　　B．共享设备　　C．虚拟设备　　D．一般物理设备

24．按________分类可将设备分为块设备和字符设备。

A．从属关系　　B．操作特性　　C．共享属性　　D．信息交换单位

25．________算法是设备分配常用的一种算法。

A．短作业优先　　B．最佳适应　　C．先来先服务　　D．首次适应

26．利用虚拟设备达到 I/O 要求的技术是指________。

A．利用外存作缓冲，将作业与外存交换信息和外存与物理设备交换信息两者独立起来，并使它们并行工作的过程

B．把 I/O 要求交给多个物理设备分散完成的过程

C．把 I/O 信息先存放在外存，然后由一台物理设备分批完成 I/O 要求的过程

D．把共享设备改为某个作业的独享设备，集中完成 I/O 要求的过程

27. 将系统中的每一台设备按某种原则进行统一的编号，这些编号作为区分硬件和识别设备的代号，该编号称为设备的________。

A. 绝对号　　B. 相对号　　C. 类型号　　D. 符号名

28. 等待当前磁道上的某指定扇区移动到磁头下所需的时间称为________。

A. 寻道时间　　B. 启动时间

C. 旋转延迟时间　　D. 传输时间

29. 通道是一种________。

A. I/O 端口　　B. 数据通道　　C. I/O 专用处理器　　D. 软件工具

二、填空题

1. 设备分配应保证设备具有较高的 ① 和避免 ② 。

2. 设备管理中采用的数据结构有 ① 、 ② 、 ③ 、 ④ 等四种。

3. 从资源管理（分配）的角度出发，I/O 设备又分为 ① 、 ② 和 ③ 三种类型。

4. 按所属关系对 I/O 设备分类，可分为系统设备和________两类。

5. 引起中断发生的事件称为________。

6. 常用的 I/O 控制方式有程序直接控制方式、中断控制方式、 ① 和 ② 。

7. 设备分配中的安全性是指________。

8. 通道指专门用于负责输入输出工作的处理机。通道所执行的程序称为________。

9. 通道是一个独立于 ① 的专管 ② 的处理机，它控制 ③ 与内存之间的信息交换。

10. 虚拟设备是通过 ① 技术把 ② 设备变成能为若干用户 ③ 的设备。

11. 实现 SPOOLing 系统时，必须在磁盘上开辟出称为 ① 和 ② 的专门区域以存放作业信息和作业执行结果。

12. 发生中断时，刚执行完的那条指令所在的单元号称为断点，断点的逻辑后继指令的单元号称为________。

13. 打印机是独占设备，磁盘是________设备。

14. 磁带是一种 ① 的设备。它最适合的存取方法是 ② 。

15. 磁盘是一种 ① 存取设备，磁盘在转动时经过读写磁头所形成的圆形轨迹称为 ② 。

三、综合题

1. 假定有一个具有 200 个磁道（0～199）的移动头磁盘，在完成了磁道 125 处的请求后，当前正在磁道 143 处为一个请求服务。若当前队列以 FIFO 次序存放，即 86，147，91，177，94，150，102，175，130。对下列每一个磁盘调度算法，若要满足这些要求，则总的磁头移动次数为多少？

（1）FCFS。

（2）SSTF。

（3）SCAN。

（4）CSCAN。

2. 简述设备驱动程序的处理过程。

3. 简述设备的独立性。

习题参考答案

第 1 章

一、选择题

1B，2C，3A，4B，5D，6A，7A，8B，9C，10B，11B，12C，13A，14D，15C，16C，17B，18C，19C，20C，21C，22A

二、填空题

1．硬件，软件　2．多道程序设计　3．多路性，交互性，“独立”性，及时性

4．多道程序设计　5．文件管理　6．软硬件资源，系统软件

第 2 章

一、选择题

1B，2C，3A，4A，5C，6D，7B，8①D，②B，9C，10A，11D，12①A，②C，③B，④D，13C，14①A，②B，15B，16C，17B，18D，19A，20C，21C，22D

二、简答题

1．简述进程的概念。

进程是程序在一个数据集合上的一次运行活动，是操作系统进行资源分配和调度的一个基本单位。

2．简述程序与进程的区别。

程序是有序代码的集合，通常对应着文件，可以复制，其本身没有任何运行的含义，是一个静态的概念。而进程是程序在处理机上的一次执行过程，它是一个动态的概念。

进程有被创建、生存到退出消亡的过程，是有一定生命周期的。程序是静态的，可以作为一种软件资料长期存在，是永久的，无生命的。

进程更能真实地描述并发，而程序不能；

进程是由进程控制块、程序段、数据段三部分组成；

进程具有创建其他进程的功能，而程序没有。

同一程序同时运行于若干个数据集合上，它将属于若干个不同的进程。也就是说同一程序可以对应多个进程。通过调用关系，一个进程也可以包括多个程序。

在操作系统中，程序并不能独立运行，作为资源分配、调度和独立运行的基本单位都是进程。

3．简述 PCB 的作用。

描述进程的全部信息的数据结构；进程存在的标志。

4．画图说明带挂起的进程状态及其转换。

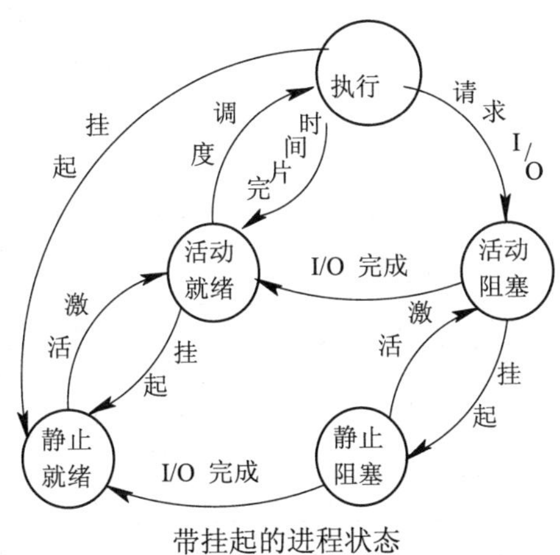

带挂起的进程状态

进程的状态反映进程执行进程的变化。这些状态随着进程的执行和外界条件发生变化和转换。进程被创建后，已经具备了运行的条件，只要获得 CPU 就可以运行，所以它首先进入的是就绪状态，等待进程调度程序调度；一旦被调度到（分配 CPU），获得 CPU，就可以执行程序，处于执行状态；在执行状态，如果时间片用完，操作系统通过一个时钟中断，使它停下来，把它的状态再置为就绪状态；在执行状态如果进程申请 I/O 操作，或者需要等待某种资源，才能继续向前推进，这时进程转入阻塞状态，不参与进程的调度；当它等待的 I/O 操作完成了，或者它等待的资源具备了，进程转入就绪状态。

被挂起的进程可能是就绪的，也可能是阻塞的。处于就绪状态，并且是挂起的，我们称为静止就绪状态，原来就绪状态可以称为动态就绪状态；处于阻塞状态，并且是挂起的，称为静止阻塞状态，原来阻塞状态可以称为动态阻塞状态。

5．简述处理机调度的三个层次。

作业的提交首先进入操作系统的后备作业队列，操作系统的作业调度进程负责处理、判断，符合条件的作业被接纳，为其创建进程，送入进程的就绪队列；进程在就绪队列中等待进程调度程序的调度，经历进程的各种状态转换，获得了足够的 CPU 时间后，程序执行完毕，进程管理程序收回分配给它的所有资源，注销进程，程序运行结束退出。

高级调度指的是作业调度，即根据作业控制块中的信息，审查系统能否满足用户作业的资源需求，以及按照一定的策略、算法，从后备队列中选取某些作业调入内存，并为它们创建进程、分配必要的资源。然后再将新创建的进程插入就绪队列，准备执行。在每次执行作业调度时，都须考虑接纳多少个作业，接纳哪些作业。

低级调度是指进程调度。被作业调度所接纳的进程，宏观上看都是处于运行状态了，但是 CPU 只有一个，这些进程是以时间片为单位轮流来使用 CPU 的。每一个时刻只能有一个进程使用 CPU，处于实际的执行状态。处于执行状态的进程怎样停下来，就绪状态的进程获得 CPU。

中级调度又称中程调度（Medium-Term Scheduling）。它是一种带有挂起功能的调度方式，引入中级调度的主要目的，是为了提高内存利用率和系统吞吐量，使那些暂时不能运行的进程不再占用宝贵的内存资源，而将它们调至外存上去等待（挂起进程），把此时的进程状态称为驻外存状态或挂起状态。当这些进程重新具备运行条件且内存又稍有空闲时，由中级调度来决定把外存上的哪些具备运行条件的静态就绪的进程，重新调入内存，并修改其状态为活动就绪状态，挂在活动就绪队列上等待进程调度。

三、计算题

1．

作业情况 \ 调度算法	进程名	A	B	C	D	E	平均周转时间
	到达时间	0	1	2	3	4	
	服务时间	3	5	2	6	1	
FCFS	开始时间	0	3	8	10	16	
	完成时间	3	8	10	16	17	
	周转时间	3	7	8	13	13	8.8
SJF	开始时间	0	6	3	11	5	
	完成时间	3	11	5	17	6	
	周转时间	3	10	3	12	2	6.4

2．

作业	开始时间	完成时间	周转时间
J1	0	a	a
J2	a	$a+b$	$a+b$
J3	$a+b$	$a+b+c$	$a+b+c$

平均周转时间：

$$\frac{3a+2b+c}{3}$$

3．

1）先来先服务：

作业号	开始时间（时）	结束时间（时）	周转时间（分钟）	带权周转时间
1	8:00	10:00	120	1.00
2	10:00	10:30	100	100/30(3.33)
3	10:30	10:42	102	102/12(8.5)
4	10:42	11:00	70	70/18(3.89)
平均			98	4.18

执行顺序: 1→2→3→4

2）短作业优先：

作业号	开始时间（时）	结束时间（时）	周转时间（分钟）	带权周转时间
1	8:00	10:00	120	1.00
2	10:30	11:00	130	4.33
3	10:00	10:12	72	6.00
4	10:12	10:30	40	40/18(2.22)
平均			90.5（分钟）	3.388

执行顺序: 1→3→4→2

第 3 章

一、选择题

1A，2B，3C，4B，5B，6D，7B，8D，9①B，②D，③F，10B，11A，12D，13B，14C，15B，16D，17D，18B，19B，20B

二、填空题

1．绝对值　　2．S<0　　3．4，0　　4．–（M–1）　　5．4，1

三、简答题

1．简述临界资源与临界区。

系统中同时存在有许多进程，它们共享各种资源，然而有许多资源在某一时刻只能允许一个进程使用，这种每次只允许一个进程访问的资源叫临界资源。

每个进程中访问临界资源的那段代码称为临界区（critical section）。

2．简述死锁的概念。

死锁是指两个或两个以上的进程在执行过程中，因争夺资源而造成的一种互相等待的现象，若无外力推动，它们都将无法推进下去，此时称系统处于死锁状态或者说系统产生了死锁。

3．简述死锁产生的原因。

（1）资源不够，资源的数量不是足够多，不能同时满足所有进程提出的资源申请，这就造成了资源的竞争，而且资源的使用不允许剥夺。

（2）进程的推进不当，进程的推进次序影响系统对资源的使用。

4．简述死锁产生的必要条件。

（1）互斥条件：指进程对所分配到的资源进行排他性使用，即在一段时间内某资源只由一个进程占用。如果此时还有其他进程请求资源，则请求者只能等待，直至占有资源的进程用毕释放。

（2）请求和保持条件：指进程已经保持至少一个资源，但又提出了新的资源请求，而该资源已被其他进程占有，此时请求进程阻塞，但又对自己已获得的其他资源保持不放。

（3）不剥夺条件：指进程已获得的资源，在未使用完之前，不能被剥夺，只能在使用完时由自己释放。

（4）环路等待条件：指在发生死锁时，必然存在一个进程——资源的环形链。

四、综合题

1．

A <u>wait(Mutex)</u>　B <u>singal(Mutex)</u>　C <u>wait(S)</u>　D <u>wait(Mutex)</u>　S <u>0</u>　Mutex <u>1</u>

2．

（1）说明信号量 empty，full 分别代表什么值，它们的初值是多少？

Empty 代表空的缓冲器资源个数，初值是 n，

full 代表满的缓冲器资源个数，初值是 0。

（2）解释语句 a、d 的作用。

a：是申请一个空的缓冲器资源。

d：释放一个满的缓冲器资源。

（3）语句 b、c 的作用是什么？

b、c 的作用是使想缓冲器中送产品的和调整缓冲器指针的动作不能被分开，生产者之间需要互斥。

（4）补充？？？的位置的语句。

Signal（mutex）

（5）若语句 d 换成 signal(empty)；可能会出现什么样的结果？

生产者不停地生产数据，送入缓冲器，数据会被覆盖；消费者进程无法推进。

3.

（1）说明信号量 Rmutex，Wmutex 的作用，它们的初值是多少？

读者用互斥信号量 Rmutex，写者用互斥信号量 Wmutex，初值 1，1

（2）解释变量 Readcount 的作用。

记录读者个数。

（3）解释语句 a、b。

a：判断如果是第一个读者，申请写信号量，即读者与写者互斥。

b：读者个数加 1。

（4）解释语句 c、d 的作用。

c：申请写者互斥信号量。

d：释放写者互斥信号量，实现写者之间的互斥。

（5）如果读者进程改成下列语句，请描述一下会出现什么情况。

读者之间互斥，一次只能有一个读者在读，不能使读者共享读了；

在读的同时可以有写者在写，读的信息不完整。

4.

```
semaphore a=0，b=0，c=0，d=0，e=0，f=0，g=0；
Parbegin
   Begin S1；Signal (a)；Signal (b)；End
   Begin Wait(a)；S2；Signal (c)；Signal (d)；End
   Begin Wait(b)；S3；Signal (e)；  End
   Begin Wait(c)；S4；Signal (f)；  End
   Begin Wait(d)；S5；Signal (g)；  End
   Begin Wait(e)；Wait(f)；Wait(g)；S6；  End
Parend.
```

5.

```
int S=1;int Sa=0;int Sb=0;
main()
{
   cobegin
      father();
      mather();
      son();
```

```
    daughter();
    coend
}
father()
{
  while(1)
    {
      P(S);
      将一个苹果放入盘中
      V(Sa);
    }
}
mather()
{
while(1)
{
P(S);
    将一个桔子放入盘中
    V(Sb);}
}
son()
{
  while(1)
  {
       P(Sb);
       从盘中取出桔子
       V(S);
       吃桔子;
  }
}
daughter()
{
while(1)
  {
      P(Sa);
      从盘中取出苹果
      V(S);
      吃苹果;
  }
}
```

6.

```
Mutex=1;  //A、B文件的互斥信号
Sa=1, Sb=1; //同组进程读文件计数器的互斥信号
```

```
Ac=0, Bc=0; //同组正在读文件的读者人数

A-Readers:
begin
  P(Sa);
  If Ac=0 then P(Mutex);
  Ac=Ac+1;
  V(Sa);
  Read file P;
  P(Sa);
  Ac=Ac -1;
  If Ac=0 then V(Mutex);
  V(Sa);
end;
B-Readers:
begin
  P(Sb);
  If Bc=0 then P(Mutex);
  Bc=Bc+1;
  V(Sb);
  Read file P;
  P(Sb);
  Bc=Bc-1;
  If Bc=0 then V(Mutex);
  V(Sb);
end;
```

7.

```
e1=n,e2=m,f1=0,f2=0,mutex1=1, mutex2=1
in1=0,in2=0,out1=0,out2=0
P1: 生产item1
Wait (e1)
Wait (mutex1)
     Item1=>buf1[in1]
     In1= ( in1 + 1 )  mod n
     Signal (mutex1)
     Signal (f1)
P2:  Wait (f1)
     Wait (mutex1)
     Buf1[out1]=>item
     Out1=( out1 + 1 )  mod n
     Signal (mutex1)
     Signal (e1)
     加工item=>ietm2
     Wait (e2)
```

```
      Wait (mutex2)
      Item2=>buf2[in2]
      In2= ( in2 + 1 )  mod m
      Signal (mutex2)
      Signal (f2)
 P3:  Wait (f2)
      Wait (mutex2)
      Buf2[out2]=>item
      Out2=( out2 + 1 )  mod m
      Signal (mutex2)
      Signal (e2)
      输出item
```

8.

```
Var count(顾客数):integer:=0;
   Mutex(互斥), sofa(沙发),empty(理发椅空),full(理发椅满): semaphore:=1,N,1,0;
   Cut(理发完), payment(付费), receipt(收费): semaphore:=0, 0, 0;
begin
   parbegin
         guest:begin
           wait(mutex);
           if((count>N) then;
             begin
                signal (mutex);
                exit shop;
             end
           else
             begin
                  count:=count+1;
                  if (count>1) then
                    begin
                       signal (mutex);
                       wait (sofa);
                       sit on sofa;
                       wait (empty);
                       get up from sofa;
                       signal (sofa);
                    end
                else    /*count=1 */
                   begin
                      signal (mutex);
                      wait(empty);
                   end
                signal (full);
                sit on the baber _chair;
```

```
                    signal (full);
                    wait(cut);
                    pay;
                    signal (payment);
                    wait(receipt);
                    get up form the baber_chair;
                    signal (empty);
                    wait (mutex);
                    count:=count-1;
                    signal(mutex);
                    exit shop;
                 end
      barber:begin
          repeat
            wait (full);
            cut hair;
            signal (cut);
            wait (payment);
            accept payment;
            signal (recipt);
          until false;
        end
   parend
end
```

9.

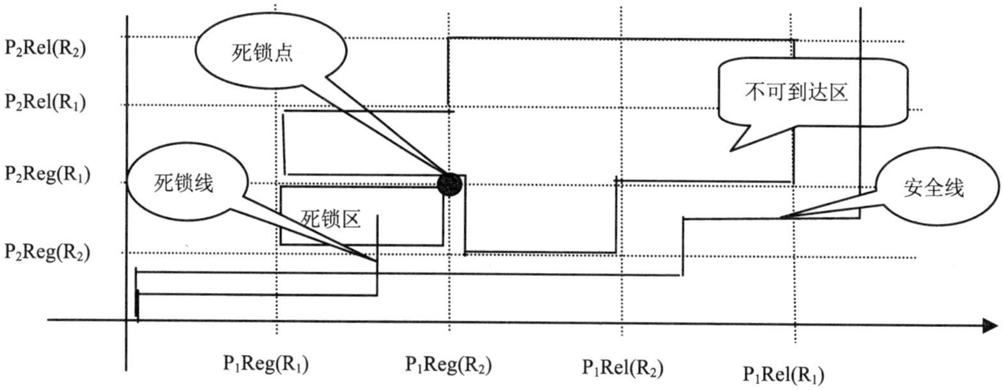

10.

（1）求系统可用资源向量。

4 2 1

（2）需求矩阵。

6 3 1
0 1 1
6 0 1
0 0 1
2 3 1

（3）问 T_0 时刻系统是否安全？如果安全，给出一个进程执行的安全序列。

系统是安全的，

安全序列 $P_2 \rightarrow P_3 \rightarrow P_4 \rightarrow P_5 \rightarrow P_1$ （不唯一）

（4）T_1 时刻 P_1 进程提出资源请求 Request（2，2，0），系统能否将资源分配给它？为什么？

T_1 时刻：假设系统将资源分配给 P_1。

资源进程	最大需求量 R_1 R_2 R_3	已分配资源数 R_1 R_2 R_3	需求矩阵 R_1 R_2 R_3	可用资源向量 R_1 R_2 R_3
P_1	6 5 2	2 4 1	4 1 1	2 0 1
P_2	2 2 1	2 1 0	0 1 1	
P_3	8 0 1	2 0 0	6 0 1	
P_4	1 2 1	1 2 0	0 0 1	
P_5	3 4 4	1 1 3	2 3 1	

资源进程	Work R_1 R_2 R_3	已分配资源数 R_1 R_2 R_3	Work+已分配资源数 R_1 R_2 R_3	状态
P_4	2 0 1	1 2 0	3 2 1	Finish
P_2	3 2 1	2 1 0	5 3 1	Finish
P_5	5 3 1	1 1 3	6 4 4	Finish
P_1	6 4 4	2 4 1	8 8 5	Finish
P_3	8 8 5	2 0 0	10 8 5	Finish

因为有安全序列 $P_4 \rightarrow P_2 \rightarrow P_5 \rightarrow P_1 \rightarrow P_3$，所以系统将资源分配给 P_1。

11．

（1）求可用资源向量。

（1，1，2）

（2）需求矩阵。

2 2 2
1 0 2
1 0 3
4 2 0

（3）当前的状态是否是安全的？如果是安全的，写出进程安全序列。

是安全的，进程安全序列 $P_2 \rightarrow P_3 \rightarrow P_4 \rightarrow P_1$

（4）如果进程 P_2 发出资源请求向量（1，0，1），系统能否将资源分配给它？为什么？

能将资源分配给它。因为若分配给进程 P_2 资源（1，0，1），则可用资源为（0，1，1）。

需求矩阵：

2 2 2
0 0 1
1 0 3
4 2 0

存在安全序列 $P_2 \rightarrow P_4 \rightarrow P_1 \rightarrow P_3$

（5）如果是进程 P_1 发出资源请求向量（1，0，1），系统能否将资源分配给它？为什么？

不能，因为若分配给进程 P_1 资源（1，0，1），则可用资源为（0，1，1）。

需求矩阵：

1　2　1

1　0　2

1　0　3

4　2　0

不存在安全序列，故不能分配。

12.

序号	*M*	*N*	*K*	是否死锁
1	6	3	2	不能死锁
2	6	3	3	可能死锁
3	7	3	3	不能死锁
4	9	4	3	不能死锁
5	13	6	3	不能死锁

第 4 章

1. B，

2. 发送，接收

3.

（1）共享存储区通信：

共享存储区通信可使若干进程共享主存中的某一个区域，且使该区域出现在多个进程的虚地址空间中。

（2）消息传递系统：

在消息传递系统中，进程间的数据交换以消息为单位，在计算机网络中，消息又称为报文。程序员直接利用系统提供的一组通信命令（原语）来实现通信。

（3）管道通信：

管道是指用于连接一个读进程和一个写进程以实现他们之间通信的一个共享文件，又名 pipe 文件。

第 5 章

一、选择题

1A，2B，3C，4A，5A，6A，7C，8B，9B，10A，11A，12D，13①A，②D，14D，15A，16B，17C，18C，19C，20B，21A，22B

二、填空题

1. 地址变换　2. 界限寄存器和存储保护键　3. ①页号及页内偏移量，②段号及段内偏移量　4. 段号、段在内存的起始地址、段长度　5. ①逻辑，②物理　6. ①静态重定位，②动态重定位　7. 地址递增　8. 页号和块号　9. 存储空间　10. ①程序装入内存，

②程序执行

三、综合题

1.

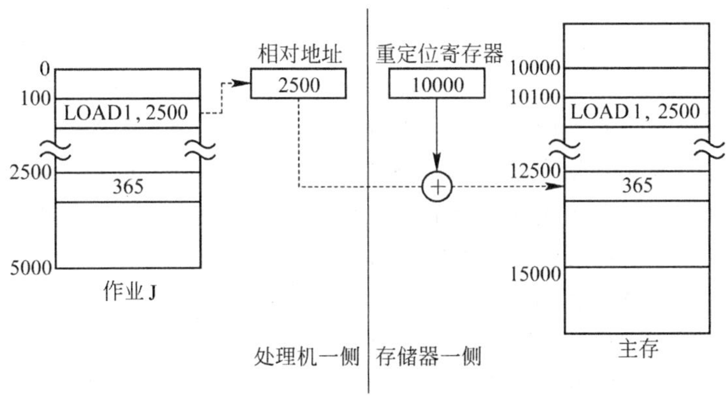

图 5-1　重定位

如图 5-1 所示，程序要想运行，必须被装入内存，装入内存的地址是由操作系统决定的。程序的逻辑地址与内存的物理地址一般是不一样的。程序中的指令可能会根据自己的逻辑地址存取数据，程序被装入内存后，地址发生了变化，所以需要重定位，调整程序中的地址才能使程序正常运行。

重定位有两种，分别是动态重定位与静态重定位。

（1）静态重定位：即在程序装入内存的过程中完成，是指在程序开始运行前，程序中的各个地址有关的项均已完成重定位，地址变换通常是在装入时一次完成的，以后不再改变，故成为静态重定位。

（2）动态重定位：它不是在程序装入内存时完成的，而是 CPU 每次访问内存时 由动态地址变换机构（硬件）自动进行把相对地址转换为绝对地址。动态重定位需要软件和硬件相互配合完成。

2.

1）首次适应算法（First Fit）

首次适应算法要求空闲分区按首址递增的次序排序（队列），当进程申请大小为 u.size 的内存时，系统从空闲区表的第一个表目开始查询，直到首次找到等于或大于 u.size 的空闲区 m。如果 m.size-u.size，大于等于系统规定的最小分区的大小 size，则从 m 中划出大小为 u.size 的分区分配给进程，余下的部分 m.size-u.size，仍作为一个空闲区留在空闲区表中，但要修改其首址和大小。

如果多出来的部分 m.size-u.size，小于系统规定的最小分区的大小 size，则把多出来的部分 m.size-u.size，看成是分区的“零头”，把整个分区 m 分配给进程，分区存在 m.size-u.size 的“零头”，在空闲分区表中把 m 分区置成空表目。

2）循环首次适应算法（Next Fit）

首次适应算法每次都是从内存的低地址部分开始查找的，所以内存的低地址部分使用比较频繁，出现小的、比较零碎的空闲分区的机会比较多，内存的使用不够均衡。在首次适应算法基础上稍加改动就是循环首次适应算法。循环首次适应算法的做法是不在每次从队首开始查找，而是从上次找到的空闲区的下一个空闲区开始查找，直到找到第一个能满足要求的空闲区，如果找到最后一个空闲区（队尾），还没找到要求的空闲区，则返回队首，从头再来查找。

3）最佳适应算法（Best Fit）

最佳适应算法是从全部空闲区中找出能满足作业要求的且大小最小的空闲分区，这种方法能使碎片尽量小。为实现此算法，空闲分区表（空闲区链）中的空闲分区要按大小从小到大进行排序，自表头开始查找到第一个满足要求的自由分区分配。每次分配时，找到的空闲区表中能够满足作业需求的最小的空闲区，尽量不分割在空闲区保留大的空闲区，便于大作业的装入。但会造成许多小的空闲区。

4）最坏适应算法（Worst Fit）

每次都找最大的空闲分区分配给作业，是这个空闲分区分配一部出去后，剩余的空闲区最大。这样做主要是考虑使剩余部分的空间不至于太小，仍可用来装入其他作业，被其他作业利用。

当回收内存分区时，应检查是否有与回收区相邻的空闲分区，若有，则应合并成一个空闲区。

相邻可能有上邻空闲区和下邻空闲区、既上邻又下邻空闲区、既无上邻又无下邻空闲区。

若有上邻空闲区，只修改上邻空闲区长度为回收的空闲分区长度与原上邻区长度之和即可；

若有下邻空闲区，修改这个下邻空闲分区的首地址为被回收分区的地址，长度为下邻空闲分区的长度和回收分区的长度即可；

若既有上邻又有下邻空闲分区，修改上邻空闲分区的长度为上邻空闲分区长度加下邻空闲分区长度再加上回收分区长度之和，再把下邻空闲分区的状态改为空表目即可；

若既无上邻区又无下邻区，那么找一个状态为空表目的一行，记下该回收分区的起始地址和长度，且改写相应的状态为空闲，表明该行指示了一个空闲分区。

3.

在分页式管理系统中，进程的每个页面离散地分配到了内存中的任意物理块中。系统为每个进程建立一张页面与物理块的映射表，简称为页表，进程通过页表能够找到每个页面对应的物理块。不需要连续存放。

4.

不对。

页面一般是 2 的整数次幂，但程序不一定正好是 2 的整数次幂。

最后一页一般存在页内零头，非彻底解决。

5.

若采用最佳适应算法：在申请 96KB 存储区时，选中的是 5 号分区刚好一样大；接着申请 20KB 时，选中的是 1 号分区，一分为二，剩下 12KB 空闲区；最后申请 200KB 时，选中 4 号分区，分配后剩下 18KB，显然采用最佳适应算法进行分配，可以满足该作业序列的需求。

若采用首次适应算法，在申请 96KB 存储区时，选中的是 4 号分区，分配后剩下 218–96=122KB；接着申请 20KB 时，选中的是 1 号分区，一分为二，剩下 12KB 空闲区；最后申请 200KB 时，现有的五个分区都无法满足要求。因此，首次适应算法不能满足该作业序列的需求。

6.

1KB 作业进 1 号分区，主存浪费为 7KB；9KB 作业进 2 号分区，主存浪费为 23KB；33KB 作业进 3 号分区，主存浪费为 87KB；121KB 作业进 4 号分区，主存浪费为 211KB；所以主存空间的浪费为 7+23+87+211=328KB。

7.

（1）1.5×2=3μs

（2）1.5×85%+1.5×2×(1-85%)=1.725μs

8.

设页号为 P，页内位移为 W，逻辑地址为 A，页面大小为 L。

（1）A=1011，P=1011/1024=0

W=1011%1024=1011

物理地址为：2×1024+1011=3059

（2）A=2148，P=2148/1024=2

W=2148%1024=100

物理地址为：1×1024+100=1124

（3）A=3000，P=3000/1024=2

W=3000%1024=952

物理地址为：1×1024+952=1976

（4）A=4000，P=4000/1024=3

W=4000%1024=928

物理地址为：6×1024+928=7072

（5）A=5012，P=5012/1024=4

W=5012%1024=916

因页号超过页表长度，该逻辑地址越界。

9.

逻辑地址 2F6AH

页号 P=2F6AH /4096=2

W=2F6AH%4096=F6AH

物理地址为：11×4096+F6AH=BF6A

10.

0 - 1　　　1086=1×1024+62,　4×1024+62=4158

1 - 4

2 - 6　　　6020=5×1024+904,　页码为 5，越界

3 - 7

4 - 9

第 6 章

一、选择题

1B，2A，3D，4A，5D，6A，7B，8D，9B，10A

二、填空题

1．①先进先出 ②最近久用末使用　2．①14 ②10 ③14 ④10　3．①物理地址空间 ②机器的地址长度　4．页面置换　5．①最佳算法 ②先进先出法 ③最近最少使用　6．地址越界中断

三、综合题

1．

FIFO 页面淘汰算法：页面引用次数为 11 次，缺页次数为 9 次，所以缺页率为 9/11；

若采用后一种页面淘汰策略：页面引用次数为 11 次，缺页次数为 8 次，所以缺页率为 8/11。

2．50 次。

3．逻辑地址：$16\times2048=2^{15}$，因此至少要 15 位。内存空间：$8\times2048=16$KB

4．

（1）最佳置换淘汰算法：$M=3$ 时，7/12；$M=4$ 时，6/12

（2）先进先出页面淘汰算法：$M=3$ 时，9/12；$M=4$ 时，10/12

（3）最近最久未使用页面淘汰算法：$M=3$ 时，10/12；$M=4$ 时，8/12

5．FIFO 为 9 次，LRU 为 7 次。

6．

（1）采用 FIFO 页面替换算法，其缺页次数是多少？（画 √ 表示缺页）

	1	2	3	4	2	1	5	6	2	1	2	3	7	6	3	2	1	2	3	6
页框	1	1	1	4	4	4	4	6	6	6	6	3	3	3	3	2	2	2	2	6
		2	2	2	2	1	1	1	2	2	2	2	7	7	7	7	1	1	1	1
			3	3	3	3	5	5	5	1	1	1	1	6	6	6	6	6	3	3
缺页	√	√	√	√		√	√	√	√	√		√	√	√		√	√		√	√

缺页次数：16 次。

（2）采用 LRU 页面替换算法，其缺页次数是多少？（画 √ 表示缺页）

	1	2	3	4	2	1	5	6	2	1	2	3	7	6	3	2	1	2	3	6
页框			3	4	2	1	5	6	2	1	2	3	7	6	3	2	1	2	3	6
		2	2	3	4	2	1	5	6	2	1	2	3	7	6	3	2	1	2	3
	1	1	1	2	3	4	2	1	5	6	6	1	2	3	7	6	3	3	1	2
缺页	√	√	√	√		√	√	√	√	√		√	√	√		√	√			√

缺页次数：15 次。

7.

(1)

(2^6)逻辑地址的页码需 6 个二进制位。(2^{10})页内地址 10 位。

(2^5)物理地址的块号需 5 个二进制位。(2^{10})块内地址 10 位。

(2)

逻辑地址 0B5CH=$(0000\ 1011\ 0101\ 1100)_2$

物理地址 $(0001\ 0011\ 0101\ 1100)_2$ = 135CH

(3)

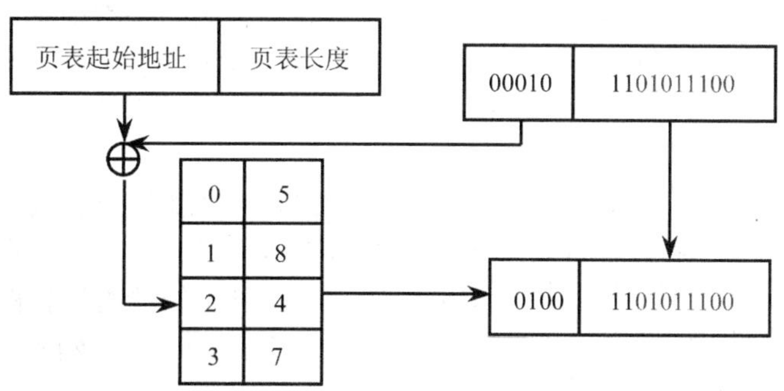

8.

(1)a 数组占据 200×200×4/256=625(个页面)

(2)根据程序,代码应占据 1 页,执行时,先发生一次代码页的缺页,将代码调入内存,执行程序,为数组赋值,需要再把 a 数组占据的每页都调入一次,所以总的缺页次数为 625+1=626.

9.

(1)虚地址有多少位? 12+9 = 21 位

(2)一个页帧有多少字节? 512 字节

(3)物理地址中有多少位表示页帧? 32 – 9 = 23 位

(4)页表有多少项(页表有多长)? 2^{12} = 4096 项

(5)页表需要多少位来存入一个页表项(页帧号及一个有效位)? 23+1 = 24 位

10.

(1)0001 01 11 1100 1010, 页号为 5。

(2)0 页,块号为 7,0001 1111 1100 1010 物理地址:1FCA。

(3)2 页,块号为 2,0000 10 11 1100 1010 物理地址:0BCA。

第 7 章

一、选择题

1B,2D,3D,4B,5C,6A,7B,8B,9A,10A,11A,12 C,13 B,14B,15A,16A,17①A,②C,③D,18B,19A,20D,21D,22D,23ADFGH,24D,25B,26A,27D

二、填空题

1. 索引,数据,索引 2. 文件 3. 文件名,文件在磁盘上的存放地址 4. 按名存取 5. 顺序、链接和索引 6. 流式文件和记录式文件 7. 存取控制表 8. 文件保护 9. 文

件控制块和文件体　10．数据块　11．顺序文件　12．索引文件　13．逻辑结构，物理结构

三、综合题

1．

系统首先从内存中划出若干个字节，为每个文件存储设备建立一张位示图。在位示图中，每个文件存储设备的物理块都对应一个比特位。如果该位为“0”，表示所对应的块是空闲块；反之，如果该位为“1”则表示所对应的块已被分配出去。

利用位示图来进行空闲块分配时，只需查找图中的“0”位，并将其置为“1”位；反之，利用位示图回收空闲块时只需把相应的比特位由“1”改为“0”即可。

当要删去一个文件，归还磁盘空间时，可根据归还块的物理地址计算出相应的块号，由块号推算出它在位示图中的对应位，把这一位的占用标志“ 1 ”清成“ 0 ”，表示该块已成为空闲块。

2．

文件系统由被管理文件、管理文件所需的数据结构（如目录表 、文件控制块 、存储分配表）和相应的管理软件以及访问文件的一组操作所组成。

从使用的角度来组织文件，用户把能观察到的且可以处理的信息根据使用要求按照一定形式构造成的文件，这种用户可见的文件外部形式称为文件的逻辑组织，也称逻辑结构。而文件系统要从文件的存储和检索等管理的角度来组织文件，文件系统根据存储设备的特性、文件的存取方式来决定以怎样的形式把文件存放到存储介质上，即内部的物理存储形式，称为文件的物理结构。

3．

（1）500/32=16，　　位示图需要 16 字

（2）第 i 字第 j 位对应的块号是 32*(i-1)+j

4．

（1）文件控制块

（2）文件分配表

（3）4,6,11

（4）9,10,5

（5）

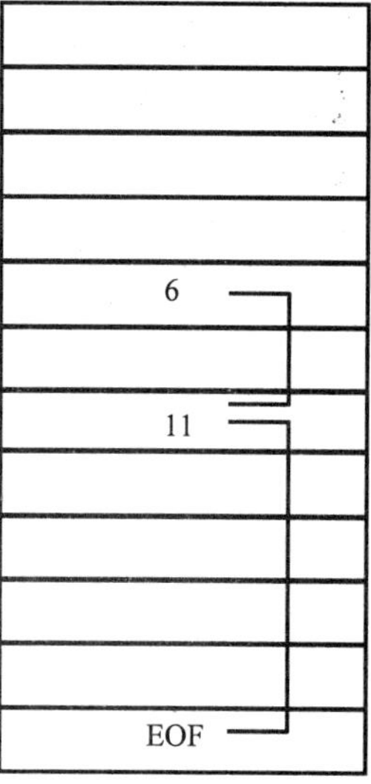

5.

（1）198,199

（2）

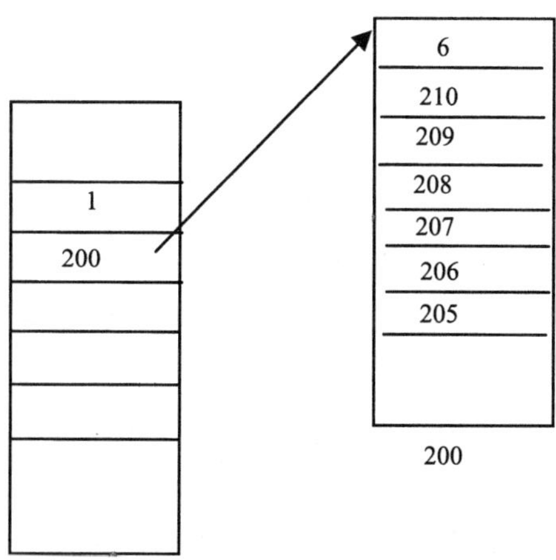

（3）200,205,206

（4）

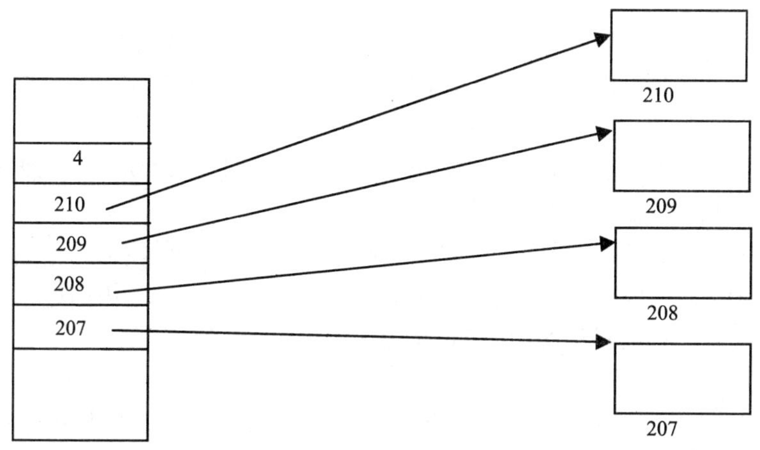

第 8 章

一、选择题

1C，2A，3D，4A，5C，6A，7D，8C，9B，10C，11A，12D，13A，14A，15B，16A，17C，18A，19B，20A，21A，22B，23C，24D，25C，26A，27A，28C，29C

二、填空题

1．①高的利用率 ②死锁问题　2．①系统设备表 ②设备控制表 ③控制器控制表 ④通道控制表　3．①独享 ②共享 ③虚拟　4．用户设备　5．中断源　6．①DMA 方式 ②通道控制方式　7．设备分配中应保证不会引起进程死锁　8．通道程序　9．①CPU ②输入输出的处理机 ③外设　10．①SPOOLing ②独占 ③共享　11．①输入井 ②输出井　12．恢复点　13．①独占 ②共享　14．①顺序存取 ②顺序存取　15．①直接 ②磁道

三、综合题

1.

（1）565。

（2）162。

（3）125。

（4）169。

2.

（1）将接收到的抽象请求转化为具体要求。

通常在每个设备控制器中都包含若干个寄存器，它们分别用于暂存数据、命令和设备的状态等信息。不同的设备在设备寄存器的数量、命令的性质等方面有着根本性的不同。因为用户和上层的设备无关性软件对设备控制器的具体情况不了解，只能发出抽象的 I/O 请求，需要设备驱动程序将这些抽象请求转化为具体要求后，再发送给设备控制器。例如当进行磁盘操作时，磁盘驱动程序要将抽象请求中的线性的盘块号转换为磁盘的盘面、磁道和扇区号。

（2）检查 I/O 请求的合法性，了解 I/O 设备的状态，传递有关参数，设置设备工作方式。

在真正启动某个设备进行 I/O 操作之前，设备驱动程序要对 I/O 请求的合法性进行验证。通常每类设备只能完成一组特定的功能，对于设备不能支持的 I/O 请求，则认为是非法的。例如对于从键盘输出数据这样的请求，很显然键盘驱动程序应予以拒绝。

（3）启动分配到的 I/O 设备，完成指定的 I/O 操作。

在完成上述各项准备工作后，设备驱动程序向控制器发出启动命令，然后把自己阻塞起来。由设备控制器控制进行基本的 I/O 操作。

设备控制器在 I/O 操作完成后，向 CPU 发出中断请求，CPU 响应由控制器发来的中断请求，并根据其中的中断类型调用相应的中断处理程序进行处理。中断处理时可能需要唤醒被阻塞的驱动程序和被阻塞的请求 I/O 的进程。

3.

设备独立性也称设备无关性，是指用户编写程序时使用的设备与实际使用的设备无关。为实现设备独立性，引入了物理设备和逻辑设备两个概念。物理设备是一个具体设备。物理设备名是系统为了能识别全部的外设，给每台外设分配的唯一不变的名字。而逻辑设备是对实际物理设备属性的抽象，并不限于某个具体设备。逻辑设备名是由用户命名且可以更改。在应用程序中，使用逻辑设备名。

参 考 文 献

[1] Tauibamum A S. Modern Operating Systems. Englewood Cliffs, N.J. Prentice Hall. 1992.

[2] William Stalling. Operating System, Internals and Design Principles (5th Edition) . Englewood Cliffs, N.J. Prentice Hall. 2005.

[3] Silberscbatz A, Galvin P. Operating System Concepts. Addison-Wesley，1994.

[4] Stallings W. Operating System. New York: MacMillan, 1992.

[5] A S Tauibamum, A S Woodhull. Operating Systems: Design and Implementation (3th Edition). Englewood Cliffs, N.J.Prentice Hall. 2006.

[6] Andrew S Tanenbaum 著. 现代操作系统. 陈向群，马洪兵，等译. 北京：机械工业出版社，2011.

[7] Gary Nutt 著. 操作系统现代观点. 孟祥由，晏益慧译. 北京：机械工业出版社，2004.

[8] 刘振鹏，张明，王煜. 操作系统（第二版）. 北京：中国铁道出版社，2007.

[9] 汤小丹，梁红兵，哲凤屏，汤子瀛. 计算机操作系统（第三版）. 西安：西安电子科技大学出版社，2007.

[10] 孟庆昌，等. 操作系统原理，北京：机械工业出版社，2010.

[11] 刘乃琦，蒲晓蓉. 操作系统原理、设计及应用. 北京：高等教育出版社，2008.

[12] 张尧学，史美林，张高. 计算机操作系统教程（第 3 版）. 北京：清华大学出版社，2006.

[13] 汤小丹，梁红兵，哲凤屏，汤子瀛. 现代操作系统. 北京：电子工业出版社，2008.

[14] 颜彬，李登实. 计算机操作系统. 北京：清华大学出版社，2007.

[15] 范策，许宪成. 计算机操作系统教程——核心与设计原理. 北京：清华大学出版社，2007.

[16] 徐甲同，方敏. 操作系统教程（第二版）. 西安：西安电子科技大学出版社，1999.

[17] 罗宇，文艳军. 操作系统. 北京：人民邮电出版社，2009.

[18] 宗大华，宗涛. 操作系统教程. 北京：人民邮电出版社，2008.

教学资源支持

敬爱的教师：

感谢您一直以来对清华版计算机教材的支持和爱护。为了配合本课程的教学需要，本教材配有配套的电子教案（素材），有需求的教师请到清华大学出版社主页（http://www.tup.com.cn）上查询和下载，也可以拨打电话或发送电子邮件咨询。

如果您在使用本教材的过程中遇到了什么问题，或者有相关教材出版计划，也请您发邮件告诉我们，以便我们更好地为您服务。

我们的联系方式：

地　　址：北京海淀区双清路学研大厦 A 座 707

邮　　编：100084

电　　话：010－62770175－4604

课件下载：http://www.tup.com.cn

电子邮件：weijj@tup.tsinghua.edu.cn

教师交流 QQ 群：136490705

教师服务微信：itbook8

教师服务 QQ：883604

（申请加入时，请写明您的学校名称和姓名）

用微信扫一扫右边的二维码，即可关注计算机教材公众号。

扫一扫
课件下载、样书申请
教材推荐、技术交流